Teaching Computing: A Practitioner's Perspective

Teaching Computing: A Practitioner's Perspective

Henry M. Walker

CRC Press
Taylor & Francis Group
Boca Raton London New York

CRC Press is an imprint of the
Taylor & Francis Group, an **informa** business

Published in 2018 by CRC Press
Taylor & Francis Group
6000 Broken Sound Parkway NW, Suite 300
Boca Raton, FL 33487-2742

Printed in the United States of America on acid-free paper
10 9 8 7 6 5 4 3 2 1

International Standard Book Number-13: 978-1-138-54978-4 (Hardback)
International Standard Book Number-13: 978-1-138-03443-3 (Paperback)

Library of Congress Cataloging-in-Publication Data
Catalog record is available from the Library of Congress

Visit the Taylor & Francis Web site at
http://www.taylorandfrancis.com

and the CRC Press Web site at
http://www.crcpress.com

To all teachers and their students,
particularly

teachers I have had through the years, including Norton Levy,
Eleanor Hoogheem, Victor Hill,
Gilbert Spencer, Neil Grabois, George Feeman,
and Franklin P. Peterson,

my teaching colleagues, including Eugene Herman,
Charles Jepsen, John David Stone, Samuel A. Rebelsky,
Janet Davis, and Marge Coahran,

my long-term mentor and guide, Nell Dale,

and my life-long partner and co-teacher/learner, Terry.

Contents

Preface

In the airport, after attending the 2016 Technical Symposium on Computer Science Education (SIGCSE 2016) in Memphis, I talked with an attendee who commented that many conference sessions seemed focused on computing-education research. Although these sessions were of some interest, he wanted practical ideas he could implement immediately in the classroom. What tips, techniques, examples, exercises, approaches, etc. can he use to address everyday challenges and to improve his teaching? In principle, he could work his way through volumes of proceedings from SIGCSE symposia, ITiCSE conferences, ICER workshops, and CCSC regional conferences, but digesting the full range of conference and research articles takes much time — even if he could acquire that range of materials.

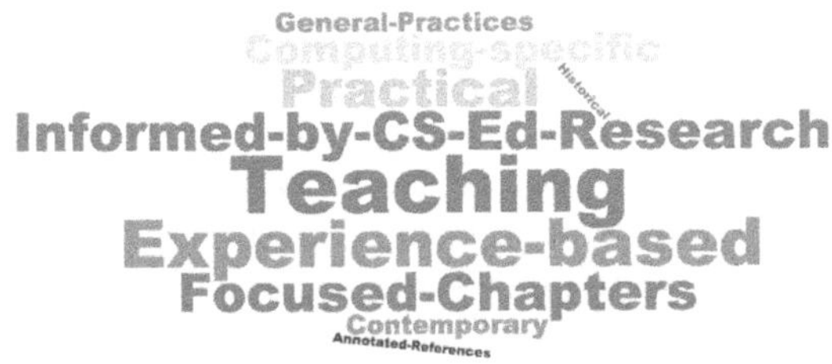

FIGURE 1 Book Elements Word Cloud

For beginning faculty, starting to teach can be even more intimidating. Some graduate schools and some computing faculty provide guidance and mentoring, but many do not. Often a new faculty member is assigned to teach a course and then left on their own to put the course together. Sometimes the new faculty member can find materials from a previous instance of the course; sometimes an experienced faculty member may be teaching another section of the same course. However, much of the time, the new faculty must construct the course with little input, experience, or feedback. Further, the new faculty member likely does not know the computing-education literature and does not know where to begin.

This book addresses these issues by providing a solid resource for both new and experienced computing faculty. The book serves as a practical, easy-to-use resource, covering a wide range of topics in a collection of focused down-to-earth articles.

Motivating Vision for this Book

New computing faculty need a way to get started in their teaching, and on-going faculty want to improve. This book addresses these needs by building upon the 50+ teaching-oriented columns I have written for the *SIGCSE Bulletin* and ACM *Inroads*[1], adding several additional articles and selected annotated references.

Over the years, I have observed that many new college faculty, recently graduated with Ph.D.s from research-oriented universities, often have rather little teaching experience. Some graduate schools provide teaching experience and feedback to their graduate students, but others do not. Thus, many new computing faculty have meager background regarding teaching. They may be eager to learn, but they may have difficulty finding a place to start. Also,

[1]According to ACM policy, the author of columns retains copyright control, and the author organizes and expands those columns in this single volume.

mid-career computer science faculty may have gained experience about effective teaching, but feedback and discussion sometimes may be lacking.

Between 1997 and 2009, I wrote columns on *Classroom Issues* for the SIGCSE *Bulletin*, with a pieces normally appearing twice a year in June and December. In 2010, the extended format of the SIGCSE *Bulletin* shifted to a new ACM magazine, called *Inroads*, at which time I wrote columns on curricula (*Curricular Syncopations*) appearing in June and December, and columns on in-class teaching (*Classroom Vignettes*) appearing in March and September. Altogether, between 1997 and March 2017, I have written over 50 featured columns. Feedback from the editors suggests that these articles are very well received by readers — computing educators.

Putting these two themes together, this book provides practical, down-to-earth suggestions, observations, and perspectives for computing teachers. Although a complete book on all aspects of teaching, including practices and a full range of cutting-edge research results, would require several volumes and likely overwhelm many practicing teachers, a modest, one-volume resource on practical approaches, tips, and techniques could provide a strong starting place for new computing faculty. The envisioned volume covers many basic elements, but follows an approach that is not intimidating for beginners.

Key Features of the Book

This book has numerous elements designed to connect with teaching practitioners.

- *Wide range of teaching topics:* A classroom teacher must address questions on many levels, such as:

    ◇ How does a course fit into the broader curriculum?

    ◇ How will topics be organized?

    ◇ What materials will be given to students (e.g., syllabi, readings, examples, textbooks, etc.)?

    ◇ What class format should be used (e.g., lectures, small/large groups, labs, etc.)?

    ◇ How might students be encouraged to engage with the material?

    ◇ What homework, exercises, tests, projects., etc. will students be given?

    ◇ How will students be assessed?

  Although a comprehensive review of all teaching topics would fill volumes, this book discusses many of these basic elements of teaching.

- *Practical tone:* The book serves as a down-to-earth practitioners guide. Although discussions are based on principles and best practices, the style will be somewhat informal and intended for general audiences.

- *Short, focused chapters:* Classroom teachers often are pressed for time. They do not have time for long and exhaustive discussions. Rather, they need practical ideas and approaches that address current needs. At 2-6 pages apiece, each column and article can provide insights and suggestions on a common problem or circumstance in a time/space-efficient format.

- *Coherent and convenient organization:* The organization of columns and articles into theme-based parts allows teachers to navigate easily to chapters that discuss specific interests and needs.

- *Mix of general educational perspectives and computing-specific elements:* Many elements of effective teaching and learning, such as active student engagement, apply to many disciplines. Other important techniques, such as pair programming, apply specifically to computing education. This book draws upon both general and subject-specific techniques and best practices in the discussion of computing education.

- *Connections between teaching in general and teaching computing:* Much literature considers elements of teaching, and many of these works consider general principles that might apply to the teaching of any subject. For example, *McKeachie's Teaching Tips* [124] discusses a wide range to tips and techniques that might apply to teaching in departments throughout the undergraduate curriculum. This book builds upon many general principles, but also applies those ideas to computing classrooms. Thus, this text provides provides specific examples involving computing labs, programming assignments, use of national curricular guidelines for undergraduate computing [10], algorithms, and other topics specifically connected to the discipline of computing.

- *Non-intimidating, but thought-provoking:* Following the lead of ACM *Inroads*, the columns and articles in this book will target the general teaching faculty — non-specialists. Articles will use little technical jargon and will not follow an overly-formal tone. For example, articles may contain a few key references from time to time, but articles will avoid long and highly technical presentations aimed at researchers.

- *Both historical and contemporary perspectives:* Some columns collected in this book date as far back as December 1997, while others appeared in 2017. In the interests of historical perspectives, each column in this book appears with only minor editing from the original (e.g., correcting of typographical errors). In some cases, a title might appear (with the original indicated as a footnote) to provide better insight to modern readers regarding content and perspective. Also, immediately after some articles, and additional commentary considers what might have changed since the article first appeared and what might be the same. Such comments are not needed in many cases, but an update is included when a modern perspective may seem interesting and worthwhile. Further, a few chapters explicitly provide an historical perspective of curricula and/or pedagogy.

- *Annotated reference sections:* Several parts within this book contain a modest list of annotated references for further reading. As separate sections, these brief reference lists do not distract new teachers from the down-to-earth, practical columns, but the lists do provide starting places for interested readers to explore topics further.

- *Informed by current understandings and research:* Although the practical style of the book will omit long, formal presentations, ideas and perspectives will reflect modern perspectives on curricula, pedagogy, student-engagement, lifelong learning goals, and course materials.

Acknowledgments

Throughout the years, I have been extremely fortunate to be able to interact with many wonderful teachers, students, and guides.

Special thanks to

- the many wonderful teachers I had throughout my life—including
  - ◇ high school teachers, especially Norton Levy and Eleanor Hoogheem,
  - ◇ college faculty, especially Victor Hill, Gilbert Spencer, Neil Grabois, and George Feeman,
  - ◇ graduate school faculty, especially my thesis advisor Franklin P. Peterson,
- my wonderful Grinnell College colleagues, especially Eugene Herman, Charles Jepsen, John David Stone, Samuel A. Rebelsky, and Janet Davis,
- my dear mentor and guide, Nell Dale,

Thanks also to the many people who have given me feedback on the columns over the years, particularly my former colleague, Marge Coahran.

With thanks, I acknowledge the word cloud tool from http://www.wordclouds.com/, with which the author developed the word clouds used in this book, and

And, my professional efforts are possible only because of the on-going encouragement, support, understanding, patience, and discussions with my wonderful wife Terry!

Front Cover

The front cover of this book illustrates the breadth of interests found among many liberal arts students. As a Grinnell College student, Lea Marolt Sonneschein[2] explored interests in both computing and art. This book's front cover shows a photograph of her oil painting, "The Julia Set," which she based on a computer-generated image representation she had seen of this mathematical fractal. Regarding her oil painting, she wrote, "this painting tries to allude to a never ending Julia Set pattern in a visibly ending, physical form." Shortly after its creation, this painting was included in an exhibit organized by Grinnell's Art Department, and the painting now resides within one of Grinnell's introductory computing teaching-labs.

Note the original program that created the fractal image that Ms. Marolt Sonneschein observed was written in the Scheme programming language, and traces of Scheme code may be seen within this painting of the Julia Set!

The author congratulates Lea Marolt Sonneschein for her creation of the painting, and extends his sincere thanks and gratitude to her for allowing this work to appear on the front cover. Thanks also are due Jennifer Weinman, who took numerous photographs of the painting.

Publishing Support

Sincere thanks to the team at Taylor and Francis for their outstanding sense of vision, insight, and professionalism. They present a wonderful sense of collaboration, as manuscripts move from initial proposals through final production, and this project has benefited greatly from all those concerned, particularly the following.

---

[2]Ms. Marolt Sonneschein graduated as a computer science major from Grinnell in May 2015.

- Randi Cohen, Senior Acquisitions Editor, who shepherded this project from initial inquiries through the development process and on through marketing and distribution.
- Shashi Kumar, Cenveo Corporation, who handled all matters LaTeX-related (e.g., style, macros, troubleshooting, consulting) throughout this writing project.
- Robin Lloyd Starkes, Project Manager, Production, who oversaw the myriad details involved in taking the completed manuscript and moving through final publication.
- Lee Baldwin, copy editor, whose keen insight, outstanding mastery of written communication, and fine sense of balance had a substantial impact in refining the manuscript that evolved to the final book.

I have thoroughly enjoyed working with each of these people, together with their staff colleagues, throughout this project.

Copyright Matters

Many of the chapters in this book have appeared previously as columns.

- Articles between 1997 and 2009 appeared in the SIGCSE *Bulletin* as columns on *Classroom Issues*, with pieces normally appearing twice a year in June and December. A typical statement of copyright in the front matter of these issues states,

  Who owns the copyright?

  All nonreprinted material in *Inroads* is the intellectual and literary property of the authors. Original articles are unrefereed (unless otherwise noted), and are not copyrighted by ACM. Permission to reprint an article must be obtained from the author. All opinions are those of the author and do not necessarily reflect the position of ACM or SIGCSE .

- In 2010, the extended format of the SIGCSE *Bulletin* shifted to a new ACM magazine, called *Inroads*. Many articles written in 2010 or later appeared as columns on curricula (*Curricular Syncopations*) in June and December, and as columns on in-class teaching (*Classroom Vignettes*) in March and September. Within ACM *Inroads*, each article ends with the statement,

  Copyright held by author.

In summary, according to ACM policy, the author retains copyright control for all columns, and each published column confirms this arrangement.

I

Introduction

FROM MY PERSPECTIVE, TEACHING EFFECTIVELY involves elements of both science and art as courses address subject matter content. Beyond principles and practices for education in general, teaching computing also must take into account

- a rapidly-evolving discipline,
- the need for lab-based activities, often involving collaboration and/or pair programming,
- frequent dependence upon hardware and software for presentations, examples, labs, and student assignments, and
- the common use of integrated environments for problem solving and programming.

As this description of teaching suggests, becoming an accomplished teacher of computing requires not only mastery of content but also extensive knowledge of educational research, effective practices, insights into student learning, and numerous other factors. *But how does a beginning faculty member gather this extensive background?*

This book provides a solid foundation, as instructors work to master this background. Since both computing education and computing faculty evolve in their understandings, no single source can hope to completely cover all components of teaching. However, this book represents a resource that organizes and presents much material, with *down-to-earth discussions informed both by educational research and by experiences of faculty throughout the teaching community*.

To further explain the motivations, roles, and structure of this book, this *Introduction* reviews common teaching-oriented paths for college teachers (including my own experiences), clarifies the context for this book, considers the role of this book as a resource, and discusses the top-down structure for the remainder of the book. More precisely, this *Introduction* utilizes the following structure:

- Motivations
  - ◇ Common experiences as instructors start teaching
  - ◇ My personal teaching development
- Context
  - ◇ The science and art of teaching
  - ◇ Variables in teaching

- This book as a resource: Balancing research and personal experiences
- The book's general top-down structure
- Using this book

Getting Started in Teaching

Many computing faculty report common experiences as they progress from graduate school through their first teaching positions. For many future computing teachers, graduate school focuses upon essential subject matter content. Graduate work immerses students within the discipline, providing both a solid breadth of background and deep, state-of-the-art knowledge in a specific subdiscipline. Masters' graduates typically have substantially more subject-matter knowledge than those with only bachelor's degrees. Doctoral degree graduates have pushed knowledge forward by completing a substantial research project.

For many, however, graduate programs provide students with relatively little background in the science and art of teaching. At least in North America, graduate students may serve as teaching assistants, with responsibilities to teach recitation sections, grade, and provide tutoring or lab assistance. But in many contexts, assistants receive little training or guidance for their teaching assignments.

In principle, as they begin teaching, new faculty could explore the computing-education literature that contains a vast array of research studies, experience reports, curricular recommendations, and other materials . For those beginning teaching careers, and even for mid-career faculty, however, the scope of this literature can be daunting. Thousands of articles discuss research studies and experiments. National and regional organizations discuss perspectives and recommendations regarding curricular choices. Eventually, a new faculty member may be able to delve into this wide-ranging literature.

However, in practice, a new faculty member must get started in teaching specific courses, and immersion into the computing-education literature may receive a secondary priority—after immediate, day-to-day demands.

My Personal Teaching Development

As I reflect upon my experiences in teaching over the years, some early elements may be typical for graduate students and beginning faculty members. From an early age, I was interested in teaching, but as a student, I focused on my course work in my classes; I did not think much about how teachers were handling courses—I accepted course structures and tried to study effectively. Then, when I began teaching, I focused on mechanics: specific content, course organization, preparing for successive classes, and interacting with students. Although I observed what happened in courses, my initial analyses largely related to details and short-term adjustments. Over the years, I also could reflect on earlier experiences, gaining insight on what I learned throughout my teaching careers. The notes that follow in this section combine some description of my experiences with subsequent reflections.

Pre-college and College Experiences

As a child, I naturally observed my mother, who truly loved teaching. Before she married, she taught both mathematics and business in high schools. After I arrived, she did some substitute teaching, and later she taught bridge or other classes to friends and groups.

Lessons Learned:

Although I never attended a class taught by my mother (at least that I remember), two qualities seemed to pervade her teaching.

- My mother prepared carefully: materials and/or handouts, lesson plans, examples.
- My mother thrived on student-teacher interactions—the process of learning within the classroom. She gained energy and experienced excitement from the process of education, especially face-to-face connections with students.

My first actual experience teaching arose as a high school student, when I taught swimming in the summers for the Town of Concord, Massachusetts (in Walden Pond). At first, I taught beginning students (under the eye of experienced Water Safety Instructors), and over several years I progressed to teach intermediate-level and pre-lifesaving classes.

Lessons Learned:

Swimming lessons for the Town of Concord followed mastery levels from the Red Cross, and each class focused on specific skills that students would work to master.

- To master skills, each class session needed to follow a progression of activities. Students did not learn everything all at once, but steady and sequenced practice allowed students to progress nicely over several weeks.
- An instructor should pay attention to each individual student: what are they doing properly, what misconceptions they might have, what refinements might be helpful? In swimming, such feedback may apply to physical motions, but the same principle of steady student interaction seems to apply in other fields as well.
- Children learning a prone float (often called a "dead man's float") can have substantial trouble if they giggle through much of a lesson—something about breath control.

The summer after my first year in college, I worked as Assistant Waterfront Director at Camp Mattatuck, a Boy Scout camp in Connecticut. As preparation, I had earned a Red Cross Water Safety Instructor certificate and several Boy Scout credentials. The challenge was to engage junior-high and high-school boys in a constructive swimming program.

Lessons Learned

- Expanding my experience from Concord, an instructor needs both specific content knowledge, but also a sense of how topics and skills might fit together into a constructive progression of lessons.
- When an instructor can connect with students (e.g., by leading cheers during meals at a Boy Scout camp), students often are energized to engage planned activities, having a substantial impact on teaching and learning.

Throughout my K-12 and college schooling, I was extremely fortunate to have many wonderful teachers. As a student, I often took course formats and activities for granted, without question, but upon reflection, a few teachers were particularly effective.

Lessons Learned

- Norton Levy (9-12 grade math) continually worked to challenge his students, and he maintained high standards. Beyond on-going assignments, he posted weekly "Thought Questions", varying greatly in difficulty. I solved some and not others, but I had enough success to keep trying. When I succeeded, a "good work" comment meant a great deal.
- Eleanor Hoogheem (9-12 grade dramatics) directed stage productions, but also helped pupils observe reactions of an audience. Dramatics does not take place in a vacuum. Ignoring the obvious joke about not being able to breath in a vacuum, dramatics seeks to establish a connection between those on stage and those in the audience.

- Neil Grabois (college math) challenged students. He pushed me (and other students) to a limit—[mostly] manageable, but beyond a comfortable level. Finding an appropriate limit is hard, but pushing students close to their limit pays substantial dividends.
- Victor Hill (college abstract algebra) highlighted benefits of being remarkably prepared every day. Material was organized in great detail, with every piece (e.g., definition, theorem, example) presented at the just right time.

Graduate School

During my first three years as a graduate student, I had an N.S.F. Fellowship. In this capacity, I observed different teaching practices, but had no teaching responsibilities.

Lessons Learned

- Class preparation shows. When an instructor is well prepared, organized courses may progress well; when instructors are less preparation, courses may flounder.
- James Munkres (instructor, differential topology) demonstrated that extensive preparation can yield elegant lectures and classes. When combining high-level concepts and technical details, an entire subject can come together beautifully for students.

Toward the end of my graduate studies, I was fortunate to be a teaching assistant for M.I.T.'s Experimental Study Group, where I was able to discuss approaches to teaching with wonderful mentors, especially Professors George Valley and Robert Halfman.

Lessons Learned

- In encouraging student learning, an instructor can be available, respond to questions, provide careful feedback, take the initiative to help guide students, etc.
- An instructor cannot make students learn. If a student has other foci and does not want to engage the material at hand, an instructor cannot learn the material for the student. As professors Valley and Halfman observed, if a student works hard to fail a course by doing nothing during the semester, then it seems only proper for them to be successful in their goal to fail the course.

For several summers, while still at M.I.T., I was allowed to teach an introductory course on my own. For the most part, I had little guidance about how to teach—mostly I knew the course description and the record of textbooks used previously. Also, I had heard comments from students on other campuses that some instructors would delve into early topics at great length and then be unable to cover other topics later—either overwhelming students with work at the end of a semester or leaving students unprepared for later courses.

Lessons Learned

- A careful syllabus and day-to-day schedule, set up before a semester begins, can have a dramatic impact in successfully moving through a course.
- With course organization [largely] completed before a semester begins, an instructor can focus on day-to-day details through the semester without being overwhelmed
- Modest adjustments can be made in a schedule, in response to student progress and challenges, but an overall schedule provides a fine overall plan.

For the most part, courses went reasonably well; with ongoing observation of class sessions, I was able to refine some teaching basics. Interestingly, conversations with the head of the undergraduate program about a decade later indicated that the day-to-day schedules I developed as a graduate student were still being used a decade or so later.

Grinnell College and Initial Outside Contacts, 1974-1985

When I came to Grinnell College, I needed to teach courses from the moment when I arrived on campus, and my development included elements of trial-and-error. Also, my colleagues at Grinnell, particularly Eugene Herman and Charles Jepsen, were actively engaged in national discussions of the teaching of mathematics, and they served as wonderful colleagues and mentors, providing extensive background and insight.

Lessons Learned

- Each math faculty member at Grinnell was very effective, but each had a different personality and style. Faculty member used contrasting techniques that worked well for them, but likely not have succeeded if tried by others.
    - ◇ Charles Jepsen utilized a quiet wit and subtle manner to connect with students.
    - ◇ Eugene Herman emphasized graphics and promoted discussions.
    - ◇ My style evolved to include elements of demonstrations, theater, and humor.
- Collaborating on a course (e.g., choosing a textbook or planning a day-by-day schedule for multiple sections of a course) can provide wonderful insights regarding both what content might be covered and how that material might be taught.

In addition to on-going conversations with Grinnell faculty, I started connecting with national groups, with early contacts with both the Advanced Placement Program (AP) and the Special Interest Group on Computer Science Education (SIGCSE) of the Association for Computing Machinery (ACM).

Starting in 1975, I became involved with grading Advanced Placement (AP) examinations, first in mathematics (for 6 years) and subsequently in computer science (now for 30+ years). Participants at the AP readings were all experienced teachers who were teaching the same introductory courses (e.g., calculus or CS1), but the courses were offered in different settings (e.g., public high schools, private high schools, colleges, and universities).

Lessons Learned

- The development and utilization of rubrics in the grading process can yield consistent, fast, and fair grading results for either small classes or large-scale exams.
- Different experienced teachers may have remarkably different views regarding appropriate scores for specific student work. Although such differences are resolved in the grading of individual AP problems, experienced teachers (when grading on their own) may differ in the scores they assign to student work by a factor of four or more!
- Both high school and college faculty address similar problems. Content for AP courses is quite similar to introductory college courses, and challenges for teaching that content often are similar at the two levels.
- Environments different among private and public secondary schools and among types of colleges and universities. Each environment has its own constraints and opportunities. Thus, approaches in one setting may require adjustment in another environment, but ideas to address challenges may apply broadly (with adjustments in details).

Starting in 1979, I joined the Special Interest Group on Computer Science Education (SIGCSE) of the Association for Computing Machinery (ACM), attending numerous conferences, listening to new ideas and reports of educational research, and interacting with computing faculty around the country and world.

Lessons Learned

- Sessions at conferences are valuable in identifying priorities for the discipline and for courses—both related to content and pedagogy.
- Conferences and meetings are at least as important in promoting networking and encouraging interactions as in organizing formal sessions and presentations. Faculty can share challenges and brainstorm solutions.
- Involvement with SIGCSE (as a national organization) provides wonderful opportunities to connect with colleagues around one's region, the nation, and the world.
- Although working on committees and in organizations is very rewarding, a faculty member also should be realistic; saying "no" from time to time is not a bad thing (but I'm a terrible role model)!
- When serving in some official capacity for an organization or conference, organization members or conference attendees may take any statement as official policy. Thus, as a leader (e.g., SIGCSE Chair), I learned I could not make a joke or cute off-hand aside; anything I said was considered a SIGCSE pronouncement.

Connecting with the Broad Computing-education Community

Many academics have observed that a faculty member's career often proceeds through several phases. Typically, early stages focus upon basic components of teaching and scholarship. Commitments to service and the broader community frequently expand in later phases. Certainly, this pattern is clearly seen in my career path.

At Grinnell, as expected, I became progressively more involved with college service, including numerous committee assignments. Outside Grinnell, service included work with AP CS and with SIGCSE, among other activities.

Turning to outside activities involving teaching and learning, two groups became particularly important in my career and development. Starting in 1985, I was invited to join the Liberal Arts Computer Science Consortium, a focused group of faculty interested in curricula, pedagogy, laboratory experiences, facilities, and related matters within small, liberal arts colleges (e.g., see [70, 283, 109]). Also, in 1994, I organized the Iowa Undergraduate Computer Science Consortium, an informal group for computing faculty within Iowa that has continued to meet on various Iowa campuses almost every year since. Each group is relatively small (e.g.,15-30 attendees at a meeting), and that allows substantial conversation, networking, brainstorming, and collaboration.

Lessons Learned

- When teaching day after day, semester after semester, a faculty member can become focused upon course details and pay relatively little attention to high-level developments in content, pedagogy, laboratory activities, etc. Regular interactions with colleagues in groups outside one's campus can help an instructor maintain perspective and currency. (Periodic meetings can act as an interrupt to one's daily grind.)
- When schools are spread throughout a relatively large geographical area, faculty may feel isolated, and travel to existing conferences (sometimes in urban centers) can be time consuming and expensive. In such cases, when nearby meetings do not seem to exist, an interested faculty member can simply start a new group.
    - ◇ Pick a date, invite faculty within a few hours drive, and ask for suggestions for speakers and panels.

◇ If the meeting seems to address a need, ask what attendee(s) would like to host the next meeting, so the group can continue.

- Regional conferences by the Consortium for Computer Science in Colleges (CCSC) can address this need in some areas, but travel still may be unrealistic in other regions.

In addition to the development of groups, starting in the mid 1980s, I had the opportunity to teach (mostly during sabbaticals and in the summer) at the University of Texas at Austin (under the marvelous mentorship of Nell Dale), Nanjing University (through the Grinnell-Nanjing Faculty Exchange Program), UNITEC (in Auckland, New Zealand), and most recently at the University of the South and Williams College. Such opportunities have helped me experience some of the range of environments present in colleges and universities.

Also, since 1991, I have participated in external reviews of 40+ departments of computer science or mathematics/computer science. Since each review involves a campus visit, I have had the special opportunity to talk with computing faculty, students, administrators, faculty outside computing, and others, learning about programs, teaching approaches and opportunities, constraints, and ideas for the future.

Lessons Learned

- With each campus visit, I learn more about curricula, pedagogy, and alternative ideas related to teaching.
- Each school and campus has remarkably different missions, goals, opportunities, and challenges. Content and pedagogy that works well in one environment may or may not fit well in other settings. Assumptions from one environment often do *not* carry over to other settings.
- Even traveling from one small liberal arts college (e.g., Grinnell College) to another highlights remarkable differences in the goals of college and programs, faculty interests, and environments. For example, faculty at some schools seem very collaborative, wheres faculty at other schools may prefer work in isolation or with just a small group. Of course, levels of intramural politics also may vary greatly.
- Academic structure, funding, facilities, faculty appointments and promotions, classroom pedagogies, and faculty support vary widely—particularly among different regions of the world.
- Students at various schools have different interests, motivations, priorities, and goals. For example, I have been on campuses on which students were motivated by some combination of jobs and careers, grades, intellectual challenge, and service to others. However, in my experience, only a few of these areas seem primary motivators on any individual campus.

Overall, over 40+ years, I have learned much about educational theory, research studies, experiments within computing courses, experiences that seemed to work and not work, etc. Throughout my teaching and interactions with both students and colleagues, I find I am often most effective when the atmosphere integrates seriousness and fun. I work heard to be effective, creative, and colleagial, but I also want to enjoy the process.

Context for this Book: The Science and Art of Teaching

In considering my evolving understanding of computing education, much insight comes from conversations with computing faculty, but much additional background comes from published articles and reviewed proceedings. From my perspective, much of the literature regarding teaching in general and teaching computing may be organized into two main categories:

- *The Science of Teaching:* Many studies regarding education in general and computer science education in particular identify teaching techniques, approaches, and environments that are shown to be very effective or quite ineffective. For example,

    ◇ much research has demonstrated that active learning with strong student engagement is dramatically more effective than a lecture format; studies with different types of schools, different course formats, different student populations, and course environments consistently demonstrate that students learn much less with lectures than with activities that actively engage students.

    ◇ extensive research also has identified separate learning styles among students. For example, some students respond particularly well to words and sounds (auditory learning), others to visual images and graphics (visual learning), some to the process of reading and writing (learning through reading/writing), and some to the process of actions and active movements (kinesthetic learning). Students typically can learn effectively in a variety of settings, but they may strongly prefer one approach to others.

  Overall, the research literature can provide much insight and guidance for teaching.

- *The Art of Teaching:* Experience reports and conversations with colleagues provide insights regarding addressing local and cultural opportunities and constraints, student motivations and interests, classroom customs and instructor personalities, expectations for homework and workloads, etc. For example,

    ◇ I have used a lab-based pedagogy (a type of "flipped class") at several schools. In this approach (to be described more fully in Chapter 28), students are expected to read materials before class, delve into laboratory exercises (with a partner) during class, and finish any remaining lab questions for homework. At one school, students consistently adopted this approach: they came to class prepared, worked collaboratively and asked questions during class, and completed any unfinished work before the next class meeting. At another school, students consistently ignored the prescribed readings and expected the instructor to spoon feed them needed material at the start of each class session. At a third school, most students did some class preparation, although the level of study varied considerably. Altogether, student responses to the same pedagogy varied dramatically in different local settings.

    ◇ When teaching at Grinnell, first in mathematics and then computer science, I observed that each faculty member had remarkably different teaching styles. Some were extremely outgoing, some quiet, some reasonably formal, some used a dry humor, etc. Further, each faculty member utilized different pedagogy within their courses. Yet all were highly regarded by students. In considering the teachers involved, I noted that each drew upon their individual personality and traits: each instructor had developed a style that worked very well for them, but no instructor would likely be effective mimicking the style of another.

Variables in Teaching

Throughout my interactions with colleagues on different campuses, the interplay between the science and art of teaching continues as a recurring theme. Faculty need to understand

and adopt teaching principles, as revealed in numerous research students—both for education in general and for computing-education in particular. However, teaching also must fit within a specific context: a group of enrolled students work within a local or articulated context to study and learn identified material. Student details, available equipment, classroom and lab environments, school policies, etc. may change from one offering of the course to another. For example, students in different sections of the course will have varied backgrounds, interests, cultures, personalities, and life circumstances (e.g., family commitments). To be effective, an instructor may need to adapt to all of these circumstances; experienced faculty often report how an approach to a course was wildly successful one semester, but had only modest success another semester—even when the instructor thought that the same pedagogy and approach were used each time.

To expand upon this point, I think of teaching as an ongoing challenge of discovering how to connect with a specific collection of students in a particular time and place. From this perspective, effective teaching likely depends upon responding creatively and constructively to a remarkable range of variables, including

- the subject matter,
- both general and computing-related educational research,
- course formats, classroom activities, and other pedagogical approaches,
- teaching facilities available (e.g., classrooms, teaching labs) and room configurations (e.g., arrangement of desks, chairs, and equipment),
- class sizes,
- student academic backgrounds,
- student motivations, interests, and workloads,
- student cultural context,
- personal learning styles,
- an instructor's personality, and
- available personnel for in-class assistance, for grading, and for tutoring.

Pragmatically, an instructor can work to master the subject matter and, over time, the literature for both general educational research and computing-education research. An instructor can seek to understand local circumstances, customs, expectations, opportunities, and challenges. Preparation for teaching is important and helpful.

However, an instructor cannot control all factors within a course; in many cases, an instructor cannot either know or guess various qualities for students enrolled in a course. Rather, an instructor can proceed according to some initial assumptions and guesses, and the instructor can try to make adjustments and refinements as a course progresses.

This Book as a Resource: Balancing Research and Personal Experiences

Following the threads already presented in this *Introduction*, this book integrates three categories of primary sources in discussing the teaching of computing:

- research literature regarding educational research in general and computing-education research in particular,

- experience reports, including published articles and conversational statements, that describe challenges, approaches, successes, and learnings, and
- my own experiences and observations, based on 40+ years of teaching and interactions with others.

To highlight this mix of perspectives, the bibliography for this book identifies approximately 250 different articles and reports and almost 40 books. Of these articles and reports,

- about 55 represent my perspectives and experiences (e.g., my columns),
- about 40 represent papers I have authored or co-authored within a substantial peer-review process at conferences,
- about 15 represent curricular reports or recommendations from professional societies, and
- the remaining 140 or so papers include a range of research articles and experience reports.

Altogether, this book presents a wide range of different perspectives, just as my columns seek to cover multiple viewpoints and opinions. Even with this interest in a diversity of ideas, however, I am nevertheless the narrator and guide within this book. Throughout, I have worked hard to present options and to consider computing education within as broad a context as possible—using my conversations with hundreds of computing faculty to identify and clarify multiple prospectives. Ultimately, however, the emphases and conclusions presented are of necessity my own opinions.

Drawing these resources and perspectives together, this book is intended as a practical, down-to-earth resource to help beginning teachers get started and to organize discussion and reflection for experienced teachers. Throughout, discussions in the book draw upon both general educational research and specific insights and practices within the realm of computing education; studies throughout academia provide insights and perspectives for teaching, and computing faculty benefit from learning about this general scholarship. In addition, computing education also incorporates software-development methodologies, individual and group problem solving, algorithmic thinking, experimentation, interactions with hardware and software, use of [software] tools, and local equipment and facilities.

Overall, the book integrates research and experience to provide ideas, perspectives, and down-to-earth suggestions for teaching computing.

The Book's Top-down Structure

This book generally is organized following a top-down hierarchy into themed parts, which progress from general principles for an overall curriculum to considerations of an entire course and then to specifics within each course. Adding some detail to this high-level description, the parts of the book (after this introductory Part I) highlight four main themes.

- *Theme: The Overall Computing Curriculum*
  Considerations for the shape and structure of the curriculum as a whole
  - ◇ Part II: Curricular Development
  - ◇ Part III: Courses and the Computing Curricula in Context
  - ◇ Part IV: Curricular Issues
  - ◇ Part V: Computing and Mathematics

- *Theme: Course Development and Planning*
  Factors involved with initial development and planning of individual courses
  - ◇ Part VI: Course Formats: Basics, Lab-based, Active Learning, Flipped Classrooms
  - ◇ Part VII: Preparing a Course
- *Theme: Course Details and Day-by-Day Activities*
  Components that constitute the ongoing activities while a course is in progress
  - ◇ Part VIII: In the Classroom
  - ◇ Part IX: Exercises and Assignments
  - ◇ Part X: Student Progress in Courses
  - ◇ Part XI: Assessment and Grading
- *Theme: Further Considerations*
  Elements that impact relationships between computing and other academic areas or that seem separate from general curricula or specific courses
  - ◇ Part XII: Outreach and Public Relations
  - ◇ Part XIII: Additional Topics

Within each theme and part, several chapters provide background and perspective, drawing upon both educational research and experience (both from published reports and my own observations). In addition, eight Parts contain chapters that provide selected references for further exploration, with short annotations for each suggested work.

Chapters and Columns

As already noted, this book arises from the 50+ columns I have written for the *SIGCSE Bulletin* and *ACM Inroads*, starting in 1997. However, over the 2+ decades since these columns appeared, computing and computing education have evolved substantially.

With this context, this book seeks to honor historical development and also consider contemporary understandings.

- The book is organized into chapters that largely reflect the topics of my columns over the years (although a few column titles have been updated).
- The first section of most chapters presents my original column, with only minor editing (e.g., slight rewordings, adjustment of references to reflect currently available materials).
- For many chapters, the original column is followed by a subsection on *Subsequent Reflections.* This subsection updates earlier comments and provides some of my more recent thoughts on these topics.

With this combination of original columns and additional ideas, the chapters preserve the historical framework of the columns, as they appeared, and also reflects more recent understandings of teaching and learning.

Beyond this dominant composition of chapters, about ten chapters are new—covering topics that have not yet appeared in my columns, and about seven chapters provide "Selected/Annotated References" for further reading.

Using this Book

Although the body of knowledge for teaching computing is vast, teachers need to address specific questions and challenges in a day-to-day setting. Thus, the book is organized into short chapters, which target specific elements of teaching.

In addition, this organization could work well with a group of faculty interested in an ongoing seminar on computing education. Each seminar meeting could focus on one or two chapters, with the entire book serving as a reference for several months of discussions.

In summary, although the book could be read cover-to-cover from front-to-back in a systematic way, many readers, particularly new teachers, may want to focus first on immediate needs and situations.

II

Curricular Development

DEVELOPING AND EVALUATING a complete computing curriculum can seem a daunting task. A full computing curriculum must fit within the mission of the program and school, meet an extensive list of student learning outcomes, integrate numerous essential topics within cohesive courses, and package those courses into a properly structured framework. Altogether, curricular development requires extensive time and energy, as an articulated curriculum must combine many threads into a coherent and manageable tapistry that serves the needs of both an institution and its students.

Computing programs often state high-level goals and objectives, and colleges and universities typically frame their educational objectives within an over-arching mission statement. Such declarations provide an overall framework for each program within a school, informing the development of curricula and courses.

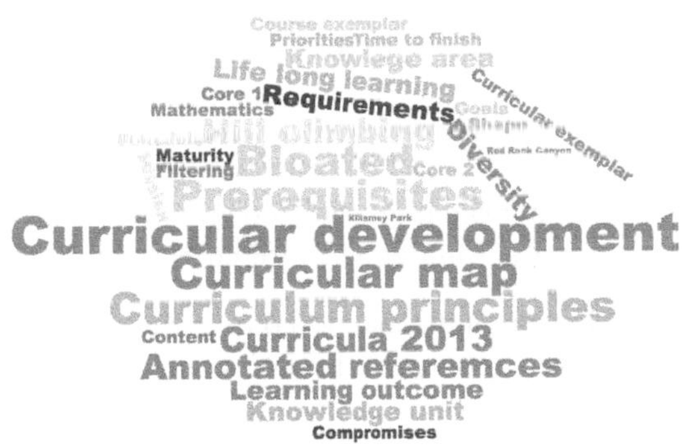

FIGURE 2 Curricular Development Word Cloud

From another perspective, the Merriam-Webster Dictionary defines *curriculum* as 1) "the courses offered by an educational institution" or 2) "a set of courses constituting an area of specialization" [125].

Interestingly, the curricular recommendations, *Computing Curricula 2013 (CS2013)* [10], of the Association for Computing Machinery (ACM) and the Computer Society of the Institute of Electrical and Electronics Engineers (IEEE-CS), follow a different approach. CS2013 frames its curriculum as a collection of topics and student learning objectives—organized into several levels. Thus, the "CS2013 Body of Knowledge is organized into a set of 18 Knowledge Areas (KAs), corresponding to topical areas of study in computing" [10, p. 14] "Each KA is further organized into a set of Knowledge Units (KUs)" [10, p. 32]. Within each Knowledge Unit, CS2013 lists a collection of topics, several learning objectives for these topics, and the number of hours recommended for the KU. Altogether, *Computing Curriculum 2013* provides recommendations at an extraordinary level of detail:

- 18 high-level subdisciplines or Knowledge Areas (KAs)
- about 85 medium-level topic categories or Knowledge Units (KUs)
- a few hundred detailed listed topics, divided into three basic categories:

  - ◇ Tier 1 Core: topics considered essential for every graduating student
  - ◇ Tier 2 Core: although not all required, graduating students should cover 80% to 90% of these topics
  - ◇ Elective: CS2013 expects some elective, but considerable flexibility is allowed

- 280-408 classroom hours, tabulating both Tier 1- and Tier 2-Core topics
- about 1050 student learning outcomes [10]

Altogether, CS2013 devotes about 227 pages to its full specification of material required for an undergraduate computer science program. In addition, CS2013 contains about 235 pages describing about 80 course exemplars that highlight innovative and successful ways to combine various topics, and about 52 pages describing five curricular exemplars that illustrate actual school curricula that cover the CS2013 curricular recommendations. [10]

Putting these strands together, a computing curriculum may be considered as an integrated collection of courses which support overall goals and objectives for both a program and a school. Courses, in turn, may be considered a collection of topics and activities that address student learning objectives at a detailed level.

Parts II, III, and IV within this book review elements of a computing curriculum and computing course content, starting generally at a conceptual level and progressing steadily to considerations for courses and then specifics within courses. Part V then connects computing curricula and courses to mathematics, as a subject both that provides support for computing and also that computing supports.

This Part (Part II) begins a discussion of curricula by examining high-level factors that relate to a curriculum as a whole. Some basic questions follow.

- What principles, beyond high-level goals and school mission statements, might help shape an undergraduate curriculum?
- How should separate courses be placed within a curriculum to allow appropriate transitions among courses or groups of courses?
- How can courses and an overall curriculum be evaluated with respect to what topics are (and are not) covered?
- What processes might be appropriate for reviewing and adjusting curricula?
- How might courses within a computing program relate to each other and to recommendations for an undergraduate degree in computing as specified by ACM/IEEE-CS in CS2013 [10]?

Chapters within Part II address all of these questions, and the final chapter also provides suggestions for further reading.

CHAPTER 1

Eight principles of an undergraduate curriculum

This column first appeared in ACM *Inroads*, Volume 1, Number 1, March 2010, pages 18–20[232] Subsequent reflections on this subject appear later in this chapter.

1.1 ORIGINAL COLUMN

THE DEBUT OF THE NEW ACM INROADS magazine seems a good time to reflect upon the principles of an undergraduate curriculum. A computing curriculum might have many high-level goals, such as capturing student imagination, covering core topics, preparing students for diverse careers, and establishing skills for life-long learning.

In my experience talking to computing faculty from many campuses, however, I have found that details, short-term conveniences, tradition, and local idiosyncrasies sometimes shape curricula in peculiar ways. In order to maintain an overall perspective, therefore, it can be useful to consider the nature of curricula and some basic perspectives for an overall computing education. This column discusses eight of those principles that I have found particularly useful when participating in numerous departmental reviews.

Principles

1.1 Typical undergraduates should be able to complete a computing program in a relatively few years.

During my conversations on one campus, both faculty and students wanted to include every useful skill and topic within their computing major. If graduates might find a topic valuable, it should be included as a requirement. This perspective led to a major with a great many requirements, and the tendency was to add more whenever another topic surfaced. With so many requirements, graduates wanted assurance that they would have all the skills required for any job.

While this viewpoint may seem appealing, the perspective often leads to curricular bloat. Further, most undergraduates may be unable to complete large numbers of courses in a relatively few years. In the United States, for example, students expect to complete a bachelor's degree in four years. Within a degree program, students study computing, mathematics, communication skills (e.g., writing, speaking, working in teams), and a breadth of other subjects (e.g., other science, humanities, and social studies). In addition, many schools have requirements regarding core requirements or the general distribution of courses.

When fitting courses and topics into a schedule that meets school expectations and major requirements, the number of topics and courses in a computing major is inherently limited. Only a modest number of topics can possibly fit into an under graduate program, and not all skills can be covered.

1.2. A curriculum sets priorities and reflects compromises.

Computing Curricula 2001 [12] and other guidelines identify many valuable topics; interesting and worthwhile material extends well beyond what can possibly fit within an undergraduate program. With so much to choose from, any curriculum must involve selection. Shaping a curriculum, therefore, sets priorities and indicates what is important.

Furthermore, any listing of topics should distinguish between what an instructor might mention and what students might realistically learn. At some conferences, for example, I have heard curricular discussions that seem to encourage pushing progressively more material into an undergraduate program, while ignoring what students can reasonably master in a semester or an undergraduate course of study. One way to view course coverage is to ask what students might be expected to remember a week, a month, or a semester after a course. If taught well, students may be comfortable with ideas and approaches, but various details will almost certainly be lost.

Pushing extreme amounts of material into a course likely will just compound this problem. With courses and topics limited, selection requires considerable care. Several points can guide choices:

- Topics should fit with the interests of students and with their long-term needs. Even if a topic is "hot" or "cool", it will take time to cover, but it may or may not have lasting value.

- Flexibility is vital, particularly at the start of a curriculum. If students in non-majors' courses become excited by the material, they may want to take more. But, if they must start over with beginning courses for majors, they may become discouraged and move to other disciplines. Enrollments in the major may depend upon how well a range of beginning courses can help channel students into later courses.

- Topic selection should consider the level for presenting that material. Each course should have a target audience, and material should be manageable for that group. It's wonderful to challenge students and help them stretch to new levels of sophistication, but the students also need to have a reasonable chance of success. With the need for strong enrollments, courses should not target just the top few.

1.3. A curriculum should meet both departmental and school mission and goals.

Schools should clarify their views of the goal of undergraduate education. One interesting perspective comes from William Wulf, former President of the National Academy of Engineering and recipient of the 1998 SIGCSE Award for Outstanding Contribution to Computer Science Education. Wulf observes, "Every other profession treats at least a Masters Degree as the first professional degree. Engineering is the only discipline that believes that the baccalaureate is a professional degree. I think the fact that we have not faced up to that causes all kinds of foolishness" [290, p.6].

Given their overall mission, colleges and universities should consider how to serve diverse populations. Some programs, however, seem to focus on specific careers. For example, some computing programs seem to assume all graduates will go to graduate school or that all will find employment at a local company. Unfortunately, such assumptions are usually false, and the resulting programs can be very limiting.

In practice, graduates pursue varying careers, so computing programs must provide a variety of experiences. For example, if some students might enter graduate school, they must learn fundamental topics in many sub-disciplines, and they would benefit greatly from a research experience. If some students plan careers in industry, they may need experience working in software-development teams. Of course, personnel, resources, and facilities limit programs. Few, if any, schools can cover all career options, and choices must reflect what can reasonably be scheduled. Ultimately, choices should reflect program priorities.

Even when a curriculum cannot do everything, it can build problem-solving skills, emphasize working a problem from an initial vision through to a fully implemented system, and provide hints of the many interdisciplinary ways computing can help people in diverse fields solve their problems.

1.4. A curriculum should provide a foundation for long-term study, professional involvement, and productivity.

The field of computing changes rapidly, and we cannot anticipate all future innovations and technologies. Further, today's practices will be replaced by new viewpoints. For example, based on a calculation from a 2000 workshop, William Wulf reported the "half-life of engineering knowledge" as between 2.5 years and 7.5 years, and Wulf concluded, "half of what we are teaching our students in some fields (computer science, by the way, was the field of 2.5 years) is obsolete by the time they [students] graduate" [290, p. 6].

If students are to be prepared to succeed in this environment and to meet future challenges, they will need the skills to remain abreast of new developments. They must be able to learn on their own, and they should be able to place new ideas within a framework of solid principles. A program in computing should provide this foundation.

1.5. A curriculum identifies core topics for all majors.

Programs typically identify some skills, principles, and perspectives as being fundamental; all students must master these ideas. For example, elements of CC 2001 [12] or LACS 2008 [109] recommend that all computing graduates have basic proficiency in such areas as discrete mathematics, programming fundamentals, systems, or algorithms.

However, if a graduate is to have certain backgrounds and experience, the curriculum should clarify where and how that material will be covered. As a contrary example, I know one mathematics program that asserts its graduates should be able to read mathematical texts and journal articles, but all courses spoon-feed students so they never need to read.

Often a curriculum specifies desired outcomes by identifying common courses that all graduates should have. These core courses provide a base for all students.

An alternative would be to require that each student take one of several restricted alternatives. In examining a curriculum, it can be helpful to review various core topics to identify how each subject is covered.

- If students are to be prepared to work in teams, where do they gain this experience?
- If students should develop the ability to learn new topics on their own, how is this skill incorporated into the curriculum?
- If students are to develop strong communication skills (e.g., writing, speaking), which courses or experiences emphasize this work?

1.6. A curriculum identifies options for students with differing interests.

One sequence of courses cannot be expected to satisfy all interests and needs. Many topics are interesting and worthwhile, but students cannot realistically take everything. In considering mathematical background, for example, multi-variable calculus might be extremely useful for artificial intelligence, statistics for systems, linear algebra for graphics, and abstract algebra (group theory) for coding theory.

When different background is needed for some areas, but not others, a curriculum can lay out various options. Each option can build upon the common core, but each option also can select elements that fit together for specific interests.

Further, different students may have distinct educational goals and priorities, so a curriculum may identify multiple levels of involvement. For example, a "standard major" might cover basics within the discipline, an "honors track" might include deeper study of foundational areas, or a "research option" might lead students through a capstone process that ends with a thesis or publication. Regardless of the details, each track should have academic integrity, provide intellectual interest, and generate excitement.

1.7. A curriculum organizes topics into a basic structure, setting prerequisites and co-requisites.

Once identifying topics and options, a curriculum should organize the pieces into a coherent and manageable framework. A curriculum identifies prerequisites and indicates the level expected for each topic and course.

Setting prerequisites can be particularly tricky. Courses with substantial prerequisites can encourage a homogeneous population; each student will have significant background, and the course can proceed at a rapid pace with considerable depth. However, long prerequisite chains limit the potential audience. When specific prerequisites are dropped as being not really needed, courses can draw strong enrollments and include a range of interesting and talented students.

Pragmatically, beginning students have varying backgrounds, and a curriculum should consider how students with different backgrounds might enter a program. Students with little prior knowledge can develop into outstanding majors, but at the start, others can intimidate these students with strong backgrounds.

1.8. A curriculum should chart how varying students might schedule courses throughout the undergraduate program.

Once courses are placed within a curriculum, planners should consider how students will flow through these courses. For example, at one school, upper-level courses (for third and fourth-year students) are offered every third year. This school's catalog indicates a lovely range of courses, but any student will find that some courses are never offered when they have the needed prerequisites. If this represented bad planning by the student, such circumstances may be acceptable. However, if the program blocks groups of students on a systematic basis, then scheduling should be reconsidered.

As another consideration, students at many schools might begin their work in computing in the second semester or even the second year of their studies. Additionally, a school may encourage students to spend a semester off campus. When such options are possible, schedules should be reviewed to ensure students can follow a curriculum in a reasonable way.

Finally, once schedules are reviewed, students need to know what options are available and how courses can be scheduled. Students and their advisors should plan when to take what courses—particularly in conjunction with off-campus study, internships, and other opportunities. The success of planning depends upon the knowledge of requirements, options, and likely course schedules.

Future Discussion

Although not comprehensive, the principles described here provide an overview of useful considerations for the design and implementation of computing curricula.

I expect future columns in this series will focus on specific principles and issues related to curricula. Throughout the series, reader feedback, comments, and suggestions will be most welcome!

Acknowledgment

Many thanks to Marge Coahran for her insightful comments on drafts of this column!

1.2 SUBSEQUENT REFLECTIONS

Developing an interesting, timely, thoughtful, flexible, and well structured curriculum is a vital component of a vibrant and successful computing program. But what other components also might be important? More generally, what characteristics and practices contribute to an outstanding undergraduate program in computing?

To my knowledge, no careful study has emerged to address such questions generally for computing. However, a thoughtful study from the Mathematical Association of America (MAA) tried to address such questions for mathematics programs in 1995 [187]. (A summary of this report appeared in the *Notices of the AMS* in November 1996 [186].) Since this work occurred over 20 years ago and focused on mathematics, specific conclusions may require interpretation and updating for the field of computing today. Perhaps a comprehensive and modern report for computing will be forthcoming, in which case definitive answers may emerge. In the meantime, my reading of the 1995 MAA report suggests several potential parallels for computing.

A natural initial step in studying success of a program would seem to be to define the notion of success. Although each school likely has its own definition, two high-level qualities seem particularly important for the success of a computing program:

- the program should attract and retain a large and diverse group of students, and
- graduates should be well prepared either for graduate school and/or for careers within business and industry.

Interestingly, the 1995 MAA report identifies similar qualities for success, with career preparation for business and industry replaced by preparation for careers in K-12 education.

In exploring qualities of mathematics programs, the 1995 MAA report examined several programs, deemed successful by these standards, and included site visits to ten two- and four-year programs. The intention was "to identify features and practices of those programs that might serve as the starting point for program improvements elsewhere" [186, p. 1356].

Both the full report [187] and its summary [186] identify several common themes for successful programs, but the study also noted there was "no single key to a successful undergraduate program in mathematics" [187]. In reviewing elements of successful programs, some common characteristics might be expected.

- Faculty care deeply about their students and devote considerable time to student-faculty interactions, both inside and outside the classroom.
- Programs focus upon students' needs rather than the faculty's interests and needs.
- Faculty utilize a wide range of pedagogical approaches to address students at all levels, not just the weakest or strongest students.
- Faculty regularly conveyed their interest and excitement about the material being taught.

Another common program characteristic involves the assignment of faculty to courses.

- In these successful programs, introductory courses were often taught by the best teachers. Beginning students may not know what to expect when they begin college courses. When they have strong, positive experiences in a first course, students may be inclined to take the next, and the next, and the next, ...
- Students may be inclined to declare a major, if they have strong teachers in early courses. Moderate or weak teaching may be forgiven at the upper level (majors already are committed to the subject), but beginning students may still be exploring their interests and may be discouraged by poor teaching.
- Good teaching and extensive student-faculty interactions also are identified as providing strong encouragement to women. Curiously, the 1995 report notes that additional factors, beyond good teaching and personal interactions, may be needed to attract students from under-represented groups.

In a recent article reporting on work by Christopher Takacs and Daniel Chambliss, Scott Jaschik writes, "Undergraduates are significantly more likely to major in a field if they have an inspiring and caring faculty member in their introduction to the field. And they are equally likely to write off a field based on a single negative experience with a professor" [93]. Altogether, both the 1995 MAA report and more recent work (e.g, by Takacs and Chambliss) indicate that effective teaching at the introductory level can have a substantial impact on the recruitment and retention of students within a discipline.

Even with these common themes among the successful programs studied, these programs also had substantial differences, and the 1995 MAA report was unable to identify a specific quality that yielded success. Rather,

- "What was a bit unexpected was the common attitude in effective programs that the faculty are not satisfied with the current program. They are constantly trying innovations and looking for improvement" [187].

The interested reader is encouraged to explore the full 1995 MAA report [187] for additional information on factors that may contribute to program success.

CHAPTER 2

Prerequisites: Shaping the computing curriculum

This column first appeared in ACM *Inroads*, Volume 1, Number 4, December 2010, pages 14–16[233] Subsequent reflections on this subject appear later in this chapter.

2.1 ORIGINAL COLUMN

A COMPUTING PROGRAM is almost always more than a collection of disjointed courses. Rather, students typically move from an introductory level to intermediate level to advanced, and each level usually builds upon background and experience obtained at earlier levels. Prerequisites often provide the structure for a curriculum, so prerequisites shape the way students progress through these courses. This column examines possible rationales for prerequisites and some consequences of prerequisite chains. The column then examines more carefully the prerequisites for two specific areas of the computing curriculum: mathematics and introductory courses.

Some Rationales for Prerequisites

Prerequisites can serve many purposes in structuring a curriculum. The following list articulates some of the most common:

- *Content prerequisites:* a course at one level may build on background a student had developed earlier.
- *Maturity prerequisites:* a course may build on general student experience and maturity beyond any specific detailed content.
- *Filtering:* prerequisites may provide evidence of student commitment or may limit enrollment in popular courses.
- *Requirement enforcement:* a program may list specific courses as requirements, and prerequisites highlight these requirements—independent of whether the topics are needed in a particular course.
- *Historical requirement:* Years ago, some faculty determined a prerequisite—possibly to meet a specific need. Now those faculty have moved on, the prerequisites remain (perhaps by inertia), and the current faculty do not really know the intended purpose.

In my experience over the years interacting with faculty at many institutions, I have encountered each of these rationales. Certainly, the first two of these rationales have been most common, and these prerequisites allow students to progress effectively to successive levels of sophistication and depth of analysis. I comment further on all of these rationales as the column progresses.

Some Consequences of Prerequisite Chains

Whatever the rationale, prerequisites have consequences. On the positive side, a prerequisite allows an instructor to assume student knowledge of specific content. As an example, some years ago, my colleagues at Grinnell and I determined that our courses in architecture, operating systems, compilers, and networks all spent several weeks introducing programming in C—largely because these courses were all offered at the same level assuming no prior C experience. Students remarked that the first couple instances of this material were useful, but four repetitions of the same material seemed clearly excessive. As a result, the faculty restructured material in several courses, added an introductory-level course involving imperative problem solving with C, and made the new course a prerequisite of the four existing courses. As a result, the material could be covered thoroughly once, later courses could eliminate their introduction of C, and each later course could cover substantially more material than had been possible earlier.

This example also illustrates a potential negative consequence of prerequisites: students cannot take upper-level courses of possible interest until prerequisites are met, so prerequisites can restrict enrollments. In the example given, the faculty were able to rearrange much material in the first several courses, so the addition of the C-based course could be taken immediately after CS1. With this adjustment, architecture, operating systems, compilers, and networks all continued to have just 2 semesters as prerequisites, as they had before the change. The nature of student background for these courses changed somewhat, but much of the new student background matched nicely with the requirements for the upper-level courses.

Another example illustrates the value of a maturity prerequisite. In particular, I have taught an upper-level algorithms course at two schools. One school specified several computing courses but little mathematics; the other required just CS1/2 in computing, but also four semesters of math (single-variable and multi-variable calculus, linear algebra and combinatorics). While both courses worked well, the nature of the classes was quite different. Analysis of algorithms was somewhat of an effort at the first school; while detailed, formal analysis was natural at the second. At the second school, the algorithms courses used little or no formal content from either calculus or linear algebra, but the mathematical sophistication of the students allowed substantial depth in describing and analyzing algorithms.

These examples demonstrate that the rationales of content and maturity can have substantial benefits. However, prerequisites also can discourage students and restrict enrollments. In times of boom, various devices might be employed to maintain manageable class sizes. However, enrollments in computing are relatively low now; further, extended prerequisite chains (with scheduling constraints) may require a full 3.5 or 4 years for a student to complete. Thus, if a student tries an introductory computing course or a course for non-majors in the second year—or even the second semester of the first year, then long prerequisite chains may require students to commit an extra semester or two to complete a major. Unfortunately, exploration of computing as a side course seems most likely for students who come to college without a prior commitment to computing; and in my experience, demographics suggest that this might impact under-represented groups strongly.

Altogether, prerequisites can serve a useful purpose as students gain background and experience. However, unneeded prerequisites also create barriers that artificially restrict enrollments, interfere with diversity, and extend graduation by a semester or two.

Mathematics Prerequisites

In his September 2010 column on "Mathematics Reasoning in Computing Education", Peter Henderson encourages offering eight mathematics courses within a computing curriculum: discrete mathematics (2 semesters), calculus I and II, linear algebra, probability theory and statistics, at least one advanced mathematics course, and theory of computation. He emphasizes, "Fundamental mathematical concepts should be introduced early and reinforced throughout the curriculum". [86, pp. 22-23] Although each of these courses can be useful, local circumstances influence what colleges and departments can offer and require for some or all students and for specific courses.

Peter Henderson notes "There seems to be consensus on the fundamental discrete math topics, but not on when/and how to fit them all in the course or curriculum" [86, p. 22]. Although there can be pragmatic local issues about when students can take discrete mathematics in their college career, it seems vital that this material be used in some computing courses. This suggests at least some upper-level computing courses have discrete mathematics as a prerequisite. Then, when the prerequisite is stated, it seems important that the upper-level course actually use this material.

Over the years, I know some computing programs have only required discrete mathematics for graduation, not for any specific course. In some other cases, a course may list discrete mathematics as a prerequisite, but that material is never used. From my perspective, both circumstances miss the opportunity to build on important topics, integrate math with computing, and encourage depth in analysis.

Further, many computing educators today argue the need to promote diversity in our field, This raises the issue of what mathematics every computing student should know and what is needed for work in specific fields. Beyond discrete mathematics, why should all computing students learn the same mathematics? Rather, why not acknowledge diversity by listing distinct mathematical prerequisites for selected upper-level computing courses? Students interested in some areas of computing might student some mathematics, while students with other interests would develop different mathematics backgrounds. A graduation requirement might provide options for mathematics, depending upon student interests.

The role of calculus highlights this point. A few years ago, a mathematician argued that all computing students should take calculus, because they would learn L'Hopital's rule, and this in turn would show them that N^2 is larger than N^3 for large values of N. Although calculus can indeed be useful for some areas of computing, this example does not seem compelling—at least to me.

Overall, students can only take a limited number of courses, and students have diverse interests. Rather than have monolithic requirements for all computing students, faculty might consider using prerequisites to help students tailor their mathematics backgrounds to their areas of interests.

Prerequisites and Introductory Computing

In recent years, computing faculty have promoted varying approaches for introductory courses. Some advocate an early emphasis on computational thinking [289], some promote a breadth-first approach with little programming, some seek to combine programming with an application (e.g., robotics, media scripting), some want to facilitate precise problem-solving skills through an in-depth study of a paradigm and associated language, etc. Also, some faculty emphasize outreach to multiple populations, some focus on efficiently moving students to a level of proficiency at which they can address extended problems and projects, etc.

Although such discussions can be worthwhile, the issue of prerequisites is sometimes forgotten.

If an introductory course is to serve as the prerequisite for a second or third course, what content or maturity will that next course be able to assume? Arguably, one problem of some breadth-first courses of the 1990s was that they covered a wide range of topics, but with minimal depth. Thus, subsequent courses could not build on the breadth-first course, but rather had to begin each topic from the beginning. With demands from ACM curricular recommendations (e.g., [7]) for subject coverage, the insertion of a breadth-first course effectively increased prerequisite chains by one course. Such a course also might increase staffing challenges and student graduation requirements, since the breadth-first course might not have replaced something else in the curriculum. For these or possibly other reasons, the inclusion of a breadth-first course in computing was not widely implemented.

Similarly a problem arises if a second course has extensive and specific expectations. For example, if CS2 assumes fluency with a large subset of the Java language, then CS1 likely is highly constrained—CS1 must cover a great deal of Java syntax and semantics.

On the other hand, suppose CS2 is designed to assume comfort with basic problem-solving constructs (e.g., conditionals, iteration, recursion), programming in some language, and application of these ideas in some application domain. In this framework, different forms of CS1 might utilize different problem-solving environments and apply basic constructs in different areas (e.g., robotics, multimedia). Alternative CS1 sections could address interests of diverse students, but all would allow students to delve into problem solving at a significant level. With this approach, care would be needed in planning CS2, so that one CS1 section was not favored over another. However, specification of desired background at a high level might allow multiple entry points to CS2 without channeling diverse students through the same monolithic starting point.

A rather extreme example from biology amplifies this approach of multiple starting points. The biology faculty at Grinnell College believe that the primary goal of the introductory biology course should be to give students significant insight about what it means to be a biologist. That is, students need to delve into a subject, learn about issues and perspectives, form a hypothesis, design an experiment to text the hypothesis, perform the experiment, collect and analyze data, refine the hypothesis, and write the results up as both a research poster and a paper in a form that might be submitted to a journal. From this perspective, introductory biology starts the process of becoming a biologist, including research and publication. As a result, the department offers nine distinct sections of biology over a year, with subjects ranging from prairie restoration to micro-biology. The sections have a little overlapping content, but mostly introductory biology emphasizes a process.

It is not clear to me whether this approach in biology might be adapted to computing. For example, at a basic level, faculty would need to brainstorm what introductory-level experiences might provide insight about what it means to be a computing professional. However, I find the idea interesting to consider.

Overall, these examples illustrate the need to articulate and clarify the nature of the prerequisites for CS2 or later courses. What, really, should CS2 assume? How much diversity in background might be allowed? Could diverse backgrounds in CS2 be used constructively to move that or later courses along?

Acknowledgment

Many thanks to Marge Coahran who reviewed a draft of this column and provided several insightful suggestions!

2.2 SUBSEQUENT REFLECTIONS

The original column appeared in December, 2010—a time of relatively low computing enrollments. In that context, extensive prerequisite chains could restrict already limited student numbers, while few prerequisites could reduce barriers and thus had some potential to increase enrollments.

In the intervening years, however, circumstances have changed. Overall computing enrollments are booming, and at the introductory level (e.g., CS1 and CS2) consideration of prerequisites can help address several challenges:

- Although computing enrollments have soared overall, much effort continues in broadening participation of diverse populations within the computing field. As diverse populations become interested in computing, the backgrounds of students also may become more diverse—particularly at the introductory level. Unless prerequisites are identified with care, students with different backgrounds (e.g., from underrepresented groups) might be discouraged or prohibited from taking beginning computing courses.
- In courses with large enrollments, prerequisites might be adjusted for different sections of the same introductory course(s), so that students in a specific section might have generally similar backgrounds. In such an environment, instructors can have a good understanding of what students know, and courses can be tailored to an audience. When each section has some homogeneity, a course might cover more material and/or provide students more support.
- If an introductory course (e.g., CS1 or CS2) includes some students with considerable background and some with very little, the true beginners can be intimidated by those who seem to know much more. Developing a CS1.5 or similarly enriched course can help challenge incoming students with experience, while it also protects those truly getting started. In this context, the CS1.5 for experienced students might specify a prerequisite, such as "1 or more years of programming in some language," whereas the traditional CS1 for those without experience might indicate "specifically designed for students with little or no programming experience." Details will depend upon local circumstances, but a "negative" prerequisite of "little programming" might help encourage beginners and avoid the undermining of confidence that can arise from those perceived to know much more.

Beyond the introductory courses, a consideration of prerequisites and course scheduling can have a substantial impact on the progress that students can make toward graduation.

- If a curriculum has relatively short prerequisite chains, students may be able to schedule required courses in parallel. If students cannot take one requirement, they might take another without losing time toward graduation. This may help students make progress through a major efficiently, even though some courses may be over-enrolled.

- Flexibility in prerequisites and requirements may allow students to take one course if another is closed due to enrollment pressures. If a specific required course is overenrolled, then students may take extra courses in a secondary area of interest. On the positive side, students can expand their background, even if required courses are closed. On the negative side, taking elective courses may be beneficial, but students taking unneeded courses may compound staffing challenges and enrollment pressures.

Additional comments regarding prerequisites may be found in the subsection on Subsequent Reflections in Chapter 13, "Staying connected with the big picture."

Overall, as the chapter title indicates, prerequisites provide an organizational structure to courses within a curriculum. Prerequisites shape where each course fits, what can be assumed as background for each course, and what material should be covered in one course to provide needed background for what comes later.

As noted in these Subsequent Reflections, prerequisites also can help shape the nature of the students enrolled within a course—particularly at the introductory level. Prerequisites can serve as a barrier to some students who may have backgrounds different from the norm (a potential issue as programs seek to broaden participation). Similarly, prerequisites (or the lack of prerequisites) may bring together true novices and those with considerable background—likely undermining confidence and overwhelming the beginners while creating relatively little challenge to those more experienced. Tailoring prerequisites to different populations for different sections of a course may yield course sections that support students with varying needs and backgrounds.

Finally, flexibility in prerequisites can help students navigate requirements for a major, by allowing them to take one required course if another is overenrolled and closed. Careful consideration of prerequisite chains can maintain needed background for courses while allowing students to make progress toward an undergraduate degree even in times of over enrollment.

CHAPTER 3

When is a computing curriculum bloated?

This column first appeared in ACM *Inroads*, Volume 2, Number 2, June 2011, pages 18–20[240] Subsequent reflections on this subject appear later in this chapter.

3.1 ORIGINAL COLUMN

QUOTES FROM CS STUDENTS AND/OR FACULTY:

- "We're one of the few, one of the proud. We're computer science majors!" [slogan for computing majors on one campus]
- "I survived CSC xyz" [campus T-shirt, for a certain course number xyz]
- "They'll never catch up now!" [instructor comment, after lecturing an hour to an empty classroom during boycott of a course that students thought was outrageously overloaded]
- "Whenever we identify a new topic that would be useful for graduates in seeking jobs, we add a requirement to the computer science major." [computing students and faculty at one undergraduate college]

Before writing this column, a search of the ACM Digital Library with the terms "curriculum" and "bloated" returned 22 hits. The reference to "bloated" largely related to software and hardware, design and interfaces, modeling, industry and government, media, and networks. However, none of these articles focused on curricula. Over the years, I have heard countless discussions about topics that should be added to an undergraduate computing program, but I have rarely heard discussions on the other side—what material is not necessary or might be dropped?

An important issue underlying the opening comments and my observations centers on an inherent conflict for curricula:

- An undergraduate computing curriculum should provide appropriate background for students seeking careers in the field—either in graduate school or in industry.
- A curriculum must be manageable; it must fit reasonably within a program that allows students to graduate in a specified length of time (e.g., four years).

Altogether, the challenge is how to accommodate new material from our discipline in an already full undergraduate computing curriculum. There are roughly four approaches:

- expand the curriculum—adding to the number of required courses or electives
- add more content to individual courses
- become more efficient in teaching and learning within existing courses
- drop other material from courses

Let's examine each of these alternatives briefly.

Expand the Curriculum—Adding to the Number of Required Courses or Electives:

At many institutions, Computing Curriculum 2001 [12] and Computer Science Curriculum 2008 [7] provide a foundation for the computing curriculum. Within these extensive reports, Principle 7 [12, pp. 12-13] seems particularly relevant:

7. The required body of knowledge must be made as small as possible. As computer science has grown, the number of topics required in the undergraduate curriculum has grown as well. Over the last decade, computer science has expanded to such an extent that it is no longer possible simply to add new topics without taking other away. We believe that the best strategic approach is to reduce the number of topics in the required core so that it consists only of those topics for which there is a broad consensus that the topic is essential to undergraduate degrees. ... At the same time, it is important to recognize that this core does not constitute a complete undergraduate curriculum, but must be supplemented by additional courses that may vary by institution, degree program, or individual student.

Although this statement recognizes that a computing curriculum cannot expand indefinitely, it leaves the specific size of a program to individual institutions. Often, a school places specific limits on the size of the computing major, but the culture at some places (among both faculty and students) may dictate that requirements expand when additional useful topics are identified.

Overall, the scope of a program follows from its basic purpose. Many computing faculty could identify worthwhile topics that could extend an undergraduate curriculum to 6 or more years. A basic question, then, is to consider what is most important. If all possible topics cannot be covered, choices must be made, a curriculum must reflect pragmatic constraints and compromise, and changes in the size of a curriculum must reflect adjustments of priorities within the overall structure of a school and its culture.

Add More Content to Individual Courses:

The discipline of computing contains many wonderful topics that might be studied profitably within a course. For example, the popular textbook, Introduction to Algorithms, by Cormen et al [49], examines an extensive range of algorithms. However, the Preface also contains this statement:

> Because we have provided considerably more material than can fit in a typical one-term course, you [the teacher] can consider this book to be a "buffet" or "smorgasbord" from which you can pick and choose the material that best supports the course you wish to teach. [49, p. xiii]

But, how does one decide specific topics? From my perspective, development of a course begins by identifying a few overall themes. What is this course attempting to do? In this regard, two examples may provide helpful warnings about what not to do.

- Traditional calculus courses of the 1970s and 1980s sometimes had the reputation of presenting a new formula each day. "If this is Wednesday, it must be L'Hopital's Rule!" In the 1980s, I reviewed one calculus textbook that contained numerous sections, each highlighting a formula and 3-5 examples that plugged expressions into the formula. The exercises at the end of the section contained 40-60 more expressions which students could plug into the formula. Following the phrasing of the time, the textbook claimed to be "lean and lively", but one of my colleagues quipped that the book was "fat and fatuous". Throughout, no formula was linked to any others, and little discussion linked formulae to underlying themes.
- Another course of the 1980s contained a remarkable smattering of topics. When I asked about this curious list of topics, the computing faculty explained that this course contained those topics that the faculty believed were important, but that did not fit anywhere else. I nicknamed this course, the "none-of-the-above" course.

Such examples highlight the need to explore and articulate the underlying motivation and themes for a course.

Turning to specific topics, my focus in selecting course topics has shifted over the years.

- No longer do I ask, "is this a nice topic?"; I can always list hundreds of nice topics?
- I regularly ask, "How does this topic support the course's main themes?"
- No longer do I focus on what I as an instructor might cover in a course, but rather what each student should learn.
- For each topic, I consider how much course time should be devoted to a topic in order for students to obtain the desired insights and technical abilities.
- Taking a long view, I ask the question, "What should students know from this course 6 months after the semester is over?"

With any new topic, therefore, an early question might be, "how does this topic support a course's overall theme(s)?" Sometimes those themes will require refinement or wholesale change to accommodate the new topic.

In some [rare] cases, a new topic can be added to a course by adjusting examples or reworking a presentation. As a example, loops are typically introduced in CS1, and students need practice constructing programs containing looping constructs. However, an instructor chooses the exercises. An important insight from recent experiments in introductory computing is the realization that careful choice of an application domain can integrate a new theme (e.g., media scripting, robotics) within an existing course by using the application for examples. Although courses sometimes require reorganization, new material could be added by adjusting examples and assignments.

Become More Efficient in Teaching and Learning within Existing Courses:

In many academic settings, faculty have been trying to pack more and more material into existing courses for years, so computing courses are reasonably efficient. Thus, for the most part, if courses could be reorganized or rethought to accommodate new material, that would have been done already.

Even with these general circumstances, however, courses sometimes can be made more efficient. Here are several suggestions for examining existing courses.

- *Examine whether a course lacks focus, energy or logical connections:* ideally, a sequence of topics should build upon each other, adding insight and leading to a satisfying conclusion. Thus, if topics wander without real direction, it seems natural to wonder about the underlying theme; what do these topics contribute to long-term education? Perhaps some topics can be combined or dropped.

- *Review the rationale for each topic:* If a course has been offered several times, the rationale may get lost. Sometimes in an external review, for example, I learn that a topic is included because "we've always done it this way". Since the discipline of computer science changes quickly, a topic may no longer provide the insight and connections that it once did—or maybe the topic is more important than ever, because of new understandings and applications.

- *Identify themes and high-level organization in a course's day-by-day schedule:* Many courses proceed by presenting a major concept or theme and then examining examples and applications. However, faculty can easily become enthralled by many lovely examples. For example, how many examples are needed to demonstrate binary search trees? Is the fifth application really needed?

- *Check for duplication of material among courses:* Even when courses are beautifully planned and coordinated, content may shift over time. New faculty with evolving interests teach the course; and textbooks may evolve. In some cases, new content in one course may duplicate material in another. For example, in a recent chat with a colleague, we discovered that RSA encryption was covered in my course on Theory of Computation and his upper-level Algorithms course. This topic was covered nicely in the textbooks for both courses, and the material could fit in either place. However, there was no need for both courses to cover the topic. As another example, I have found that exit interviews with graduating CS majors periodically identify duplicated material that can safely be removed.

- *Be careful never to introduce a topic, technique, example, etc. that will have to be unlearned later:* I learned this principle from my series editor, Gerald Weinberg, when writing my first book. With so many techniques and principles in computing Weinberg observed that students should never devote time to a technique that is inappropriate for the task at hand. For example, a bubble sort is never the algorithm of choice for any application; a bubble sort is inefficient in every setting. Instead an algorithm with about equal simplicity, such as an insertion sort, can be introduced as being useful in some contexts (e.g., nearly sorted data). Pragmatically, some beginning students try a bubble sort, and then learn their previously-known technique is awful. Similar comments apply to some introductions to recursion. If recursion is only used to compute Fibonacci numbers using multiple recursions (rather than tail recursion), why introduce recursion just to highlight that it does not work well—in that context? How much better to cover useful applications from the start! More generally, if students are not ready for a worthwhile application of a technique, why should it be covered in that course? Why not wait until students can appreciate something actually useful?

Drop Other Material from Courses:

Over the years, I have found several symptoms that suggest a course may be overloaded: high dropout rate, low enrollments, low morale, low homework completion rate, poor grades, and a campus culture that views course completion as a badge of accomplishment—like a boot camp.

More generally, if a course does not seem to be going well or if evaluations are poor, one possible cause may be excessive content. Unfortunately, students are usually not good analysts; they may feel a course is going poorly, but they often lack insight regarding the underlying cause. Thus, when asked to review a course with unfavorable student comments, I consider content, pedagogy, expectations, and many other factors as possible causes. The need to reduce content is one possibly among many.

Often when a course seems overloaded, topics often have expanded gradually over several years, and faculty must rethink the purpose of the course and its main themes. Revisions may involve discarding some topics, moving other topics to different courses (or to new courses), or reorganizing material to support high-level goals and objectives.

Overall, new developments in computing and pedagogy seem to encourage the addition of new material within the computing curriculum. However, additions in one place often require reductions elsewhere. Faculty may need to consider what to cut from a course or curriculum as much as what to add.

3.2 SUBSEQUENT REFLECTIONS

Six years after the original column appeared, conversations with faculty around the country suggest that curricular bloat remains an issue. To keep courses up to date, faculty may add new content and understandings to existing courses, but faculty may be reluctant to remove existing content. As a result, courses contain increasing amounts of material and/or new courses are added, but courses and degree requirements still have the same credit and contact hour requirements. Overall, the potential for curricular and course bloat seems high.

Interestingly, however, the ACM Digital Library still contains few articles related to curricular bloat. For example, the original column in this chapter is the only article returned with both keywords "curriculum" and "bloat" required. The term "bloat" by itself returns about 130 articles, but most relate to software development, software engineering, user-interfaces, or genetic programming. Altogether, the overloading of curricula and courses may present challenges, but apparently few faculty discuss this topic—at least in print.

An Instructive Analogy: a Team's Process in Writing a Conference Paper

In considering how to address curricular and course bloat, some approaches may be suggested by considering the process a team might follow for submitting a paper to a professional conference that places length limits and other constraints upon submissions.

To be specific, the following notes outline the approach I have used successfully several times with student-faculty teams who work with me on various projects. At the end of the research and development stage of a process, my student team and I want to write up our experiences and results, culminating with a concise, clear, and effective paper. Following a reasonably common format, the submission should cover our work in some detail, giving background, outlining what our team accomplished, presenting results (with appropriate evidence), and considering next steps.

For each of my last five or so student-faculty papers, to be presented to conferences in peer-reviewed tracks, I have followed roughly the same overall process.

- At the start, my team and I identify the primary accomplishments to be included in our paper, and we then consider what material might be needed to tell readers about background, motivation, context, approach and future directions.
- As topics emerge, we work to organize the material into primary and secondary themes, identifying a tentative structure outline for the paper.
- With an outline before us, we assign sections to team members.
- Each team member writes a draft for their sections. At this stage, the goal is to include all relevant material from the project; nothing of importance should be omitted, and we do not care about how long the draft sections might be.
- With sections written, they are combined into a single document. At this stage, the draft should contain all content that might be considered.
- Once assembled, team members perform general editing to obtain a revised draft that generally flows well.
- Since various sections have been written by different team members, some sections may overlap. Also, the writer of one section may have assumed that certain material would be covered elsewhere, so the assembled draft needs to be checked for omissions.
- At this stage a reasonable paper has begun to emerge, but it often is quite long. In particular, the current draft is compared with publication limits for the conference. For example, the current work may be 15 double-spaced pages, but the conference may limit submissions to 12. In my experience, at this stage, student-faculty papers often require a reduction in length by 25% or more.
- At this stage, each team member reviews the entire draft for wording and redundancy. Can some phrases or sentences be omitted, moved, or combined? Can wording be tightened. Can a sentence be made more effective and efficient by using active verbs? Can paragraphs and sentences be recast for greater precision? Etc.

Overall, with this type of process, the resulting paper contains the intended content, and the writing is complete tight, clear, and well organized.

In reviewing this process of the writing of a research/development paper by a team, I find numerous parallels for developing both curricula and courses.

- Initially, a faculty group or an individual faculty member identifies the student learning outcomes and the full range of content for a curriculum or course.
- The various elements are structured into courses and course segments, helping to shape prerequisites within a curriculum and the order of subjects within a course.
- Within a curriculum, faculty members may take responsibility for individual courses, and planning begins on the elements of each course.
- Often, faculty determine that more content is compressed into one or more courses, so careful paring, combining, and editing may be needed.

Eventually, the curriculum or course fits appropriately within curricular and course constraints. Redundancies are minimized, courses flow naturally, course segments are manageable and logically connected, and secondary content minimized. A curriculum and each course are tight and clear, and the material fits within the time and space provided.

CHAPTER 4

Hill-climbing with curricula and courses

This column first appeared in ACM *Inroads*, Volume 7, Number 2, June 2016, pages 36–38[266]

4.1 ORIGINAL COLUMN

SOARING IS FUN if you have a good jumping-off point, reliable equipment, and well-developed procedures, and if you are not afraid of heights.
Anonymous—at the request of the author!

It's hard to soar like eagles, when you are in a burning building.
Anonymous—no one claims credit!

Many readers of ACM *Inroads* may be familiar with the Hill-climbing Algorithm. After a quick review of the algorithm, this column notes common uses of the algorithm in curricular development and course refinement. At first, little of this discussion may appear new or different—except the jargon of hill climbing may seem unusual. However, details of hill climbing and its inherent limitations may yield insights into curricular and course development. Throughout, the column presents examples in which well-meaning faculty have applied hill-climbing to curricula and/or courses—sometimes with great success (soaring with eagles) and sometimes with less than optimal results (watching the building burn).

4.2 THE HILL-CLIMBING ALGORITHM

Challenge: In its traditional setting, we consider the problem of starting at the base of a hill or mountain and then finding a path that will lead most rapidly to the highest point.
Algorithm: Hill-climbing makes decisions locally. Over numerous iterations, we identify the direction that leads upward the fastest, and we move a small distance in that direction. (Figure 4.1 shows such a path in Red Rock Canyon National Conservation Area, Nevada.)
Note: When working with surfaces (e.g., in multivariable calculus), determining the desired direction involves computing partial derivatives and combining them into a vector, called a *gradient*.

FIGURE 4.1 Red Rock Canyon National Conservation Area, Nevada, USA. Finding a Path to the Summit, based on Hill Climbing (Photo by Theresa P. Walker; editing by the author)

Iterations within Hill-climbing for Curricula and Courses

Within the framework of curricula and courses, I find it helpful to classify iterations within hill climbing into two categories: planned refinements and crisis-motivated reactions.

Planned refinements: As a teacher, I am never satisfied; I constantly seek to improve courses. For courses, this perspective motivates me to gather feedback throughout a semester and then make adjustments. Translated to hill climbing, I believe a course is never at maximum quality; rather I need to find directions locally that will raise effectiveness. Three examples follow.

- After each class session, I may make notes about what went well and what might require attention. While the class is fresh in my mind, I may rewrite notes, record difficulties, or document ideas for new directions. Although I may not rewrite my class notes for a later course offering after a one-semester experience, I try to record enough to suggest a local adjustment for improvement.
- Many of my courses make extensive use of labs, and I may observe elements of readings, examples, or instructions that caused confusion. After class, I often revise those pieces, while I remember the problems. Typically, these class materials are passed along from

one semester to the next, so polishing one semester likely will improve effectiveness for later semesters.

- In addition to a college-mandated, end-of-course evaluation, I ask students to complete a detailed evaluation at the end of each semester. Questions ask about the pace of the course, relative time devoted to lecture or lab or class discussion, effectiveness of readings, etc. I also ask:

  ◇ "What three or four laboratory exercises did you find the most useful?"

  ◇ "Are there some laboratory exercises which you think should be revised or discontinued? If so, please identify which of these lab exercises should be given highest priority for revision or replacement, and indicate any modifications you would recommend."

  Effectively, such questions serve to identify gradients for hill climbing. When asked at the end of a semester, I can use the feedback to make adjustments for the next semester.

A similar approach of listening and facilitating feedback can be applied to curricular refinement. For example, here are two approaches used at Grinnell and elsewhere.

- Some departments organize a weekly gathering at lunch time—a CS Table, a practice that has been used for years by foreign-language departments to encourage language fluency. Although such meetings may advertise discussion of a topic-of-the-day or readings on a theme, gatherings also provide a mechanism to receive on-going feedback on courses and curricula.
- Shortly before graduation, faculty may reach out to graduating students for exit interviews, again soliciting feedback.

When questions are open ended and a climate of trust has been established, themes may emerge that suggest curricular refinements. For example, if CS faculty discover that RSA encryption is discussed in each of several courses, modest adjustments may reduce coverage to one or two courses—avoiding unnecessary repetition and freeing time in other classes for additional material. Similarly, if students remark that discrete mathematics is a prerequisite for an algorithms course, but several topics in the math course are never used, then the math-algorithms sequence might be examined to the flow of topics from one semester to the next.

Throughout planned refinement, a key is to listen, seek feedback, and adjust. Often small changes, when made in response to identified observations, can make a difference: calculating a gradient identifies the direction for moving a course or a part of a curriculum to a higher level.

Crisis-motivated reactions: Over the years, often at conferences, I have talked with faculty who have made quick adjustments to a course in response to pressures (frequently time-related)—only to discover that addressing short-term challenges created further difficulties later in the same course or in later courses. Here are five examples:

- Algorithmic analysis is downplayed in CS2, only to find that students cannot do even basic analysis in an upper-level algorithms course.

- Staffing constraints (e.g., time) motivate a decision not to grade program documentation in one course, and students then seem ill-prepared to approach projects in a later course.
- When the skill of writing out logical arguments is dropped from a discrete structures course, students cannot produce even simple proofs in subsequent automata or other theory courses.
- A few topics are cut at the end of several courses in the interest of time, only to discover that graduating students have never seen such concepts as refactoring, classes P and NP, or non-solvability.
- Staff teaching assignments may be shifted from one semester to another to address immediate student demand, without a consideration of long-term needs: a fix now in the short term may create a larger crunch later.

Often, time is a motivating factor to make an immediate change, although sometimes it may be tempting to avoid a difficult or burdensome topic (e.g., grading proofs). Placed within the context of hill-climbing, a direction for change may entail numerous dimensions, but immediate demands narrow the focus to a single factor. For hill climbing, one must consider the full landscape (north, east, south, and west), and considering only east-west factors may be risky. (At points A and B in Figure 4.2, left-right-only motion takes one right with hill climbing, when consideration of all directions take one back left (A) or back (B).

Translating back to courses and curricula, it is important to consider the full range of goals and objectives for a course when making a change—even in reaction to a crisis. A simple course adjustment may address an immediate time crunch, but faculty must be careful that the change does not contradict the principles of the course or curriculum. Over the years, I have seen this conflict play out badly in too many settings—particularly with overworked faculty!

When Hill Climbing Fails

The Hill-climbing Algorithm searches for local maxima. If one location is higher than those nearby, no gradient indicates a direction for further improvement—even if global maxima exist elsewhere (Figure 4.2). Thus, refinements of this algorithm include mechanisms to jump to another (non-nearby) place—at least from time to time. Applied to curricula and course development, tinkering can only go so far; sometimes dramatic changes are needed. Again, five examples illustrate limitations with tinkering.

- As technology evolves, a common tendency is to add topics to courses. Over time, prescribed content may extend far beyond what is reasonable for a single course or course sequence. Sometimes refinements in course organization and pedagogy may allow modest gains in efficiency, but adjusting details in an overloaded course typically still yield an overloaded result. Instead, either substantial cuts in content are needed or a new course may be needed.
- In some curricula, several courses may have a common, minimal prerequisite. Although students may express enthusiasm for each course separately, feedback may highlight several topics repeated in each course. No course could assume the common material, so all began with similar units—making students repeat material multiple times, dampening student interest, and taking time that could be used for other exciting content. Such situations may suggest the need for curricular reorganization. Perhaps

FIGURE 4.2 Picture from Killarney Park, Ireland Numerous local maximums without an upward path to the summit at back in clouds
At both A and B, applying hill climbing for just left/right moves one right, whereas the full algorithm goes back left (A) or back (B)
(Photo and editing by the author)

the common material could be gathered into a new course that serves as a prerequisite to others. Perhaps material could be redistributed among existing courses with adjustments in prerequisites. Perhaps the common material could be inserted into the existing prerequisite, with some content from the earlier course moved to one of the later courses.

- In mathematics curricula over the years, faculty often complained that students had difficulty writing proofs, and this hindered many upper-level courses in abstract algebra, real or complex analysis, topology, and other theory-oriented courses. In response, mathematics curricula now commonly include what is called a "bridge course" that serves as an intermediate stage (or bridge) between lower-level courses (e.g., calculus) and more advanced courses. Often specific content for the "bridge course" takes a secondary role to the process of mathematical thinking: logical deduction, proof techniques, and practice in writing proofs. If common deficiencies are generating problems in upper-level computing courses, faculty may consider what type of course might be the computing equivalent of math?s "bridge course". (For example, when do students learn to structure programs, write meaningful documentation, develop strong test suites, etc.)

- From my experience teaching mathematics, multiple upper-level mathematics courses may implicitly assume that certain topics are covered in that "other course". In fact, one of the many outstanding qualities of the textbook *Topology* by James Munkres was the inclusion of a chapter zero that explicitly brought those topics together in a manageable context. Similarly, in computing curricula and courses, faculty might consider what tools of the trade and topics (e.g., debuggers, Linux tools, productivity suites, code analysis) might be assumed in advanced courses. Although such background may be known to some, others may never be exposed to such material—possibly with disproportionate impact on students from under-represented groups. If such assumptions seem minimal, tinkering with a course (e.g., hill climbing) might yield a satisfactory solution. However, when many such assumptions seem to arise, a more dramatic type of rethinking or reorganization may be required.

- At times, student feedback suggests that all courses are overloaded, and the overall curriculum seems bulging with a large number of courses and requirements. In such circumstances, moving topics among courses may change places of stress, but overloads will remain. Further, adding requirements may be impractical on an already full curriculum. Instead, the CS faculty might consider priorities—what really is important, and what (although nice) can be cut—sometimes in climbing, one may want to jettison unnecessary weight that gets in the way of overall progress.

The discussion and examples suggest that modest refinement of courses and curricula may work very well, when thought through in advance to reflect observations and feedback. However, when tinkering, faculty should be careful to keep goals and objectives firmly in mind and not just respond to immediate crises. While modest changes can address small-scale difficulties, a hill-climbing approach cannot address some large-scale difficulties. Sometimes a fresh start is needed by rethinking of courses or curricula.

Acknowledgment

Many thanks to the reviewer of this column for several suggestions in specific wording. Thanks also to Theresa P. Walker for the photo in Figure 4.1.

CHAPTER 5

Developing a useful curricular map

This column first appeared in ACM *Inroads*, Volume 3, Number 4, December 2012, pages 14–16[243].

5.1 ORIGINAL COLUMN

SOME BASIC QUESTIONS for any curriculum are:

- Does the curriculum cover the right material?
- Do course prerequisites provide the appropriate background for the next course?
- Throughout a curriculum, are topics covered efficiently and in a reasonable order?

When courses are developed individually, each course may have cohesion and present worthwhile material in an appropriate way, but the overall curriculum may include significant redundancies or omissions, and prerequisites may not provide needed background for an individual course.

National curricular recommendations (e.g., from the ACM/IEEE-CS [7, 9, 12] or the Liberal Arts Computer Science Consortium [109]) can provide a helpful framework for answering basic curricular questions and reviewing the flow of content from one course to the next.

This column discusses a curricular map as a helpful tool for comparing a curriculum with programmatic goals and objectives and with national curricular recommendations.

Form of a Curricular Map

I think of a curricular map as a table. Labels in one dimension (e.g., rows) specify goals, objectives, topics, or other content. Labels in the second dimension (e.g., columns) specify courses. A cell within the table provides information about goals, objectives, topics, etc. for a course. In my experience, curricular maps describe various types of information, such as:

- the extent to which an objective or topic is covered in a course (e.g., introductory, moderate, or advanced level)
- the depth of coverage of a topic, according to Bloom's taxonomy [25]

- the number of in-class hours a course devotes to a topic comments about how an objective is met or a description of how topics are covered

Practically, the amount of information stored in a cell may be limited by the number of rows and columns. For example, if there are many objectives or topics, and if there are several courses, a curricular map table can become unwieldy if much information is stored in each cell.

Motivations

Although some may consider the development of a curricular map as a formal exercise to satisfy an external agency (e.g., a dean or assessment office), my own experience suggests this exercise can have several helpful benefits that I truly care about:

- If a curricular map charts goals and objectives, then tabulation of coverage in courses allows a useful mechanism to track what work happens in the overall curriculum, as well as in individual courses.
- A topic-level curricular map encourages faculty to examine an overall curriculum—both analyzing the detail of individual courses and reviewing the overall priorities of a curriculum.
- A curricular map can help identify gaps or redundancy within a curriculum. Are some topics covered many times, while other topics apparently skipped everywhere?
- Development of a curricular map can bring faculty together.
  - ◇ Faculty develop a shared view of courses, prerequisites
  - ◇ Faculty review the depth of coverage at introductory, intermediate, and advanced levels
  - ◇ Faculty clarify where a course fits within an overall curriculum—what can it assume, and what foundation should it provide for later work

Even if faculty have worked together for many years, courses tend to evolve, and faculty may lose track of what really is covered in various courses. The development of a curricular map provides a focus for reviewing developments in the discipline and in courses, leading to a renewed and shared vision of a program's curriculum.

Level of Detail Makes All the Difference

In my experience, high-level curricular maps provide little insight. At one extreme, I have seen a department's curricular map with only high-level topics (e.g., "programming", "communication skills", "problem solving", etc.). A cell for the course entry identified the level of proficiency students were expected to obtain for each high-level topic—options were "introductory level", "moderate proficiency", "advanced level". Then, in filling out the table, "introductory level" was specified for all topics for each introductory course; "moderate proficiency" was listed for all topics for second level courses; and "advanced level" was used for all advanced courses. From my perspective, this table avoided the issue of what really was covered in each course.

A more useful curricular map is given in Section 9 of the *2007 Model Curriculum for a Liberal Arts Degree in Computer Science* [109], in which each recommended course is connected with the various subjects (or Knowledge Areas) in Computing Curricula 2001 [12].

This curricular map indicates not only coverage of individual topics, but also a high-level review of what areas are emphasized in the *2007 Model Curriculum* (e.g., algorithms and complexity, theory of computation, and programming languages) and what areas are de-emphasized somewhat (e.g., intelligent systems, information management, operating systems, and software engineering). Any curriculum reflects priorities and choices, and the curricular map in [109] clarifies the perspectives of the Liberal Arts Computer Science Consortium.

Implementation Suggestions

Over the years, I have had occasion to develop reasonably detailed curricular maps that related at least three different curricula to ACM/IEEE-CS curricular recommendations [7, 9, 12]. In each case, the task required substantial time (perhaps 1-2 hours per course), suggesting that this work may require considerable motivation, dedication, and careful analysis. (The high-level curricular map, described at the beginning of the previous section, might require rather little time, but it also seems to have little consequence—why bother?)

A Naive Algorithm for Producing a Curricular Map On the surface, development of a curricular map might utilize the following naive algorithm:

```
For each Knowledge Unit
    For each course
        For each class session determine how much time or what content is devoted to
          this topic in this course for this session
```

Computationally, Computing Curriculum 2013 identifies about 81 Knowledge Units that contain required topics and another 74 Knowledge Units with only elective material. Also, a computing major likely involves 10 or more courses, each of which might meet 3 hours per week for 15 weeks. This suggests that the inner loop of the curricular-map-development algorithm might be executed 81 * 10 * (3 * 15) = 36,450 times if only Knowledge Units with required topics are reviewed or (81+74) 10 * (3 * 15) = 69,750 times if all Knowledge Units are reviewed. Such a level of work for a faculty member might seem excessive. At just 2 seconds per class session per course per Knowledge Unit, this might suggest 20.25 hours of work to cover just the Knowledge Units with required topics or 38.75 hours of work if all Knowledge Units are reviewed.

Further, even this algorithm can be inadequate, because some course material might be categorized in various ways. For example, the "divide-and-conquer" problem-solving strategy arises with different variations in "PD/Parallel Algorithms, Analysis, and Programming", "SDF/Analysis and Design", and "AL/Algorithmic Strategies" in Computer Science Curriculum 2013. If one classification is used repeatedly, the curricular map may suggest that one Knowledge Unit is covered extensively (more than a recommended amount), while another Knowledge Unit is not covered at all. Thus, even when topics are classified for a class session, some reclassification may be needed so that the shape of hours for a course reflect all priorities and perspectives.

With these difficulties, I do not believe the above naive algorithm is practical for developing a curricular map. The naive algorithm might produce a reasonable result, but the work for the nested loops followed by the review of priorities and perspectives likely requires more time than seems realistic.

A Better Approach for Developing a Curricular Map

Rather than the naive algorithm, I suggest the following approach. In my experience, this requires about 1 hour per course (working collaboratively with a colleague) or 1.5-2 hours per course (working individually).

- Use a spreadsheet to record hours allocated to topics within courses. However, rather than develop your own spreadsheet, use an existing template to reduce your start-up time. For example, Computer Science Curriculum 2013 contains a pre-formatted spreadsheet to allow hours to be mapped to courses.
- Focus on one course at a time
- Consult a day-by-day course schedule, if available
- Work collaboratively, if possible, with a faculty member who has taught the course often
- Methodically go through the high-level Knowledge Areas in *Computer Science Curriculum 2013.*
    - ◇ If a general Knowledge Area does not seem relevant to the course, scan the Knowledge Units quickly (sometimes topics are inserted in places that may be unexpected). Typically, one can skip over a significant fraction of the Knowledge Areas in very little time.
    - ◇ If a general Knowledge Area seems relevant to a course, make an initial estimate of time spent, based upon faculty expertise. Then scan the day-by-day course schedule to confirm or adjust the time for the Knowledge Units within the overall Knowledge Area.
    - ◇ Write brief notes on your allocation of time and coverage of topics, as this can be helpful later on.
- After a first pass through all Knowledge Units for a course, check the total number of course hours (available from the spreadsheet). Often this number will not match the actual amount of time the course meets (even adjusting for tests, end-of-course evaluations, etc.).
    - ◇ Review your notes on time allocation and make adjustments as necessary so the spreadsheet matches the amount of time in the course
    - ◇ If some topics are covered a little here and a little there, add the topic hour(s) in the spreadsheet as needed.
- Check that the overall shape of assigned hours reflects the priorities and topic coverage. Adjustments may be needed if some topics could reasonably be listed in several Knowledge Units or Knowledge Areas.

In my experience, working course-by-course helps a curricular map reflect both the details and the overall complexion of a curriculum. Thus, a curricular map can be extremely helpful in reviewing an overall curriculum, as well as the progression of topics from one course to the next.

However it is created, the development of a curricular map requires dedicated focus, analysis, and review. Working with a colleague is much more fun and insightful than working alone in this process!

Acknowledgment

The development of this column benefited greatly from insightful feedback obtained from Marge Coahran. The author is deeply grateful to her for her comments!

CHAPTER 6

Selected/Annotated references for curricular development

A VIBRANT AND SUCCESSFUL COMPUTING PROGRAM incorporates a well-designed curriculum (content) within a creative and supportive environment. This chapter suggests several basic references for further reading within four main categories:

- national curricular recommendations
- characteristics of successful academic programs
- program review
- programs for broadening participation

National Curricular Recommendations

Approximately every ten years, professional societies in both computing and mathematics publish curricular recommendations for undergraduate programs.

- ACM/IEEE-CS Joint Task Force on Computing Curricula, *Computer Science Curricula 2013*, ACM Press and IEEE Computer Society Press, December, 2013 [10]
  The Association for Computing Machinery (ACM) collaborates with the Computer Society of the Institute of Electrical and Electronics Engineers (IEEE-CS) to develop and publish curricular recommendations for computing science about every ten years. The most recent recommendations appeared in December 2013.

- ACM/IEEE-CS Joint Task Force on Computing Curricula, *CS2013 Curricular (Learning Outcome) Spreadsheet*, 2013 [11].
  This spreadsheet includes several different tables. Rows for one table include all student learning outcomes and rows in another table identify all Knowledge Units. Courses can be assigned to columns, and the spreadsheet tables allow mapping of local courses to the detailed CS2013 recommendations.

- Committee on the Undergraduate Program in Mathematics (CUPM), Carol S. Schumacher and Martha J. Siegel, Co-Chairs, and Paul Zorn, Editor, *2015 CUPM Curriculum Guide to Majors in the Mathematical Sciences*, Mathematical Association of America, 2015 [44].
  These recommendations from the Mathematical Association of America provide guidance regarding connections between computing and mathematics.

- Henry M. Walker, Daniel Kaplan, and Douglas Baldwin, *Reports of the MAA Program Study Group on Computing and Computational Science*, 2015 [280, 281]
  In developing recommendations for undergraduate mathematics, the Mathematical Association of America created several Program Study Groups to examine the mathematical needs of client departments. The Summary and Supplemental Reports of the Program Study Group on Computing and Computational Science maps topics within mathematics to computing subdisciplines, including computer engineering, computer science, information systems, information technology, software engineering, computational science and big data.

- Liberal Arts Computer Science Consortium , "A 2007 model curriculum for a liberal arts degree in computer science", *Journal on Educational Resources in Computing (JERIC)*, Volume 7, Number 2, June 2007, Article 2 [109].
  Although a response to the ACM/IEEE-CS *Computing Curricula 2001* (not *Computing Curricula 2013*), this model curriculum provides insights about a strong computing program building upon the liberal arts. In addition to suggesting courses within a structure, the article discusses staffing, student scheduling, and other matters of a computing program.

Characteristics of Successful Undergraduate Programs

- Alan C. Tucker, *Models That Work: Case Studies in Effective Undergraduate Mathematics Programs*,

  - Full report: MAA Notes Number 38, Mathematical Association of America, 1995. [187].
  - Summary report: Notices of the AMS, Volume 43, Number 11, November 1996, pages 1356-1358. [186].

  A study by a committee of the Mathematical Association of America that explores what makes a successful mathematics program. The study investigates in depth several programs, including ten for which the committee held site visits.

- Scott Jaschik, "Majoring in a Professor," *Inside Higher Ed*, August 12, 2013 [93].
  A report of work by Christopher Takacs and Daniel Chambliss, showing that students may choose a major based on the quality of instruction in the first course rather than by the nature of the subject matter, job possibilities, or expected earnings potential.

Program Review

- Accreditation Board for Engineering and Technology (ABET), *Accreditation*, 2017 [2].
  The Accreditation Board for Engineering and Technology (ABET) is recognized by the Council for Higher Education Accreditation to formally review and assess programs

in several fields of engineering and technology. Accreditation criteria and supporting documents, updated annually, provide considerable information about the process and criteria for programs in the field of computing.

- MAA Committee on Department Reviews, *Program Review*, Mathematical Association of America [42].
  The Mathematical Association of America (MAA) does a particularly fine job helping mathematics departments review their programs. Their Committee on Department Reviews coordinates the posting of resources for these reviews, and the Committee organizes a listing of consultants which can be accessed by contacting the Committee Chair. Although these resources focus on the mathematical sciences, many of the processes and materials may be adapted for computing as well.

- Henry M. Walker and Joan Krone, "Conducting Departmental/Program Reviews and Serving as a Reviewer," 2014 Midwest Conference of the Consortium for Computer Science in Colleges (CCSC), September 19, 2014 [199].
  Colleges and universities, particularly at small colleges, often conduct periodic reviews of their various academic programs. Typically, these reviews are somewhat informal (in contrast to ABET accreditation). After a self study by program faculty, external reviewers visit campus and write a report, and the program faculty consider appropriate next steps. This reference provides slides for a workshop/conference presentation that outlines the external-review process.

Programs for Broadening Participation

In recent years, several programs have emerged to encourage K-12 students to gain experience with computing, with a hope that these students might consider computing careers in the long term. The following programs include curricula and other supporting activities.

- code.org, *Teach Computer Science*, [39], 2017.
  The vision of Code.org is that "Every student in every school should have the opportunity to learn computer science" [39]. Toward this end, the organization provides resources for teaching elements of computing at the elementary level, middle-school level, and high-school level. Additional activities, aimed at reaching out to students at all levels, include an annual Hour of Code.

- The College Board, *Advanced Placement (AP) Computer Science Principles*, 2017 [26].
  The College Board, with substantial funding from the National Science Foundation, has developed a computing curriculum/course designed for pre-college students to encourage the exploration of computing and technology and to understand connections between computer science and real-world problems. This effort includes substantial resource material and has provided numerous opportunities for teacher training. The AP Computer Science Principles examination premiered in 2017, providing a mechanism for students to receive college credit for work completed in high school. With about 49,000 examinations taken in its first year, this course set a record for the most number of AP exams taken in the first year of an AP test.

- Office of Science and Technology Policy (OSTP), *CS for All*, led by the National Science Foundation (NSF) and the U.S. Department of Education (ED), 2017 [142]

This nationwide effort "will leverage and expand NSF and ED investments in CS education and STEM education more broadly. The effort accelerates ongoing efforts to build the knowledge base and capacity for rigorous and engaging CS education in schools across the nation. It will also bring together NSF and other federal agencies with private partners to support professional development for educators to teach CS." [142]

III

Courses and the Computing Curricula in Context

CURRICULAR AND COURSE PLANNING AND DEVELOPMENT extend well beyond specific content (although content certainly is important). Both curricula and courses must contain content to meet articulated student learning goals (e.g., as discussed in Part II of this book), but curricula and courses also are shaped by the backgrounds, interests, and needs of the enrolled students. Some broad considerations include the following elements.

- Computing curricula and courses exist within an overall context.
  - ◇ Prospective or declared computing majors constitute one group of students.
  - ◇ Students with primary interests in other fields may utilize computational thinking as part of their problem solving.
  - ◇ Work in some disciplines may require programming or knowledge of other key elements within a computing curriculum.
- Students have varying backgrounds and abilities, particularly within introductory courses; students with limited prior experience may not mix well in introductory courses (e.g., CS1) with more experienced students.
- Students with varying interests (e.g., both majors and non-majors) may have different motivations and thus respond to different incentives, policies, and procedures.

Part II examines computing curricula and courses within this broad context and also explores design factors that go beyond considerations involving specific content.

CHAPTER 7

An historical view of computing curricula

THIS CHAPTER PROVIDES AN OVERALL PERSPECTIVE on curricular recommendations for undergraduate computing curricula. Initially, a committee of the Association for Computing Machinery (ACM) developed recommendations in 1968 [5] and 1978 [17]. After that, ACM partnered with the Computer Society of the Institute of Electrical and Electronics Engineers (IEEE-CS), to publish recommendations in 1991 [8], 2001 [12], and 2013 [10]. Each of these curricular recommendations targeted a full range of schools, and thus they naturally represent a compromise position. For example, engineering-oriented programs often place more emphasis on computer architecture and organization than might be found at non-engineering-oriented schools.

In particular, liberal arts programs generally emphasize algorithms, the theory of computation, and programming languages. Thus, in response, to the ACM-based recommendations, the Liberal Arts Computer Science Consortium (LACS), established in 1985 with initial funding from the Sloan Foundation, has published "model curricula for a liberal arts degree in computer science" in 1986 [70], 1996 [283], and 2007 [109].

This new chapter reviews the evolution of these curricula over roughly 50 years, highlighting both the development of computing as a discipline and the interactions of national recommendations from ACM [and IEEE-CS] with LACS' perspective of the liberal arts.

7.1 NEW ARTICLE

The ACM [and IEEE-CS] curricular recommendations for undergraduate computing programs appeared in 1968 [5], 1978 [17], 1991 [8], 2001 [12], and 2013 [10]. Subsequent responses by the Liberal Arts Computer Science Consortium (LACS) appeared in 1986 [70], 1996 [283], and 2007 [109]. Both sets of guidelines reflect substantial thought, insight, and discussion, and thus computing faculty consistently have taken these recommendations seriously.

Naturally, the discipline of computing has developed and matured substantially over the years, and both the ACM [/IEEE-CS] and LACS recommendations reflect much of this evolution. However, various faculty have different perspectives on changes in both the discipline and these curricular recommendations; sometimes reactions to new or emerging guidelines have led to much debate and controversy.

Also, different schools have different priorities, goals, and perspectives, so any single collection of recommendations will fit some programs better than others. Such differences also have yielded a range of responses.

The comments that follow reflect both some historical developments and some reactions. Although this discussion is intended to provide perspective and background, space limitations require the discussion to be reasonably brief. In what follows, observations and notes are organized into these sections.

- General Context and Notes
- The Role of Mathematics
- Architecture, Organization, Operating Systems, [Uni-processor] Systems
- Parallel and Distributed Computing, Networking, Security
- Programming Languages
- Theory of Computation
- Files and Databases
- Additional Notes

General Context and Notes

The curricular recommendations from ACM [later with IEEE-CS] have appeared about every 10–13 years, and the LACS curricula appeared about 5–6 years after corresponding ACM recommendations. With over a decade between versions, both the computing discipline and computing education changed substantially from one version to the next. In addition, the initial ACM recommendations suggested courses, but when ACM partnered with IEEE-CS, the recommendations shifted to topics and, eventually, student learning objectives.

This chapter begins an historical review of these recommendations with my perspective on the environment for each recommendation. Table 7.1 presents some overall perspective chronologically; to allow reasonable formatting, Table 7.1a highlights some basic statistics for curricula for 1991 and before; Table 7.1b provides parallel statistics for curricula beginning in 1996. Subsequent subsections discuss the evolution of content within general subject areas.

Curriculum 1968 broke new ground for an emerging discipline. Work for the report started in 1962, with preliminary recommendations distributed in 1965. Major issues involved whether the discipline should be called "computer science" or "information science", whether degrees should be at the undergraduate or graduate levels, what background was needed for all computing students, what mathematics should be part of computing programs, and how a curriculum could provide appropriate background in some focused areas.

Curriculum 1968 reported a growing number of bachelor's degrees in computing—148 were projected for the following year with most or all offered at Ph.D.-granting universities, and most of the report focused upon undergraduate curricula. However, Curriculum 1968 also reported somewhat more computing programs existing or projected at the the master's level, and discussion provided guidelines for graduate programs, including descriptions of numerous advanced courses.

Curriculum 1978 evolved during a time of nation-wide debate about the nature of the computing discipline, the relationship between computer science as a discipline and professional computing (e.g., in industry), the role of mathematics and theory, the scope and content of material essential for all computing students, and an appropriate range of electives. This report mandated 5 required mathematics courses, with 2 more deemed essential for some students. This requirement was lower than in Curriculum 1968, but both the scope and content of the mathematics requirement were controversial. For example, as discussed later in this chapter, Ralston and Shaw argued that Curriculum 1978 required more continuous mathematics than was needed for all students, and they proposed a 2-year required sequence in discrete mathematics. [154].

Table 7.1a: Computing Curricula Overview 1968—1991				
	Curr.1968	Curr.1978	LACS 1986	Curr.1991
Sponsoring Org.	ACM	ACM	LACS	ACM/IEEE-CS
Min. CS Faculty FTE	5	6	3	—
Total pages	47	20	9	160
Background structure (pgs.)	9	6	4	33
Required content (pgs.)	10	5	3	33
Electives content (pgs.)	20	5	< 1	5
Service courses (pgs.)	1	1	—	< 1
Graduate courses (pgs.)	2	—	—	—
support material (pgs.)	5	4	2	7
Required courses				
computing	6	8	6	283 hours ($\geq$ 6–7 courses)
math	6	4 (+2 for some)	1 (discrete)	3–4
Elective courses				
computing	2 of 5	4 of 10	1 (locally)	NA
math	2 of 4	TBD locally	advised)	1 of 4

Table 7.1b: Computing Curricula Overview 1996–2013				
	LACS 1996	Curr.2001	LACS 2007	Curr.2013
Sponsoring Org.	LACS	ACM/IEEE-CS	LACS	ACM/IEEE-CS
Min. CS Faculty FTE	3	—	3	—
Total pages	11	240	34	518
Background structure (pgs.)	4	43	7	37
Required content (pgs.)	4	34	22	~89
Electives content (pgs.)	1	40	< 1	~72
Service courses (pgs.)	—	7	—	—
Graduate courses (pgs.)	—	—	—	—
Support material (pgs.)	2	36	5	32
Required courses				
computing	6	~8–12 (280 hours)		~7–11 (239-267 hours)
math	1	3		41 hours
Elective courses				
computing	3	~2–5	—	—
math	2 (local)	—	1 of 4	—

Note: All page counts depend upon judgement calls and estimates, and thus should be considered rough approximations.

LACS 1986 evolved in response to Curricula 1968 and 1978, which largely reflected computing programs in large, Ph.D-granting universities and which explicitly required at least

5 or 6 computing faculty. Indeed, at a session at SIGCSE 1985 on computing programs for small colleges, the panelists suggested (at least from the perspective of myself and others from small, liberal arts schools), that small and/or liberal arts colleges could not construct legitimate computing programs but would have to water down the ACM recommendations.

The SIGCSE 1985 session led to the creation of the Liberal Arts Computer Science Consortium (LACS), with initial Sloan Foundation, to construct a strong computing curriculum starting with the strengths of the liberal arts rather than the framework of large universities. In effect, LACS 1986 legitimized computing programs at liberal arts colleges, showing a limited curriculum that covered a solid core of computing while being practical; students could complete the major within a 4-year undergraduate program, and 3 computing faculty could staff the full program.

Computing Curricula 1991 recognized computing as an interdisciplinary discipline, drawing from mathematics (e.g., theory, formal methods), science (e.g., the scientific method), and engineering (e.g., problem-solving methodologies). The Report also discussed "recurring concepts, the social and professional context, mathematics and science, language and communications" and an "integrated laboratory experience" [8, p. 35].

To encourage experimentation and flexibility, Computing Curricula 1991 specified 9 categories of topics (Subject Areas) and material within those categories (Knowledge Units or KUs). The Report specified minimum lecture hours for each KU, with a total of 283 hours of lecture specified. In addition, the Report suggested some ideas for lab exercises, some of which were identified as open (e.g., unstructured, students working on their own) and some as closed (e.g., "a scheduled, structured, and supervised assignment" [8, p. 27]). To further encourage experimentation, Appendix A of the recommendations (about half the Report) outlines 12 different hypothetical curricula and their courses, each covering the needed topics but with different emphases. The recommendations also highlighted alternatives for a programming-first versus a breadth-first approach . Since these recommendations did not mandate specific courses, however, publishers and some faculty still used the course structure of Curriculum 1978 well into the 1990s.

LACS 1996 observed that Curricula 1991 "requires faculty to analyze their current course offerings with regard to these 11 areas and 56 modules", but notes that "the review and redesign of an entire curriculum is a tremendous job, and it seems that few colleges and universities have been willing to undertake this task." [283, p. 86]. LACS 1996 argued that Curricula 1991 had at least three difficulties for liberal arts colleges:

- Curricula 1991 over-emphasized architecture and software engineering for a liberal arts program,
- Curricula 1991 under-emphasized algorithms, data structures, and theory, and
- The 283 required lecture hours must be organized into coherent courses, so covering this material would yield almost an entire major at many liberal arts colleges and leave little time for upper-level electives and other important topics.

LACS 1996 also noted updates from LACS 1986 regarding multiple problem-solving paradigms and languages, coverage of parallel and distributed computing, formal labs for introductory courses, an additional course in mathematics and theory, "the explicit inclusion of social and ethical issues," and "a required senior project" [283, p. 88]. As with LACS 1986, these recommendations build upon the strengths of liberal arts and fit realistically within staffing and student constraints for small, liberal arts colleges.

Computing Curricula 2001 sought to update curricular content through a time of rapidly evolving technical advances, including the World Wide Web, networking, graphics, multi-media, embedded systems, relational databases, object-oriented programming, APIs, human-computer interactions, issues of safety and security, and applications. [12, pp. 9–10]. Further, computing subdisciplines had come into their own, and Computing Curricula 2001 represented one of several volumes of curricular recommendations. (Others included an Overview, Computer Engineering, Software Engineering, and Informations Systems, with separate curricular guidelines for each subdiscipline [12, pp. 1–2].)

To address the breadth of the discipline, Curricula 2001 organized its Body of Knowledge into 14 Knowledge Areas, each with a collection of core and elective Knowledge Units (KUs). Required content involved 280 class-hours of "core" material, identified as essential for all undergraduate programs, together with an extensive discussion of elective topics.

As a separate matter, over the previous decade, schools had experimented greatly with different curricular structures and course organizations. Thus, in addition to a listing of core topics, Computing Curricula 2001 provided several alternative and hypothetical approaches for course sequences to help guide schools. For example, at the introductory level, Curricula 2001 outlined possible approaches, including "imperative first," "objects first," "functional first," "breadth first," "algorithms first," and "hardware first." For each of these, the introductory sequence might involve either 2 or 3 courses, with a 3-course introductory sequence often integrating some discrete mathematics with computing topics. At the intermediate level, approaches involved "topic-based," "compressed," "systems-based," and "Web-based" approaches, each with 8–9 core courses. The advanced level identified various electives, with the specification that a team-based project be included somewhere. The scope of electives, however, was left to the local environment. Overall, Curricula 1991 emphasized a substantial listing of core topics, indicated possibilities for numerous electives, and identified some implementation models that might be considered at different types of schools. However, all courses and curricular examples were hypothetical (not actually offered at existing schools). In practice, therefore, each school was left with the time-consuming and intimidating task of developing its own entire curriculum.

LACS 2007 evolved "from three basic motivations: (1) Changes in emphasis between Computing Curricula 1991 and Computing Curricula 2001; (2) A renewed commitment to the inclusion of multiple paradigms for problem solving within the curriculum; and (3) The need to update" LACS 1996.

As with LACS 1986 and 1996, the 2007 Model Curriculum is limited to 12 courses that can fit practically within a liberal arts context. Software engineering, multiple views of problem solving (with multiple languages), and discrete mathematics are highlighted, while some theory is reorganized or deemphasized. In comparing LACS 2007 with Curricula 2001, priorities of earlier LACS curricula highlight liberal arts perspectives. For example, LACS 2007 contains noticeably more coverage in areas of algorithms, theory, and programming languages than the ACM/IEEE-CS recommendations.

Pragmatically, LACS 2007 provides reasonable detail regarding each of its recommended courses, largely accounting for the increased page count (from 11 to 34) from LACS 1996 to LACS 2007. Also, to illustrate some possible options at the introductory level, two different introductory sequences are presented, one integrating discrete mathematics with functional programming, and the second with more traditional introductory computing and separate discrete math courses.

Interestingly, these recommendations combine to yield 303 class-hours. Although this total is somewhat more than the 280 class-hours prescribed by Curricula 2001, the LACS

2007 curriculum has material divided into courses, whereas the Curricula 2001 hours requires additional class time to organize topics into coherent courses.

Computing Curricula 2013 again updated curricular recommendations to reflect the evolution of the computing discipline, following an interim update in 2008 [7]. The recommendations added a new Information Assurance and Security Knowledge Area, including computer and network security, and reorganized several topics, such as Software Development Fundamentals and Systems Fundamentals. As with Curricula 1991 and 2001, topics are divided into broad Knowledge Areas and more detailed Knowledge Units, which include both in-class hours and student learning objectives. Required material is divided into Core-1 Topics (content which all schools must cover) and Core-2 Topics (worthwhile content, of which schools should strive to cover 80% or 90%). Overall, schools covering all Tier-1 and Tier-2 topics would require about 308 in-class hours of instruction, whereas schools covering all Tier-1 material and 80% of the Tier-2 material would include 279.4 in-class hours.

By listing content through Knowledge Units, Curricula 2013 continued the approach of 1991 and 2001 of encouraging schools to experiment with curricular structures, course organization, class formats, and pedagogy. However, rather than outline hypothetical courses that combine some topics in interesting (but untested) ways as done in Curricula 2001, Curricula 2013 presents "Course Exemplars" which are courses actually offered at various schools and are judged to be interesting, innovative, and successful. Further, Curricula 2013 identifies five "Curricular Exemplars" that outline in moderate detail curricula developed at identified schools (2 two-year programs and 3 four-year programs) that illustrate how Knowledge Units might be combined within an overall curriculum.

Interestingly, the task force for Curricula 2013 included at least two faculty from liberal arts colleges, and the published recommendations include two curricular exemplars from undergraduate, liberal arts colleges. Altogether, Curricula 2013 includes many fundamental liberal arts perspectives, and the Liberal Arts Computer Science Consortium does *not* plan another Model Curriculum in response to the Curricula 2013 recommendations.

The Role of Mathematics

The role of mathematics within an undergraduate computing curriculum has been an ongoing subject of discussion for at least half a century.

Initially, in the 1960s, many computing curricula evolved from departments and programs that had mathematical connections. As a consequence, the mathematics requirements within Curriculum 1968 are extensive [5, pp. 160-161]:

- *Required:* all of the following
    - ◇ Introduction to Discrete Structures
    - ◇ Introductory Calculus
    - ◇ Mathematics Analysis I
    - ◇ Probability
    - ◇ Linear Algebra
- *Electives:* at least 2 of the following
    - ◇ Mathematical Analysis II
    - ◇ Advanced Multivariable Calculus
    - ◇ Algebraic Structures
    - ◇ Probability and Statistics

In this listing, all courses except Discrete Structures are described by references to 1965 course guidelines from the Committee on the Undergraduate Program in Mathematics

(CUPM) of the Mathematical Association of America (MAA). At that time, discrete structures was not discussed by the CUPM guidelines, and Curriculum 1968 placed Introduction to Discrete Structures among the computing courses rather than with the mathematics requirements. Altogether, Curriculum 1968 required exactly the same number of required and elective courses in mathematics as it does for computing (6 and 2 respectively).

Between 1968 and 1978, extensive debate focused upon the appropriate level of mathematics for an undergraduate computing degree. Curriculum 1978 expanded the number of required and elective computing courses to 8 and 4, respectively, while reducing the number of required mathematics courses to 4 (sometimes 6) and allowing mathematics electives to be determined by local schools. With this approach, a computing degree includes an expanded range of computing topics, and the amount of mathematics was reduced correspondingly.

In contrast, Ralston and Shaw argued [154] that

- the nature of computing should require substantially more mathematics than required in Curriculum 1978, but also
- the required mathematics should emphasize discrete mathematics rather than traditional, continuous mathematics.

Thus, Ralston and Shaw promoted a full 2-year sequence in discrete mathematics. Further, they noted that much of the math recommended in Curriculum 1978 still referred to courses in the 1965 CUPM recommendations, and new emphases were needed (e.g., a good background in statistics, but only a semester or two of calculus). [154, p. 70]

Although no curricular recommendations after 1978 recommended two full years of discrete mathematics, several narratives supported expanded study of discrete mathematics beyond a single semester. For example, additional discrete mathematics was identified as an option for LACS 1996. Also, Pedagogy Focus Group 2, studying supporting courses for Curriculum 2001 recommended a 2-semester sequence for Discrete Mathematics [140], although the final Curricula 2001 recommendations only indicated discrete math beyond one semester as an option. Within LACS 2007, however, a second-semester foundational math course, including much discrete math, was required.

Beyond discrete mathematics, computing has evolved, so that various subjects within computing utilize a wide range of mathematical subjects. For example, cryptography and security may draw upon abstract algebra, artificial intelligence upon multi-variable calculus, and graphics upon linear algebra. Altogether, most courses within a mathematics major might support one or another area within computing. Pragmatically, an undergraduate computing major cannot also be required to fulfill a mathematics major, so the selection of mathematics courses beyond discrete mathematics likely must depend upon a student's interests and the advice of a faculty advisor.

Architecture, Organization, Operating Systems, [Uni-processor] Systems

Problem solving with computing, of course, involves running programs on hardware, with operating systems providing helpful services. With this natural interaction of hardware, software, and operating systems, all curricular guidelines include significant components covering architecture, organization, and operating systems. Differences among curricular recommendations involve both the scope and the emphases of this required material. Coverage of both architecture and operating systems was relatively high (1.5 courses each) in Curriculum 1968, and required courses in these areas have declined in ACM-connected recommendations as associated topics have evolved. The three LACS curricula consistently have included 1 course for architecture, but the recommendations for operating systems have steadily declined as concurrency and networking have become prominent.

In the 1960s, much computing-application development worked reasonably close to the machine, computer memory was limited, and operating systems supported only moderate levels of abstraction. In this environment, issues of hardware (e.g., registers, data representation, memory layout, etc.) arose frequently in applications, and such topics touched many courses. For example, the first two courses in Curriculum 1968 ("Introduction to Computing" and "Computers and Programming") included elements of data representation, machine instructions and their execution, addressing, "Computer systems organization, logic design, micro-programming, ..., Symbolic coding and assembly systems, ..., and program segmentation and linkage" [5, p. 156, 158]. Intermediate-level courses included "Computer Organization" and "Systems Programming." Altogether, the equivalent of about 3 beginning and intermediate courses covered architecture, organization, and [operating] systems. Electives might extend discussion in these areas further.

By the last 1970s, some of these low-level elements had yielded basic virtual machines (although that term was not used), and the two specified introductory courses emphasized program syntax and semantics, algorithmic development, structured programming, testing and debugging, and a few standard algorithms, with only a rather small amount of background in areas of architecture and systems.

Interestingly, from the late 1970s though the early 2000s, the ACM curricula required about 1.5 intermediate-level courses related to architecture and organization, while the LACS curricula reduced this coverage to about 1 course. Before 2001, the audience for ACM-related recommendations included schools with engineering components, and the recommendations partially reflect this perspective. With Curricula 2001, however, ACM/IEEE-CS issued separate curricular recommendations for computer science and computer engineering, and the computer science guidelines showed a modest reduction for architecture. Following a liberal-arts perspective, the LACS curricula have consistently deemphasized (but not eliminated) architecture and organization in favor of algorithms and theory.

Through this period, coverage of operating systems, by themselves, generally decreased from about 1.5 courses in Curriculum 1978 to only about half a course by Curriculum 2001. The LACS curricula also showed a steady reduction, starting at about 1 course in 1986 to 0.5 course in 1996 to 0.3 course in 2007. However, as a narrow focus on operating systems decreased, interest in parallelism, distributed computing, networking and security expanded, so required content related to operating systems and concurrency might be considered as generally constant with required content equivalent to about 1 course. In Curricula 2013, this trend continued, with the Operating Systems Knowledge Area (KA) requiring only 15 core hours, but with other KAs for "Information Assurance and Security," "Networking and Communication," and"Parallel and Distributed Computing". In earlier curricula, some topics now included in these KAs might have been considered part of operating systems.

Parallel and Distributed Computing, Networking, Security

As might be expected from the evolution of computing hardware, both Curricula 1968 and 1978 largely omitted discussion of parallel and distributed algorithms. However, gradually from the mid 1980s, parallel and distributed computing and networks entered undergraduate curricula. For example, LACS 1986 identified networks as a possible elective, but neither parallel nor distributed computing were explicitly discussed in those recommendations.

As the gradual expansion of multi-processing continued, Curricula 1991 included discussions of parallelism within Knowledge Units within the Knowledge Areas for "Algorithms and Data Structures," "Architecture," and "Programming Languages" . Similarly, both Algorithms and Programming Languages included elements of distributed computing, and elements of networking appeared both in required Knowledge Units and in descriptions of possible advanced electives. Further, security was identified as one of twelve "recurring concepts" for consideration in many settings. [8, p. 13].

Following this evolution, LACS 1996 included a core course on Sequential and Parallel Algorithms [283, p. 91], and parallelism also was discussed in the course on programming languages. Elements of security, however, were not included in undergraduate study, and these might be considered as part of graduate study.

In the 2000s, parallelism, networking, and security became well established as important topics, although computing faculty discussed whether this content belonged at the undergraduate versus graduate level. For example, "Net-Centric Computing" emerged as its own Knowledge Area in Curricula 2001, including required content in both networking and security. Parallel algorithms were discussed, but as elective content rather than requirements.

LACS 2007 highlighted parallelism, networking, and security in some core courses, although only "Elementary concepts in security" were discussed [109, p. 31].

Finally, Curricula 2013 included "Information Assurance and Security," "Networking and Communications," and "Parallel and Distributed Computing" as Knowledge Areas, each with several required Knowledge Units.

As this narrative suggests, undergraduate curricular recommendations largely have followed the development of computer systems and technology, with increasing coverage of parallelism, networking, and security at the undergraduate level over the years.

Programming Languages

Introductory courses of each computing curriculum [almost] always involve programming. In some cases (e.g., CS 1 and 2, Computer Programming I and II in Curriculum 1978), the development of programming skills may have high priority; in some other cases (e.g., Curricula 2001's "breadth-first" approach), high priority is given to an overview of computing as a discipline, with programming present but of somewhat lower priority. Whatever the approach, students develop reasonable facility with programming over their first few courses, as they develop problem-solving skills and learn software-engineering methodologies.

Beyond the coverage of introductory programming, each curriculum includes discussion of programming languages in courses at the intermediate-level. As suggested in Table 7.2, Curricula 1968, 1978, and 2001 recommend a full semester course on programming languages, as do all three LACS curricula, although the emphases of these courses vary considerably. With the need to include other emerging material in recent years, both Curricula 2001 and 2013 recommend the equivalent of only half a course.

Table 7.2: Programming Languages

	Curr. 1968	Curr. 1978	LACS 1986	Curr. 1991	LACS 1996	Curr. 2001	LACS 2007	Curr. 2013
Programming Languages	40	40	40	46	48	21	47	28
Notes:	• 1 [three-credit] course equated to 40 semester hours • all hour counts are rough estimates only							

- Curriculum 1968 examines capabilities and implementation of several languages. Some topics covered include use of blocks in FORTRAN and Algol, list processing in LISP,

IPL-V, and SLIP, and string processing in SNOBOL. Modest coverage also includes elements of the formal specification of language syntax.

- Within Curriculum 1978, CS8, "Organization of Programming Languages," "is an applied course in programming language constructs emphasizing the run-time behavior of programs" and provides some additional programming experience but also explores underlying language issues and their impact on run-time behavior [17, p. 154].
- LACS 1986 "emphasizes the principles and programming styles that govern the design and implementation of contemporary programming languages" [70, p. 208].
- Curricula 1991 provides modest coverage of four different paradigms: functional, procedural/imperative, object-oriented, and logic/declarative. Additional language-related recommendations cover concepts and theory, such as virtual machines, representation of data types, sequence and data control, dynamic storage, and formal grammars, syntax, and semantics.
- LACS 1996 discusses the four different problem-solving paradigms described in Curriculum 1991, but in noticeably more detail.
- LACS 2007 updates both LACS 1986 and LACS 1996. The core course on programming languages largely focuses on language principles, formal syntax and semantics, language support for parallel and distributed computing, and run-time considerations. Recommendations for several courses show "A renewed commitment to the inclusion of multiple paradigms for problem solving within the curriculum" [109, p. 4].
- Curricula 2013 touches upon many language elements of LACS 2007, but with substantially reduced coverage. Programming paradigms include object-oriented, functional, event-driven, and reactive. Concepts include type systems, program representation, language translation and run-time considerations—all required at modest levels.

Theory of Computation

Topics within the category, "Theory of Computation," typically include finite automata, regular expressions, Turing machines, classes P and NP, and undecidability.

As Table 7.3 suggests, Curriculum 1991 specified the equivalent of about half a course on topics within the Theory of Computation. However, Curriculum 1969, 1978, and 2001 all mention this material, but for consideration within elective or advanced courses.

Table 7.3: Arch., Org., OS, Uni-proc. Systems

	Curr. 1968	Curr. 1978	LACS 1986	Curr. 1991	LACS 1996	Curr. 2001	LACS 2007	Curr. 2013
Theory of Computation	electives outlined	electives outlined	40	20	40	topics outlined	26	6
Notes:	• 1 [three-credit] course equated to 40 semester hours • all hour counts are rough estimates only							

Table 7.3 also suggests that these topics typically are included within liberal arts curricula, as these schools often emphasize underlying principles and theory as part of the liberal arts. Both LACS 1986 and 1996 describe a full course with this material as part of the intermediate core. LACS 2007 includes this material as part of two required courses:

- Course C.1 *Principles of Algorithm Analysis* includes solid coverage of algorithmic complexity.
- Course FC2: *Theoretical Foundations of Computer Science* includes substantial coverage of finite state machines and computability.

Interestingly, CS2013 identifies 3 Core-Tier1 hours to "Finite-state machines, Regular expressions, [and] The halting problem" and 3 Core-Tier2 hours to discussion of classes P and NP [10, p. 59]. This requirement provides a partial merge of a liberal arts perspective on theory with perspectives from other types of schools.

Files and Databases

The general topic of permanent data storage and data processing illustrates well how the field of computing has evolved in its understandings and approaches over the years.

Through the 1960s and 1970s, file processing became progressively more important: About 20%-25% of the core courses in Curriculum 1968 was devoted to the study of file design and storage, and Curriculum 1978 expanded expands the treatment of files to the entire Course CS5, "Introduction to File Processing," including an "Overview of Database Management System (5%)" [17, p. 153]. By Curriculum 1991, however, Knowledge Unit OS7, File Systems and Naming, identified some fundamental material on files, but the scope of coverage was identified as only 4 required hours. However, Curriculum 1991 promoted "Database and Information Retrieval," to a full Knowledge Area, with Units for "DB1: Overview, Models, and Applications of Database Systems" (4 hours) and "DB2: The Relational Data Model" (5 hours), and the recommendations outlined an elective course, CS11, "Database Management Systems Design," with coverage of "methods for data manipulation in the environment of a database management system" [17, p. 154] the "network, relational, and hierarchical data models" [17, p. 155].

By 2001, the ACM/IEEE-CS recommendations indicated increased importance of relational databases, and one of the sample courses describes combining artificial intelligence and database. Altogether, files have become somewhat secondary, but databases have become more prominent and more sophisticated.

Within the liberal arts models, LACS 1986 does not mention either files or databases, while LACS 1996 mentions databases as a possible upper-level elective. By LACS 2007, files were combined with streams as a type of input and output. Relational databases were mentioned as a possible application of areas of discrete mathematics and as examples for principles of software development.

Additional Notes

Beyond much discussion of specific technical skills, curricula discussions over the years seem increasingly interested in perspectives beyond computing itself. For example, Curricula 1991 included a Knowledge Area on "Social and Professional Issues", and all ACM and LACS curricula since then have included substantive requirements in this area.

Other areas outside specific technical content include both communication skills and the ability to work in teams. For example, both LACS 1996 and 2007 specify a substantial team project as part of a liberal arts degree in computer science, and both highlight the need for strong communication skills (which often are associated with the liberal arts).

Concluding Thoughts

The historical development of computing curricula over the past half century reflects to a considerable extent the evolution of the computing discipline itself. Some curricular discussions, such as the role of mathematics, reflect divergent perspectives within the computing community overall. Also, new technical discoveries and understandings within the discipline often are reflected within curricula within a rather short time interval.

In considering the curricular recommendations themselves, the national computing curricula from ACM and IEEE-CS have developed within a context to strongly encourage experimentation, flexibility, and innovation. Also in the spirit of engineering specification, curricula have become extremely detailed regarding content and learning outcomes. Specific courses are identified as examples, but faculty are encouraged to organize material to fit local priorities and opportunities.

In contrast, the LACS curricula in 1986, 1996 and 2007 provided more specific course recommendations, specifically for a liberal arts context.

Overall, the recent ACM/IEEE-CS recommendations since 1991 provide faculty with great flexibility, but also require faculty to devote substantial time to organizing topics into courses and then developing them. Putting a full curriculum together in this context can be very rewarding, but also requires a massive commitment of time and energy.

CHAPTER 8

Connecting computer science with other disciplines and the wider community

This column first appeared in ACM *Inroads*, Volume 8, Number 2, June 2017, pages 29–32[269].

8.1 ORIGINAL ARTICLE

MUCH CONTEMPORARY ACADEMIC DISCOURSE involves the term "interdisciplinary" (or an equivalent term). The word commonly appears in course syllabi, curricular descriptions, public-relations pieces, advertising for prospective students, grant proposals, etc. However, recently I have heard several computing faculty ask why they should prioritize interdisciplinary activities, when course enrollments already are stretching staffing and resources to the limit. Recently, a computing educator, outside Grinnell, my home institution, asked me to reflect on the importance of teaching computing within a broad (e.g., liberal arts) context. Although highly-skilled, but narrowly-focused, specialists can make worthwhile contributions, much work in both industry and research requires teams with individuals of diverse backgrounds and perspectives. This column suggests a framework for connecting computing with other disciplines and the wider community. Discussion naturally considers opportunities (philosophical, strategic, practical) and challenges (strategic, practical). The column also suggests some approaches for expanding interdisciplinary connections while respecting practical realities.

Opportunities

From my perspective, motivations for engaging with other departments and the wider community fall into at least three categories: educational philosophy, strategic/forward-looking education, and practical considerations.

FIGURE 8.1 An Interdisciplinary Word Cloud

Educational Philosophy

Grinnell College's mission statement seems reasonably typical for liberal arts schools, when it states that the college seeks to educate students "for the different professions and for the honorable discharge of the duties of life." [77] In today's technological society, citizens need a general understanding of computing—its opportunities, its limitations, its impact, etc. Further, Jeannette Wing [289] and many other computing educators have highlighted computational thinking as a fundamental problem-solving perspective—not just in computing but across disciplines and throughout curricula. In addition, various disciplines require specific technical skills and perspectives related to computing.

Beyond specific computing topics, a liberal arts education typically promotes broad perspectives and highlights multiple points of view. Many disciplines approach problems in different ways, and a liberally-educated person should be comfortable with many ways of addressing problems. Of course, within traditional computing, problem-solving approaches include imperative, functional, object-oriented, and declarative; some would add event-driven as well.

Altogether, college graduates and informed citizens need general background regarding computing, experience with computational thinking and problem solving, and often additional technical knowledge. Within a liberal-arts context, computing can provide important views of problem solving. And, as educators with appropriate expertise, computing faculty have important roles to play in meeting this need. Faculty in other disciplines (e.g., mathematics, other sciences) may be able to help (e.g., in developing skills for computational thinking), but, as experts, computing faculty likely should be contributing. Overall, computing is important throughout education as a subject and perspective that can expand horizons for all concerned.

Strategic/Forward-looking Education (Career and Strategic Planning)

Separate from philosophical priorities, especially within the liberal arts, undergraduate education is promoted as providing a solid foundation to students for their long-term careers. Increasingly, many positions require both algorithmic problem-solving skills and technical knowledge. For example, sciences today require computers to collect and analyze data, social sciences often require the structuring and processing of very large data sets, and the arts utilize computing techniques. In a recent conversation, for example, a visual artist observed that existing packages support some image manipulation, but innovative work requires substantial algorithmic development.

On a broad level, many [most?] research/development efforts today in both industry and graduate school involve teams—often including people with multiple interests and expertise. Although team members have their own perspectives and backgrounds, each person also must be able to communicate effectively with people of varying backgrounds. Pragmatically, this means non-computing people need to know basics of technology, and computing people need to know enough about applications and the broader world to be able to interact effectively with team members and clients.

Beyond specific uses of technology in future careers, today's talented undergraduates can be expected to become the policy makers and managers of the future. People in decision making roles may not need extensive technical details, but they should understand basic principles concerning opportunities, limitations, approaches, and cost trade-offs. Solid decision making in a today's world requires technological insights.

In summary, interdisciplinary undergraduate work, including computing, often is essential as students prepare for future careers and as educators provide a framework for future managers and policy makers.

Practical Considerations

Even with these educational motivations for considering computing within an interdisciplinary framework, departments and computing faculty might reasonably ask, "what's in it for us?"

Even in the short term, when computing enrollments often are high, at least two answers come to mind:

- Academia often places a premium upon scholarship, grants, and research. In recent years, I have heard progressively more opportunities for support of interdisciplinary endeavors. Projects involving connections with the broader community seem to have strong potential for faculty (and student) collaboration.

- In a time of strong student demand, interdisciplinary connections can provide options to support projects in disciplines throughout the curriculum. At Grinnell, for example, computing students are actively recruited for research projects in the fine arts (e.g., art), languages (e.g., English), the social sciences (e.g., anthropology, economics), the sciences (e.g., physics, biology), and mathematics. These opportunities allow computing students to be involved in serious scholarship—well beyond what the computing faculty itself can support.

Challenges

Although many motivations may encourage existing and expanded interdisciplinary ties between computing and other disciplines, substantial difficulties can arise on at least two levels: programmatic and individual faculty.

Programmatic Challenges

At many colleges and universities, student demand within computing is already high and growing. With overwhelming enrollments, computing faculty and resources are often spread thin, course sizes may increase substantially, and work loads may push teachers to a breaking point. In today's environment, proposals for additional outreach for further interdisciplinary connections may seem completely unrealistic. Of course, interdisciplinary initiatives may help bolster enrollments in times of meager demand, but such considerations likely do not resonate in todays' boom time.

Another programmatic difficulty can involve contrasting interests and perspectives between computing programs and other disciplines. For example, when I have talked with faculty of client programs during some external reviews, initial discussions may highlight "how to" issues, such as which buttons to push with a specific software package to accomplish certain tasks. Although such mechanics are far from computational thinking, creative course development may be able to infuse approaches to problem solving, algorithms, and data structures with the use of specific software. For example, a course might utilize a spreadsheet in response to a client program, but also provide practice with such notions as algorithms, conditionals, and elementary programming. Although such development can yield wonderful courses, course planning and delivery can require extensive faculty time and, perhaps, resources.

Challenges to Individual Faculty

For individual faculty, interdisciplinary initiatives may have several important consequences. Development of new or revised courses can require substantial effort, and energy devoted to course development likely means less progress in other teaching and scholarship. Further, student evaluations for a new course typically are lower than courses which have been offered several times (over time, schedules can be adjusted, details can be refined, etc.). For established faculty (e.g., tenured faculty), a focus on new and interdisciplinary initiatives may result in weak personnel reviews and relatively low salary increases. For untenured faculty, lowered evaluations and reduced scholarly productivity may result in contracts not being renewed and in denials of tenure. Altogether, development of connections, courses, and programs takes substantial resources, and such work may impact faculty careers negatively.

Further, for individual faculty members who might be interested in expanding their background to join interdisciplinary scholarship or to develop new courses, retraining in companion disciplines is likely to be time consuming and reduce scholarly production. In the long run, retraining of computing faculty may yield new opportunities and collaborations, but the short term outcomes may not fit well with the reward system at many colleges and universities.

Approaches

Weighing opportunities and challenges, some computing faculties may want to expand connections with other disciplines and the wider community—even with current enrollment pressures. Although specific ways forward almost surely depend upon local circumstances, here are some general thoughts.

Taking Stock

A natural first step in programmatic evolution and faculty development involves listening to client programs and understanding individual agendas. Perhaps surprisingly, in my experience with external program reviews, many computing programs have little contact with potential clients and collaborators. What are their needs, interests, and expectations? Which individuals might want to collaborate, what areas might have potential for interdisciplinary projects, and what might other faculty be able to contribute?

Interestingly, on the national level, some reports provide insights regarding interdisciplinary connections. For example, the ACM/IEEE-CS Task Force invited a background report on "The roles of mathematics in computer science" in its preparation for CS 2013. Although I am not aware of its public release, a summary piece appeared in ACM Inroads [20]. More broadly, the Mathematical Association of America (MAA) has a strong track record of identifying needs and opportunities of client departments. Thus, the 2015 Curriculum Guide by the MAA Committee on the Undergraduate Program includes "Program Reports" that discuss needs and perspectives of 14 client departments (including "Computing and Computational Science") [44].

Although national reports may provide insights, my discussions during on-campus interviews indicate considerable variation regarding opportunities from one campus to another. Overall, pursuit of specific opportunities largely seems dependent upon serendipity and individual interests. Although not present on every campus, reaching out to specific faculty outside computing can be particularly effective in the right setting.

Introductory Level Courses

At the introductory level, course offerings typically need to address three audiences: students who already have decided to major in another area, students who have committed to a computing major, and students who are exploring several options as majors. In practice, interests of undergraduates often evolve over several years, so students in one of these groups might well shift their foci as they gain experience.

When a school can offer numerous introductory courses, separate courses might be developed for each audience. Here are three examples.

- A broad non-majors course might provide a high-level view of computing topics (e.g., architecture, operating systems, networking), highlight applications, review social and ethical issues, and provide practice with problem solving and computational thinking.
- A discipline-oriented course for majors in a specific area might combine algorithmic problem solving with coverage of a programming language and use of identified software tools.
- A cross-listed course might integrate introductory topics in both computing and another discipline—providing students with academic credit in both disciplines.

Each of these approaches can be very successful in meeting general educational needs for an informed citizenry and for supporting students in other disciplines. High enrollments in these courses also can provide strong justification for expanded computing faculty who not only support these non-majors courses but also might contribute generally to computing programs. However, often, these courses have little long-term payoff for the computing programs themselves. The courses require staffing, but few students in these courses may progress to later courses in computing.

As an alternative to separate courses for non-majors, introductory computing courses might include breadth that may serve multiple audiences.

- A CS1 course might integrate fundamental algorithms, data structures, and algorithmic problem solving with strategic examples and enrichment lessons that may connect with needs from client departments.
- In one case (that eventually yielded my first two books), in-class discussions and common readings may cover many common topics for CS1 (e.g., conditionals, iteration and recursion, arrays, common algorithms). Then separate problem sets and programming assignments connect with separate disciplines and backgrounds (e.g., pre-calculus, calculus, statistics, chemistry, physics).
- When a computing faculty can agree upon common learning goals and problem-solving techniques for CS1, then different courses (perhaps utilizing different languages and applications) can address different audiences, but each course provides appropriate background for the next course, CS2. These multiple tracks in CS1 require substantial discussion and planning, but have been quite effective in meeting diverse needs while bringing a range of students to computing.

With these alternatives, applications and techniques within CS1 may provide desired background for a client's program, but the computing background and algorithmic problem solving also allows students to progress within the mainstream computing program.

Cooperation and Collaboration with Other Faculty

Beyond the introductory level, connections with faculty outside computing may take many forms. The following list just begins to identify some options.

- *Sitting in on courses:* Over the years, faculty from several departments (e.g., physics, anthropology, French, library, psychology) have sat in on my CS1 or CS2 courses. These faculty had specific needs for their scholarship, and attending my class regularly seemed an efficient means to gain needed experience. This contact also can open wonderful possibilities for publicity and collaboration.
- *Team teaching:* Reports from several campuses describe course offerings that include faculty from two [or more] departments. For example, a bioinformatics course might include both computing and biology faculty, and students might be recruited from both fields. In the humanities, examples might include faculty and students from music or art as well as computing. With two faculty for one course, staffing costs can be expensive, but students see contrasting perspectives on the subject at hand.
- *Faculty-faculty tutorials:* On some campuses, programs (sometimes funded by grant(s) or a dean's office) may provide support through which one faculty member can study another's field (or two faculty can methodically explore each other's subjects). Such tutorials often promote interdisciplinary scholarship, grants, and publications.

- *Research/development opportunities:* Discussions between computing faculty and other scholars may uncover computing needs. At Grinnell, such conversations often identify projects, in which computing students can contribute to research teams throughout a college or university. When student demand for computing-based projects exceeds what computing faculty can support, interdisciplinary connections may help provide other opportunities.

In summary, compelling motivations (philosophical, strategic, practical) exist for computing programs to reach out to outside disciplines and engage in interdisciplinary projects. However, today's circumstances also may present substantial obstacles for expanded engagement. A range of approaches for collaboration may be available for collaboration, although details almost certainly depend upon local circumstances.

CHAPTER 9

The role of programming in introductory computing courses

This column first appeared in ACM *Inroads*, Volume 1, Number 2, June 2010, pages 12–15[234] Subsequent reflections on this subject appear later in this chapter.

9.1 ORIGINAL COLUMN

SHOULD PROGRAMMING BE PART of introductory computing courses, and if so, what role should programming play?

Many [most?] introductory computing courses, both for non-majors and for potential majors, expect students to write computer programs. Some educators, however, criticize these courses as overemphasizing details and ignoring broad themes. More generally, computing faculty regularly debate the appropriate content for introductory computing courses. The question opening this column addresses one important component of this more general discussion.

Context: SIGCSE 2010 Panel

At SIGCSE2010, Michael Goldweber organized a panel on "What Everyone Needs to Know About Computation" [72]. My presentation raised issues related to problem solving, the need for precision of exposition, and the choice of a language in the development and specification of algorithms and data structures. This column expands my brief comments at that session to the more general issue of the role of programming in introductory computing courses.

Some Themes

Over the years, I have heard numerous discussions regarding programming at the introductory level. From my perspective, a significant fraction of the comments and arguments fall into three categories: problem solving, laying a foundation for later courses, and job preparation. The next subsections examine these themes in some detail.

Problem Solving

From my perspective, problem solving is central to the discipline of computing. Computers have become important in contemporary life, because computers help people solve problems. Computers assist people in their work or aid in tasks that people want to do. Further, the popular phrase "computational thinking" [289] highlights the importance of using ideas of computation to help solve problems.

In introductory computing courses, problem solving also motivates many students. Some students particularly like the intellectual challenge of tackling hard and tantalizing problems, and some students find motivation when contemplating how computing can help people.

Of course, problem solving is not unique to computing, but computing does allow practice and insights for many applications. Further, computing can integrate experimentation, mathematical rigor, and engineering methodologies.

With this importance of problem solving in computing, a fundamental question arises: how can our courses help students develop their problem-solving skills? A natural answer would seem to be: Practice! Courses at all levels should ask students to tackle a wide variety of problems.

Unfortunately, practice in problem solving can be hindered by a basic difficulty: many environments allow people to be sloppy. Reasoning can be inconsistent, approaches can be incomplete, plans may be ill-defined, suggested solutions may represent more wishful thinking than careful analysis, etc. These difficulties are compounded because natural language (e.g., English) is imprecise, ambiguous, and contradictory. High-level rhetoric can mask problematic details. Vague generalizations can allow people to avoid facing hard problems.

Introductory courses, therefore, must address several vital questions:

- How should solutions be expressed?
- How can the correctness of proposed solutions be evaluated?
- How can alternative solutions be compared?

The fundamental answer is that students must be able to write solutions with sufficient care that they can be analyzed. Solutions cannot involve just high-level abstractions, but also must extend down to various details. Thus, the subtitle of my presentation at the SIGCSE 2010 Panel was "the Devil is in the Details".

But if English or another natural language is too vague for expressing solutions, how should students express their work? Historically (particularly in the 1980s), one approach used pseudo-code—a stylized language may seem more natural than a formal programming language. To obtain adequate precision, however, an acceptable pseudo-code had to have a tight syntax and semantics. Thus, in practice, a sufficiently formal pseudo-code needed the same level of specification as an actual programming language. As a result, pseudo-code had few advantages over programming languages for problem solving, but pseudo-code solutions had the disadvantage of being unable to execute on machines as a check to correctness. Perhaps these limitations on pseudo-code have been part of the reason why pseudo-code has not been widely accepted and used over the years in introductory courses.

In summary, an emphasis on "problem solving" in introductory computing courses leads naturally to writing solutions in a programming language. To be helpful to students, solutions must be stated with reasonable precision, and students must be able to analyze and compare solutions. Programming provides this framework, but most other approaches(e.g., written English) do not.

Laying a Foundation for Later Courses

Introductory courses often provide a framework for later courses. In computing, many upper-level courses require students to program and test solutions. Thus, rather than cover the same programming topics in each course, upper-level courses commonly expect students to have mastered the basics of programming at the beginning level.

At my own institution, for example, we were able to upgrade and streamline the curriculum substantially by collecting several overlapping topics together into a single course, positioned reasonably early in the curriculum. Previously, courses in computer architecture, operating systems, networks, and compilers all introduced programming in C. Students commented that seeing this material once or twice was indeed useful; but the repetition over four courses was excessive. Thus, my department developed a course on imperative problem solving and C, typically taken in a student's second year of study. With this background collected in a single course, later courses can cover substantially more material than was previously possible, and discussions in those later courses can focus more upon concepts and approaches than low-level programming constructs.

In planning the level of programming at the introductory level, two observations may help provide parameters for topic coverage and selection.

- If later courses will build on beginning courses, then those first courses must provide a reasonable level of mastery. If students from a prerequisite course do not have adequate background, the later course will likely have to repeat the material. Such repetition takes time and undermines the prerequisite structure.

  As a possible example, some proposals in the 1980s and early 1990s for a "breadth-first introductory CS course" often touched upon a wide range of interesting topics. However, after some experimentation, this approach was not widely adopted. Although many reasons have been cited, one hypothesis suggests that later courses had to repeat all subjects in the "breadth-first" course. Effectively, the "breadth-first" introduction added an extra course to the major. With pressure to insert many topics into the curriculum, it could be difficult to fit a broad survey into an already full major. This experience suggests that whatever is covered at the beginning level must be substantive enough to allow later courses to build upon them.

- If introductory courses over emphasize extensive details and language idiosyncrasies, then students may not see the wealth of concepts and applications that have made computers vital in contemporary life.

  For over 20 years, many computing educators have promoted a "breadth-first" approach for introductory courses. Computing involves algorithms, data structures, abstraction, hardware, problem solving, applications that help people, and much more. However, students often have little idea of this breadth, and misconceptions abound. For example, informal conversations suggest that many students believe that games comprise 50% or more of the revenue generated by IT, when in fact gross revenues for games account for under 2% of IT revenue in North America [225]. Correcting such misconceptions and publicizing the breadth of computing can have a significant impact in recruiting the many prospective computing students who have diverse interests and want to contribute to society.

  However, some programming-oriented courses focus upon syntax, language semantics, library functions, system calls, coding tricks, and the like. At the extreme, coding details might displace any emphasis on problem solving at all. Syntax and precision are necessary for describing solutions to problems, but it is reasonable to question to what extent people should be asked to be compilers in search of syntax errors.

Overall, a moderate coverage of programming at the introductory level seems necessary to provide a base for later courses, while excessive attention to details can detract from the higher goal of practice with problem solving.

Job Preparation

Computing regularly ranks high among fields for interesting and well-paying jobs; and the possibility of strong job prospects resonates with many students. Often job preparation motivates the incorporation of practical programming experiences throughout a computing curriculum. Possible internships, co-op programs, summer jobs, part-time employment, or work-study options may provide additional motivation for extensive programming in introductory courses.

Of course, each computing faculty must determine its response to these pressures to prepare for both short- and long-term career preparation. Each undergraduate program has its own priorities, and these will influence how programming might fit within introductory computing courses.

Even when high-level perspectives have been resolved, however, several factors can complicate practical decisions regarding which problem-solving approach(es) will be supported by what language(s).

- CS Curriculum 2008 articulates renewed emphasis for programming in the computing curriculum. For example, Section 3.2.1 states, "Attention is drawn to the need for increased levels of care and attention in the teaching of basic programming" [7].

- Faculty might choose a language for an introductory course in order to illustrate problem-solving techniques and provide a gentle beginning to programming. For example, some environments (e.g., Alice, Dr. Scheme, Greenfoot, Logo, Scratch) may be designed primarily for beginners. However, the important pedagogical and learning goals of introductory courses may yield different language choices from languages used by professional programmers in industry.

- Since different languages support different views of problem solving, Section 9.1.5 of CS Curriculum 2008 adds a new requirement, "Exposure to different programming paradigms", and Section3.2.4 states, "Computer science professionals frequently use different programming languages for different purposes and must be able to learn new languages over their careers as the field evolves. As a result, students must recognize the benefits of learning and applying new programming languages. It is also important for students to recognize that the choice of programming paradigm can significantly influence the way one thinks about problems and expresses solutions of these problems. To this end, we believe that all students must learn to program in more than one paradigm." [7]

- Different jobs likely require experience with different languages. For example, an April, 2010, TIOBE study reports the relative popularity of various programming languages. In examining articles indexed by several search engines, no language commanded more than 18% of the references examined, only two languages (C and Java) had ratings over 10%, and only five languages (C, Java, C++, PHP, and BASIC or Visual Basic) had ratings over 5% [178]. Language popularity by other measures almost certainly would yield variations in specifics, but the main conclusion is clear: software developers use a wide range of languages in different settings. When schools seek to prepare students for varying careers, therefore, it can be difficult to know just what language(s) should be covered.

- Programming languages evolve; sometimes new languages supplant earlier ones. For example, TIOBE reports [178] track changes in language popularity over time. Also, based on a calculation from a 2000 workshop, William Wulf reported the "half-life of engineering knowledge" as between 2.5 years and 7.5 years, and Wulf concluded, "half of what we are teaching our students in some fields (computer science, by the way, was the field of 2.5 years) is obsolete by the time they [students] graduate" [290, p. 6]. Whatever language might be taught in an introductory course, different languages and perspectives likely will be needed for many careers four or five years later.

Conclusions

Problem solving is a central component of introductory computing and computational thinking [289]. Further, the language used to articulate solutions matters by shaping our thinking. Thus, Herbert and Stuart Dreyfus observe:

> The precision essential to a computer's way of manipulating symbols constitute both a great advantage and a severe limitation. Since what the symbols in a computer represent must be absolutely precise, and the programming must be absolutely clear in defining each symbol, the attempt to write a computer program inevitably exposes hand-waving, fuzzy thinking, and implicit appeals to what everyone takes for granted. Submitting to this rigor is an immensely valuable discipline [60, p. 54].

All students (including both non-majors and potential majors) benefit from sharpening their problem-solving skills; and precision in exposition is particularly helpful in presenting, analyzing, and testing proposed solutions. Programming provides an essential vehicle for this needed precision.

Programming in introductory courses also can provide a base for work in later courses, allowing upper-level courses to emphasize concepts and high-level thinking skills. Programming also can provide experience that can help prepare students for jobs and careers.

Too little emphasis on programming may lead to repetition of coverage later on—taking valuable time out of subsequent courses. Too much emphasis, however, can overwhelm students with technical details and distract students from the connection between computing and its use in helping people.

Further, faculty face significant challenges in selecting any specific language, since different types of problems encourage different problem-solving paradigms, and different employers and jobs utilize different languages . Altogether, many factors suggest that programming has a role in many introductory computing courses, but great care is needed in clarifying that role. For each program and school, faculty will need to chart a path among conflicting demands and choices.

Acknowledgment

Many thanks to Marge Coahran who reviewed a draft of this column and provided several insightful suggestions!

9.2 SUBSEQUENT REFLECTIONS

The original column discussed three themes related to the role of programming in introductory courses: problem solving laying a foundation for later courses, and job preparation.

The Subsequent Reflections section at the end of Chapter 47, "The balance between programming and other assignments," identifies three audiences for introductory courses: non-computing majors, general audiences, and computing majors or potential majors.

Chapter 47's Subsequent Reflections section also contains an extended discussion of the role of programming for those various audiences. Beyond that discussion, two additional thoughts come to mind: Faculty and courses must consider whether

1. programming in a specific language is an end in itself (e.g., to get a job or as a tool for another discipline), or
2. programming is a vehicle for high-level goals (e.g., to sharpen problem solving skills).

Programming as an end or a vehicle

In some settings, student proficiency with details of a particular language may address the needs of a local client (e.g., an employer or academic program): local employers may want interns and employees to have experience with Java, Python, Ruby on Rails, or other language, engineering departments may require background in C or MatLab, the social sciences may expect knowledge of R, MySQL, or other tools. In meeting these needs, however, computing faculty also must consider underlying principles: details of a language evolve from one version to another, and language preferences change (e.g., many engineering departments used to require FORTRAN, but now C or MatLab seems quite common).

In other environments, programming may serve as a vehicle to meet other objectives, such as enhancement of problem-solving skills, practice using contrasting problem-solving paradigms, development and analysis of algorithms, or experience structuring large-scale software projects. With such high-level priorities, language selection and details may be of secondary importance. At some schools, for example, different CS1 sections may utilize different languages, although all cover identified core topics (e.g., conditional statements, iteration, recursion, basic data structures). Students enter CS2 with a common understanding of approaches for problem solving, but syntax details are not considered significant.

Paradigms and Languages

Over the years, I have heard many energetic discussions regarding the choice of language for an introductory course. In a few cases, the underlying issue seemed to be preparation for jobs or opportunities in the local area, in which case specifics of a language had a high priority. However, in many discussions, the underlying issue seemed less about a specific language than about a choice of problem-solving paradigm.

For example, one group might argue the need for organization of programs according to object-oriented principles, while another the need for a top-down (e.g., imperative) methodology, and a third group the use of functions in higher-order procedures.

Upon reflection, many past debates seemed focused upon language details and whether a language supported specific constructs. Perhaps a better approach would consider what paradigm(s) might be emphasized in introductory courses; specific languages can be chosen later. Certainly languages have different capabilities, advantages, and disadvantages, but discussions of language may not be productive when the underlying issue involves the choice of paradigm (e.g., functional, imperative, object-oriented, declarative, or event-driven).

Summary: Do not confuse language wars with paradigm discussions.

CHAPTER 10

Motivating students and working with gifted students

This article appears in this book for the first time, demonstrating that new topics might be considered as subjects for future columns.

10.1 NEW ARTICLE

WHEN ENROLLED STUDENTS ARE DIVERSE, providing an appropriate level of content, challenge, and support for each student can require substantial creativity and effort. In shaping a course to meet student needs, at least three major variables should be considered: students' current background and skills, students' motivations, and students' level of confidence.

Students' current background and skills

From the perspective of course content and basic format, a student's learning within a course depends on such factors as the student's

- prior technical background,
- problem-solving experience and ability,
- communication skills, including the student's proficiency in reading, writing, oral communication, and
- ability to work in groups (e.g., pairs).

In my experience, students' ability in these areas need not be connected. Students with extensive technical backgrounds may know details and tricks-of-the-trade, but may not be creative problem solvers. Students who seem to be gifted problem solvers (with or without strong computing backgrounds) may have poor communication and/or inter-personal skills.

Further, students' levels of competence in each of these areas evolve over time. For example, some of my best graduating computing majors started with very little computing background. Also, since they entered college with little exposure to computing, they started with limited problem-solving experience. Thus, at the start of their college careers, few, if any. of these students would be identified as being gifted or knowledgeable.

On the other hand, some incoming students with considerable technical background may enjoy wallowing in the complexity of their programs and hacking on details—largely ignoring higher-level problems solving methodologies, data structures, and algorithms.

Altogether, some experienced incoming students flourish and remain strong through graduation, but other experienced students do not progress as their initial background might have predicted. Similarly, some students who enter college as true novices become truly outstanding computing majors, while others become capable but undistinguished majors.

Students' Motivations

Just as students different in their backgrounds and abilities, they also may respond to dramatically different environments. The following list suggests some common motivations:

- intellectual challenge: some students truly enjoy tackling difficult or thought-provoking problems,
- technical involvement: some students want to delve into elements of computing to explore technical complexities and nuances,
- potential to help others: some students observe social needs, locally or globally, and want to make a positive difference in the lives of others,
- employment and/or a career: some students focus upon obtaining jobs to support families, gain financial independence, enter a rewarding career, etc.,
- money: some students seek wealth—perhaps hoping to become millionaires before they reach age 25 or 30,
- grades: some students will do what is required to achieve a targeted grade level (perhaps to get A's, but perhaps just to get by with B's or C's),
- stature: some students want to gain recognition among their peers,
- video games: some students seem focused on the design, creation, and and interaction with video games, and
- please parents: some students seem externally driven to satisfy parents or families.

In the latter category, I know one student whose family literally sold their farm, so the student could attend college and then have the means to support all of the student's siblings as they trained for possible careers.

Undoubtedly, many more motivations could be added to this type of list of motivations. Although students at any specific campus have different motivations, faculty at various school suggest that one or a few of these perspectives seem to predominate in local environments. For example, I know one school where most studentw respond to intellectual challenges and the potential to help others, while at another school students will expend great effort if they think an activity will lead to a good job.

Students' Levels of Confidence

Just as preparation and motivation vary dramatically among students, so does the level of self confidence. Some students (often white males) may have extremely high regard for their abilities and skills. For some of these students, a low grade on an exercise or test is taken to indicate a poorly prepared assignment or misguided exam rather than an indication of the students' knowledge.

On the other side, some students (often women and those from underrepresented ethic minorities) may have little confidence, perhaps wondering if they belong in the course at all. For some of these students, an outstanding performance on an exercise or test may be attributed to luck, or perhaps the grader missed the crucial flaw. (For more discussion of this perspective, see the Subsequent Reflections for Chapter 38—especially the comments concerning the Imposter Phenomenon.)

To illustrate some in-class implications of students who are over- or under-confident, consider the well-meaning instructor who, in an introductory course, asks, "are there any students who have not compiled a program?" or "does everyone know about file protection codes in Linux?" In a course for beginners, many students may lack this background. However, an over-confident student may claim understanding, because such matters may seem minor and thus not worth worrying about. In contrast, insecure students may believe they are the only ones who lack this knowledge, supporting their nervousness that they do not belong and reinforcing their feeling that they should consider dropping the course.

Approaches for Courses

In attempting to handle diversity (e.g., differing backgrounds, motivations, and confidence levels) within a classroom, several different approaches might be considered, including the course structure, student activities, and mechanisms to motivate students. Throughout, however, an instructor might want to remember several cautions.

Structural Approaches

Some ways to respond to diverse populations, backgrounds, and interests are structural: students might be separated into different sections with different outlooks.

- Students with a moderate level of experience might be placed into one section, and true novices into another. Initial placement of students into these sections may present a challenge (how much background does each student actually have?), but placement tests, questionnaires, or personal interviews might help. Separate sections may not completely resolve diversity challenges, but narrowing the range of students in each section may help focus content and the type of support offered—at least somewhat.

- One or more CS1 or CS2 courses might highlight an application area that may be different for many incoming students. Three examples include image processing [78], the control of robots [102], and connections with music [84, 117]. Such applications can provide new and interesting opportunities to apply computing techniques, even if some data structures and algorithms may be familiar to experienced students.

- Focusing upon topic coverage or presentation, one or more introductory courses might be organized, so that the content may be equally new to most students. For example, utilizing functional problem solving in CS1 may seem unfamiliar, so both beginners and experienced students may start on an equal footing.

- Different introductory sections might adopt different pedagogical approaches. For example, on one campus (for physics), one semester the introductory course utilizes a workshop or lab-based format (e.g., following Chapter 28 or [184], and the next semester, the course utilizes a somewhat more-traditional format with lectures and small-group discussions. To communicate the alternatives to students, course descriptions are circulated with registration materials each semester.

- On one campus, students with substantial background (e.g., Advanced Placement or International Baccalaureate computer science) may have strong backgrounds in some areas, but may not fit easily into the school's second or third computing course. Rather than mixing this group with true beginners, a special CS1.5 course builds upon the students' background in an accelerated framework, so these students can jump ahead in the curriculum while not intimidating the novices.

Student Activities

Some ideas for meeting students at their own levels and backgrounds follow.

- In my introductory courses, I typically present a range of problems—some required and some possibly for extra credit. Particularly for the optional problems, the exercises include problems of varying difficulty—from easy to moderate to challenging. In this framework, motivated novices might solve some problems, while experienced students might be challenged by other problems—but all students can have success, without stigma about which problems they attempt and solve.

- I also present students with a range of labs—some required and some for extra credit. For all labs, motivated novices should be able to work through required steps, although some optional parts might be included as enrichment (possibly for experienced students, but also for those with considerable motivation).

- In-class student mentors and evening lab assistants can be very helpful, particularly for helping novices in learning details of a lab environment, interpreting compiler/operating system messages, and making suggestions for students who become stuck on steps for a laboratory exercise. When utilizing students as mentors and lab assistants, care is needed to clarify parameters: what types of activities can assistants help with, how much help is appropriate, to what extent can assistants provide answers rather than suggest directions, etc. Overall, the goal is to provide guidance on activities for which collaboration is allowed, but to clarify what, if any, help is appropriate when an assignment calls for a student to work without collaboration.

  When designed well, the use of student assistants can have at least three benefits:

  1. Assistants can provide assistance and reassurance to students having trouble, particularly if the students are just getting started.
  2. Assistants can serve as role models—upper-level students who may have started with little background, but who now are clearly successful.
  3. Assistants can provide upper-level students (even second-year students) with reassurance and a sense of self confidence; even if assistants started their careers with little background, they are recognized as being knowledgeable with fine communication skills.

- In creating tests, I tell my students that 1/3 to 1/2 of each test will consist of problems from labs or other homework. In my courses, such material covers most of the course content, so such coverage parallels the course well. Further, drawing test questions from labs and other homework highlights the need for students to work through homework material. Also, students who feel insecure can gain some confidence that they will have seen many test questions, if they work labs and homework regularly.

- After grading and returning student tests, I often allow students to redo questions for which they have lost more than 20% of the possible points. (Realistically, if students lose 20% of the points or fewer, it seems unlikely that students will learn much from revising a question, and I do not want to have to grade that material again. On the other hand, if a student loses more than 20%, there likely is something the student should look at again for that question.) If a question requires identifying output from a program or otherwise producing answers that are easily checked (e.g., online), then in addition to a revision of a question I also require a paragraph or more explanation.

  When students submit the revision, I often average their original test score with the revision, so the original counts a substantial amount but they also have an incentive to look again at material which may have been unclear. In my teaching, the possibility of a test revision is received very well by students, and I have found students often take the opportunity to look back over topics than need further work.

- Some faculty around the country report framing a grading rubric, so that students doing solid, but basic, work on an assignment will receive B's. Work beyond the basics is aimed at encouraging students to extend the foundational assignment—perhaps adding their own creative embellishments or added features. Although some faculty indicate that this approach provides a reasonably safe foundation for novices while providing potential challenges to more experienced students, care must be needed to avoid at least three potential pitfalls.

  1. For true novices, writing programs that meet the basic assignment may require substantial development of problem-solving and technical skills. When A's require doing even more (likely requiring dramatically more time and effort), such a grading policy may imply that novices at the start of a course cannot expect to achieve grades higher than B's. In summary, this type of policy may frustrate novices.
  2. Experienced students, on the other hand, may be able to develop programs for a basic assignment with relatively little time and effort. For these students, extending assignments may yield A's with only modest work. In summary, this type of policy may cause experienced students to think of the course as an "easy A".
  3. In many real-world applications, programs must be written to clients' needs; adding extra features can introduce errors, increase code complexity, and complicate code maintenance—all without helping with the actual needs of a client. With this in mind, students can (and should) be encouraged to be creative in addressing clients' needs, but students should not be given the message that added elements are always beneficial.

     To partially address such potential difficulties, an assignment might provide both a collection of required features and some optional additional capabilities. Students can be creative in developing software to address these features and capabilities, but they also will be reminded that the eventual goal of software is often to meet clients' needs.

Student Motivation

In my experience, determining how to motivate students seems largely a local matter. For example, when I have used the lab-based approach discussed in Chapter 28,

- On some campuses, students feel pressure to do [at least some of] the reading ahead of time, because they do not want to be embarrassed before their peers. If they come to class unprepared, they would display their lack of knowledge to their lab partner, and would feel awkward if they had to acknowledge they had not done the reading.
- On at least one campus, students seem to have little interest in coming to class prepared. When they talk to their lab partner, they might readily indicate they are unprepared. Sometimes, students even seem to boast about how unprepared they are.

As this experience suggests, there seems little reason to believe that one approach or collecting of techniques is likely to motivate students across all schools.

However, each student likely is motivated by something. In talking to faculty at different schools, each may have its own culture, and tapping into that environment may provide hints for motivating students. Perhaps one can highlight intellectual challenge, describe applications that connect to addressing human needs, include many quizzes or exercises that impact a final grade, utilize projects that relate to job opportunities in the area, etc.

Cautions

While being pro-active in addressing the needs of one group, an instructor also is advised to be careful not to discourage others.

- In talking to students, be careful NOT to say, "this idea is easy" or "of course, you know this." Since students' background and experience vary, what may be straightforward for some may also be challenging for others. If an instructor indicates a concept or exercise will be simple, and if students have difficulty (perhaps because of a weak background), then the students may feel that they do not belong—even if they are doing well and making good progress. Altogether, instructors are advised not to indicate that some topic is easy or straightforward.
- Be careful about telling a student that their situation is hopeless, since such messages may be self fulfilling. As an example, years ago one student in my introductory programming course started the semester with remarkably little background and problem-solving ability. Each week, the student faced a new challenge, but consistently expanded their background and worked through required content. By the end of the semester, the student's background, abilities, and problem-solving skills had developed substantially. In this case, I was initially skeptical about the student's chances of success. However, I kept working with the student, and the student never gave up. By graduation, the student had earned a prestigious award for further study. Of course, such success stories do not always occur, but they arise with sufficient frequency that they should not be ignored. The basic lesson is that students with weak backgrounds can excel over time—if both they and their instructor persist!
- In setting course expectations, try not to frame the grading process, so that only experienced students can achieve strong grades. Although instructors may be tempted to set standards to reward efforts by experienced students, this approach also may discourage novices and prevent them from flourishing and achieving long-term success.

CHAPTER 11

Capstone-, research-, and project-experiences

This article appears in this book for the first time, demonstrating that new topics might be considered as subjects for future columns.

11.1 NEW ARTICLE

MANY DISCUSSIONS AMONG COMPUTING FACULTY FOCUS upon introductory courses, such as CS1, CS2 and breadth courses (e.g., CS Principles). This article describes approaches and experiences at the other end of the computing curriculum. Often, courses at the senior, undergraduate level seek to build upon material learned earlier and to encourage students to integrate disparate themes within a new and creative context.

In describing a special learning experience near the end of a computing program, this chapter follows common practice in using the term "capstone" to reflect an activity that

- brings ideas from [some] previous courses together and
- asks students to utilize that material in some new context or new environment.

Although many types of senior courses may be appropriate, focusing on only one of these elements, the term "capstone" here requires both; endeavors with just one of these elements are not considered "capstones" here. To clarify further, a "capstone", as used here, need not draw upon every prior course in a curriculum, but a capstone should build upon several ideas covered previously—likely in several different courses.

In describing capstone experiences, Boyer's discussion of scholarship, involving four main categories, provides a helpful starting place. A second, independent consideration focuses upon whether a capstone experience should be done individually or within a group framework. After reviewing these perspectives, this article provides four contrasting examples of capstone courses for each of the four types of creative scholarship identified by Boyer.

Boyer's Categories for Scholarship also may Apply to Capstones

Within the parameters of a capstone experience, breadth and flexibility seem important when evaluating the notion of "something new." One approach draws from insights from Ernest Boyer, who identifies four types of scholarship: "the scholarship of discovery", "the scholarship of integration", "the scholarship of application", and the "scholarship of teaching and learning" [27]. Traditional research fits within the "scholarship of discovery," but scholarship (creating new understandings, insights, and artifacts) encompasses substantially more breadth than just traditional research. In addition, Boyer's categorization acknowledges that faculty's scholarly endeavors may evolve. One project may involve the scholarship of discovery, but the next may involve integration, application, or teaching and learning.

For computing faculty, some activities may relate to new algorithms or data structures (e.g., discovery), but the development of software packages may draw upon database principles, algorithms, human-computer interfaces, and other matters (perhaps integration), and the creation of tools may involve the creative combining of existing technologies (perhaps application). And sometimes evolving software packages, hardware units, or re-thought course materials may relate to course formats and pedagogy (e.g., teaching and learning). Altogether, some computing faculty may focus upon a traditional form of research (e.g., the scholarship of discovery), but many faculty work on a range of projects—perhaps focusing on other types of scholarship for all or part of their careers.

Similarly, upper-level students have a range of interests and motivations. Some may have interests in new discoveries, but many want to develop tools or write software packages to address local needs of people or organizations. In other cases, upper-level computing students may want to reach out to K-12 students or to beginning college students to provide encouragement and support as individuals first encounter the world of computing. In considering these varying perspectives, Boyer's insightful framework for faculty scholarship also can provide a helpful framework for a consideration of student capstone experiences.

Just as one type of scholarship does not match all intellectual endeavors of computing faculty, one type of capstone experience will not likely meet the needs and interests of all students. Some students may be prepared and motivated by a capstone of discovery, but others may find such work unmotivating and unrelated to their goals of helping others or gaining career-oriented experience in developing large-scale software projects.

Teams or Individual Work?

In some academic work, individuals are expected to work individually. For example, many courses require in-class tests and final course examinations to be done without collaboration—often without consultation of notes or outside sources (sometimes called "closed-book assessments"). Further, some schools require an honors thesis to reflect the work of one student; a thesis may not represent a collaborative effort by two or more students. Perhaps in these contexts, schools want to be able to attest to an individual student's achievements and abilities.

Although individual work certainly has a place within the discipline of computing, most contemporary software development represents team efforts. Software often is too large and complex to be created and maintained by one person. Also, many applications require expert knowledge of the application domain, algorithms, data structures, human-computer interfaces, etc. With such a wide range of backgrounds needed for a single product, a development team often requires multiple contributors with separate specialities and insights.

Putting Boyer's categorization of scholarship together with the nature of work within the field of computing, it seems natural and important that some capstone experience support work by an individual, but other experiences utilize work by teams. For example, capstone experiences involving integration and/or application likely are well served by collaboration. In many cases, a software package developed by one individual will not have the scope, depth, and range of insight that is likely obtained through a team effort.

As a separate matter beyond practices of individual or team work, faculty widely observe that supporting and monitoring student projects often is time consuming. An instructor must review objectives, specifications, design decisions, data structures, algorithms, interfaces, etc.. Further, an instructor must Interact regularly with students involved in a project, providing feedback, offering alternatives, following progress, etc.

With substantial time commitments involved with faculty oversight of capstone experiences, large student demand can yield severe time commitments for supervising faculty. One school of which I am aware has a graduation requirement that every student must complete a senior project under the guidance of a faculty member. With enrollments soaring, the computing faculty now require most such senior projects to entail collaboration. Occasionally, an honors student is allowed to delve into a discovery-based project individually, but most students are strongly encouraged (required?) to form teams for projects involving the scholarship of integration, application, or teaching and learning.

In summary, team-based projects and collaborative activities seem preferable for many computing-oriented capstone exercises over work by individuals. Collaborative, group-based work matches the development teams found in many computing enterprises, but also has the practical advantage that a relatively modest number of faculty may be able to accommodate large numbers of students.

Capstone Experiences Highlighting the Scholarship of Discovery

Using Boyer's categorization of scholarship as a helpful way to organize thinking about capstone experiences, this section and the next three sections each provide four contrasting examples for each of Boyer's four categories. As its title suggests, this section focuses upon contrasting approaches for capstone experiences involving the scholarship of discovery.

Disappointing Observations

Over the years, I have attended many student poster sessions at numerous regional conferences. In many cases, I appreciate the student effort, but I am disappointed about the quality of the work.

- In some cases, the instructor told the students to do a particular task (e.g., write a specific program, collect particular data), but the students seem unaware of how their work fits into a larger picture.
- In some cases, the students have expended considerable time and energy, but the results largely seem to duplicate parts of studies done elsewhere.
- In some cases, students are writing programs that, while new to them, were explored a decade earlier, and the field has move well beyond that point now.

In each of these settings, students may have learned from their labors, but the efforts do not seem to lead to new understandings for the discipline. From this perspective, these projects seem to be formal exercises to meet requirements rather than an effort that might be considered scholarship of discovery.

Final Products from Capstones of Discovery

Rather than repeating past experiments and well-known results as mechanisms to satisfy some type of capstone or project requirement, it seems appropriate to consider the possible impact of each project as it begins. Commonly, when a capstone involves some type of project, a school or course requires students to prepare a proposal that includes a description of the project's goal, a review of the history or literature related to the project, the approach to be followed, and a possible timeline for completing the work. That is, at the start of a project, students must consider the goal of the project and how the project will begin.

In addition, when proposing an independent project, some colleges require the proposal to identify where the project write-up might be published or presented—assuming all goes well. Some end-of-semester possibilities might include:

- *the ACM Student Research Competition:* In this competition, students initially enter their work within a conference framework. (At this writing, competitions were included within about two dozen different conferences.) Winners at these conferences then compete in a "Grand Finals". (See [6] for current news and a list of conferences involved with this program.)
- *a formal paper:* At the end of a project, students prepare a formal paper in a format that could be submitted for publication.
  - ◇ If the students (perhaps in collaboration with a faculty member) achieve a substantial result, then students can submit the paper to a national or regional conference. Further, papers accepted and presented at international or national (ACM) or regional (CCSC) conferences are indexed in the ACM Digital Library, so students can build their resumes while they are gaining experience and developing as scholars.
  - ◇ If a project's results provide some insights for future work, but do not merit publication, then students might write a paper to the next team, with the general theme, "here are techniques and insights we gained, which we wish we had known when we started." If well constructed, one team's paper can help jump-start their successors.
- *a carefully-prepared poster:* Since student project typically entail substantial work on a worthwhile endeavor, the project likely can be described in a professional-style poster and presented at a national or regional conference. Certainly students who present a paper also could develop a poster (two conferences activities add richness to a student's conference experience), but even students whose work is incomplete or whose results are lackluster can organize and synthesize what they have done. Presenting at regional conferences typically requires only a modest time commitment, but still gives students a worthwhile experience, provides visibility for their work, and allows them to gain feedback from a broad audience.

The key point here is that student projects, with the potential for scholarship of discovery, should include not only a proposal and work to obtain software or results, but also the write-up of a polished paper (in a publishable format) and a well-constructed poster. Work, even outstanding work, is not helpful if no one knows about it, so students need the experience of finishing their projects by writing them up.

In addition, all posters (and perhaps papers) whether posted or discussed at regional or national events, can be presented and displayed within a home department. For example, a computing department might organize one or more presentations about recent projects, as a means to advertise activities and options to prospective or incoming students. Also, displaying posters within a computing department presents visible evidence that the program, faculty, and students are active and engaged in interesting and worthwhile activities!

Four Contexts for Capstones of Discovery

Capstone experiences highlighting the scholarship of discovery may arise in many ways. Descriptions of four contrasting contexts for these experiences follow.

- *individual student-developed:* For some projects, a student may propose a subject for study or research, or perhaps a student works individually on a topic suggested by a faculty member. In this setting, a student may have a vision for new research, and the project allows the student to pursue her or his interests. Interestingly, each of the winners of the 2017 Grand Finals for the ACM Student Research Competition for both the graduate and undergraduate divisions was reported as involving an individual student. [6]. This type of scholarship seems consistent with traditional forms of research, with a student researcher working alone on a project, resulting with a concluding thesis or paper. One challenge with this type of project is that an instructor may or may not have expertise with the chosen topic, so supervising this type of project may require substantial time and effort.

- *self-contained team project:* Some projects may be proposed by either an individual student, a group of students, or an instructor, but once described, a group of students forms a team to move the project forward. Overall, this type of project expects students to brainstorm, explore, and investigate as a collaborative effort, with discussions and group interactions – perhaps including a faculty mentor at times to help guide the progress of the scholarship by discovery. In my experience, many posters at international and regional conferences represent efforts by such collaborative teams. I also know several faculty members who work regularly with student teams to obtain results that can be submitted to research-oriented conferences and journals.

- *long-term instructor-guided project:* Numerous faculty have long-term research agendas, which involve multiple components that are expected to combine into an eventual final product. As an example, a chemistry faculty member may identify a dozen experiments that are likely to provide evidence for an overall result. Sometimes, a faculty member may identify components and experiments that fit well within the capabilities of undergraduates. Within this framework, one project for a student or student team may provide one important piece of a larger body of work. Over time, the faculty member can work with recent students to prepare a substantial paper that combines these pieces into a coherent whole. In my experience, this model seems particularly common among computing faculty with long-term scholarly plans, and these faculty may have a consistent record of publications involving numerous student contributors.

- *Off-campus research experiences, such as REUs or work at independent labs:* Beyond on-campus efforts involving capstones of discovery, students may engage in projects off campus, such as national (governmental) labs, industrial research labs, or academic research labs (e.g., Research Experiences for Undergraduates or REUs in the United States). Unfortunately, in some cases, projects off campus may be subject to confidentiality agreements, so that students are not allowed to talk about the details of their work when they return to their home campus.

In many settings in which students engage in capstone experiences of discovery, students will likely require on-going and extensive faculty involvement for brainstorming, feedback, mentoring, and guidance.

Capstone Experiences Highlighting the Scholarship of Integration

This section reviews four contrasting approaches for capstone experiences involving the scholarship of integration.

- At Towson University, O'Leary describes a capstone laboratory course, in which students focus upon defensive and administrative elements of cyber security, building upon past courses on "operating systems, operating systems security, networking, and network security." In the course, "Students implement and configure defensive measures like log servers, intrusion detection systems and firewalls. They perform forensic analysis of compromised systems, and write extensive reports" [143, p. 429] Throughout, students apply background from earlier courses to a new and different framework.
- At Southern Utah University, Kesar describes a senior capstone course that examines "five real business capstone projects that were designed to provide an educational experiential learning to include ethics and professionalism in the pedagogy" [97, p. 432]. This course, designed for students from both computer science and computer information systems, explicitly builds upon background they have developed previously, but then extends their perspectives to an interdisciplinary framework involving ethical practices within business. Pedagogy includes both experiential learning and team-based activities.
- Some software development projects build upon many concepts, techniques, and skills developed in earlier courses. For example, a recent project of mine utilized student teams to develop a Bluetooth-based infrastructure, with which C programs, running on a workstation issue commands over Bluetooth to Scribbler 2 robots. Students in this effort drew upon past study of machine-level byte instructions, Bluetooth communications, blocking and non-blocking I/O and commands, threads and processes, mutex semaphores, message passing, and synchronization through pipes. [261].
- At several colleges, many projects engage students with on-going Free and Open Source (FOSS) projects, with some projects related to Humanitarian Free and Open Source Software (HFOSS). In this work, students build upon foundational work taken earlier, but then delve into a new project, understand current capabilities and needs, and then become involved in make contributions. Initial work on one of these FOSS projects may be modest. However, as students become more involved within a FOSS community, they may have expanded opportunities to contribute. Several team-based projects in the context of HFOSS efforts are nicely described by [64].

Capstone Experiences Highlighting the Scholarship of Application

Capstone experiences involving the scholarship of application typically involve team-based projects to build a software application. Two substantial issues involve the amount of time students can spend upon the project and the target audience for the completed package.

Timing and Credits

Computing curricula at the undergraduate level often are filled with numerous courses that endeavor to cover the full range of topics recommended by national professional societies, such as the Association for Computer Machinery (ACM) and the Computer Society of the Institute for Electrical and Electronics Engineers (IEEE-CS) [10]. With much material to cover, many programs have limited time to devote to a software project. In response, some schools must limit their requirement for a team-based project to a single semester, some organize the project into a two-semester sequence, and sometimes the overall project experience may extend to two years.

Target Audience/Client

Students often demonstrate considerable motivation when they work on projects that they envision and initiate. Thus, some courses allow students to describe and build their own applications—students are their own clients. Although quite common in the past, such project-oriented courses seem to be progressively less popular, as more schools seek clients beyond the students themselves. Sometimes, a student team may build a tool that will help the instructor, department, or program. In other cases, projects may target needs from another department or program within the school. In recent years, however, many team-based project courses try to identify clients outside of the school. Pragmatically, meeting the needs of an outside client requires students to consider external factors and identify client requirements; students should not simply add features that they might consider nifty, but rather must clarify and build capabilities that the client will actually find useful. This change to a client-oriented perspective often represents a significant outcome for students developing software packages that meet real needs. Williams et al [286] provide useful perspectives on selecting and utilizing outside clients for a team-based project.

Four Models

Four models suggest the range of possibilities for team-based capstone experiences of application, in which software development seeks to meet the needs of an actual client.

- *An independent project for a summer or one semester utilizing a largely internal client:* Many of my student-faculty projects over the years have focused upon software development to address opportunities and needs identified by either a colleague or me. For example, in Summer 1999, the Co-chairs for the SIGCSE 2000 Symposium asked me to develop an online system for the submission and review of papers to the conference. Although the overall goal of the endeavor seemed clear, few details were available—the SIGCSE organization had used only paper-based submissions previously, and online submission systems were rare. For this project, I worked with two students and several computing professionals to design and implement an appropriate system, with the ultimate test being whether the system met the needs of the SIGCSE 2000 conference. Over the next 15 years, this system expanded to include about 60,000 lines of code and served SIGCSE symposia and several other conferences—but the entire concept for this work began as a student-faculty project. [278, 282]

  Overall, this example illustrates two points. First, students working as an independent team can tackle challenging problems that meet specific needs within the context of a team-based project under the guidance of a mentor. Second, for a successful project, the supported activity may continue and expand long after the student development team has graduated. From a faculty member's perspective, part of the initial consideration of a proposal also should include planning for project maintenance and development in the years following initial development.

- *One-semester course divided into two separate components: project methodology and a team-based project:* Within local scheduling constraints at Grinnell College, the curriculum could support only one, semester-long, four-credit course involving software development with a team-based project. Over several iterations of the course, students were able to learn many principles, and they could make a solid start with the project itself. However, often students were unable to reach a satisfying final software package, and many students wanted to continue their development the following semester as a guided independent project.

In response, faculty member, Janet Davis (now at Whitman College), reorganized the single four-credit course into two two-credit, semester-long courses. The first of these focused upon software development methodology and principles, and the second organized students into teams for a project. Further, the project for the second course represented a multi-semester effort to support an application for a local community organization, so students gained direct experience with meeting expectations for an external client.

In this environment, students could take the methodology/principles component as a pre- or co-requisite for the project component, and students could take the project component multiple times for credit. When repeating the project component, students might take a leadership role after their first time through the course.

- *A one- or two-semester team-based project for an external client:* Variations of one- or two-semester capstone projects are becoming reasonably common.
  - ◇ *One-semester variant:* For over 10 years, computer science students at the Universidad de Chile have worked on a one-semester "real world project with a real client, external to the university, during 15 weeks, devoting 16 hours a week of effective work in the client?s organization" [24, p. 137].
  - ◇ *Two-semester variant:* Since the 1960s, Harvey Mudd College has offered a clinic, first within engineering and more recently within computer science, in which "students provide products of genuine value to external clients based on the clients' needs." Each project involves a team of approximately four students plus a faculty advisor and a representative of the sponsoring organization, and projects extend for a full academic year. [82].
- *A two-year team-based project for an external client:* At Rose-Hulman Institute of Technology, Mohan, Chenoweth and Bohner describe a two-semester senior project that builds upon preparatory material covered in a "Junior Design Sequence". In this framework, effective progress on the senior project requires a substantial level of background; students cannot begin their study of large-scale software development at the same time they begin their senior project. Rather the Junior Design Sequence provides a rich framework, and students then are well prepared for a meaningful year-long senior project. [131]

Capstone Experiences Highlighting the Scholarship of Teaching and Learning

The scholarship of teaching and learning involves organizing and synthesizing material utilizing a new class format or pedagogy to connect in an innovative way to a new or existing audience. Four, sharply contrasting examples follow.

- Suppose an instructor wants to develop a new, lab-based course, either moving from a traditionally-formatted course with separate classroom meetings and lab or changing the content of the course in a fundamental way. The lab-based format requires that students work through a different lab almost every day. Thus, if the course meets four days a week for 14 weeks, the lab-based format requires 50-56 separate labs. Although development of such labs can be intimidating for an instructor, upper-level students can draw upon their background and experience to help organize and write the readings and labs. In this context, developers can draw upon their previous knowledge, but they also must identify subject-matter priorities and consider the organization and presentation of topics to the target audience. Further, the developers will need to develop examples and lab-based exercises, as well as identify and/or write supporting readings. In this work, a team-based approach is particularly helpful, as one team can draft readings and labs, another group can review readings and work through labs to give feedback. The original team then can revise each piece, based on feedback. Reference [194] provides details for this type of a student-faculty collaboration involving the incorporation of robots in an introductory course.

- Many contemporary efforts seek to interest K-12 students in computing and to provide some experience in computational thinking. A long-term hope is that K-12 students will find elements of computing of sufficient interest that they will take computing-related courses later on (e.g., in high school or college).

  Frequently, K-12 outreach efforts are led by faculty (e.g., see [145]), and such projects can be extremely successful. However, college students also are having an impact, with regular sessions through the year (e.g., on Saturdays), with tutoring or interactions with local schools, and with involvement with "kid's camps" in the summer. Sometimes a group of college students takes an initiative, sometimes an ACM Student Chapter identifies outreach as an important project, and sometimes college students partner with one or more faculty members to develop and publicize programs.

  As these efforts continue and expand, I know of at least one instance during which college students organized outreach activities into a credit-bearing independent project under the direction of a faculty member. In another instance, college students became involved with tutoring programs at a local high school. Overall, this type of activity combines students' backgrounds in computing with interests in working with kids. Such activity may or may not be part of a combined program of the education and computing departments. In any case, carefully-constructed activities for K-12 outreach by college students will help synthesize concepts of computing with approaches to K-12 education—directly supporting the scholarship of teaching and learning.

- At one school, faculty experimented with the combination of oral comprehensive exams (required by the college) with an interest in obtaining student insights about individual computing courses and about the computing curriculum overall. In this setting, students were organized into teams of four with the following task:

  ◇ Given two current courses, develop a replacement course that would cover the key elements in one semester. (As examples, a operating systems course and a computer organization course might be compressed to single course on systems, or courses on data structures and the analysis of algorithms might be combined.)

  ◇ With the two courses now collapsed to one, staffing and the curricular schedule will be freed to allow a new course to be offered.

  For both the replacement course and the new course, students are asked to identify a proposed listing of topics, suggest a day-by-day sequence that would allow material to flow naturally for a semester, and suggest one or more textbooks or other resources.

Overall, this task requires students to review several courses, decide what material is fundamental, and determine what topics will be needed in later courses. In addition to this analysis of existing courses, students must identify the most critical gaps in the current program and how a new course might be structured to address those gaps.

- In an upper-level seminar, students may rotate identifying and presenting material, based on their explorations and study. In one simple version, a textbook might be identified that covers the desired topic. This resource is then divided into logical sections, and students present sections to the others—perhaps taking control of the entire class period every week or two; in this format, the instructor plays little or no role during a class session. In a more complex variant, students agree upon a topic, but each student then is responsible for identifying and presenting material (with any relevant handouts) every couple of weeks.

  In this approach, students might work individually or in pairs to learn, synthesize, organize, and present material to others during the class. (The instructor serves as resource, mentor, coach, and guide, but the students are responsible for class sessions—not the instructor.) Throughout, the exploration of new material builds upon students' past understandings, but students also must consider elements of teaching and learning as they master and present new material to their peers.

In summary, capstone experiences can reflect any of Boyer's categories of scholarship: discovery, integration, application, and teaching and learning. Within each category, faculty have developed contrasting approaches, each with its own motivations and opportunities. All of these approaches, however, have common features of building upon past student experiences and providing students with new perspectives and understandings.

CHAPTER 12

Selected/annotated references for courses and curricula in context

THIS CHAPTER SUGGESTS additional directions for the consideration of curricula, with sections for introductory courses, capstone experiences, student projects, and international curricula. Brief annotations provide context or commentary for each reference.

Introductory Courses

Introductory computing courses have multiple audiences. Also, some courses relate to non-majors, or to general students with different interests, or to connections between computing and goals for general education.

Courses for Non-majors

Several approaches are widely used to reach out to non-majors at various levels of schooling.

- Alice.org, *Tell Stories, Build Games, Learn to Program*, http://www.alice.org, 2017 (accessed July 7, 2017 [15].
"Alice is an innovative block-based programming environment that makes it easy to create animations, build interactive narratives, or program simple games in 3D" [15]. This environment has been extremely successful in CS0 courses and at the start of many CS1 courses throughout North America.
The alice.org Web site contains a wealth of resources, including "How Tos," "Lessons," "Exercises & Projects," "Textbooks," and an "Audio Library" [15, Resources link]
- Scratch, *Create stories, games, and animations*, https://scratch.mit.edu/, 2017 (accessed July 7, 2017 [163]
Scratch is widely used with young students, especially middle-school and high-school students, with computing through the use of stories, block-based programming, and simple animations.
Two useful sites for getting starting with Scratch include code.org [39] (elementary school [38], middle school, and high school) and a guide from the ScratchEd research team at the Harvard Graduate School of Education [19].

- NCWIT, the National Center for Women and Information Technology, *Computer Science-in-a-Box: Unplug Your Curriculum*, https://www.ncwit.org/resources/computer-science-box-unplug-your-curriculum, 2017 (accessed July 7, 2017) [137].
This material "introduces fundamental building blocks of computer science – without using computers. Use it with students ages 9 to 14 to teach lessons about how computers work, while addressing critical mathematics and science concepts such as number systems, algorithms, and manipulating variables and logic" [137].

- College Board, *AP Computer Science Principles*, https://apstudent.collegeboard.org/apcourse/ap-computer-science-principles, 2017 (accessed: February 28, 2017) [26]
A carefully-planned course, aimed at encouraging high school students and general college students to consider taking CS1 and other introductory computing courses.

- Richard Kick and Frances P. Trees, "AP CS principles: engaging, challenging, and rewarding," *ACM Inroads*, Volume 6, Issue 1, March 2015, pp. 42-45 [98].
An overview of the AP CS principles course for general students. The course includes a consideration of the role of programming at the very beginning levels.

- Henry M. Walker, *The Tao of Computing, Second Edition*, Taylor & Francis Group, CRC Press, 2013 [250].
A CS0 textbook, designed to address common questions from the general citizen. The book provides reasonable depth to address these questions in a meaningful way.

Multiple tracks for CS1

Several colleges utilize multiple, independent tracks for beginning students with various interests and to connect with a diverse student population.

- Williams College utilizes two versions of their Java-based CS1 course. Both versions cover the same core areas, but each focuses upon its own perspectives and applications.
    - ◇ Andrea Danyluk and Stephen Freund, *CSCI 134: Introduction to Computer Science*, <u>Events Version:</u> http://dept.cs.williams.edu/~freund/cs134-171/, 2017 (accessed July 8, 2017) [54]
    - ◇ Tom Murtaugh, *CSCI 134 - Digital Communication and Computation: An Introduction to Computer Science*, <u>Networking Version:</u> http://www.cs.williams.edu/~cs134/, 2017 (accessed July 8, 2017) [134]

- Union College Computer Science Department, *CSC-10X: Union College's Unique Introduction to CS*, http://cs.union.edu/intro/, 2017 (accessed July 8, 2017) [192].
Union College offers six distinct introductory courses, each of which "covers the same fundamental principles and the same core concepts in computing. All the courses introduce computers and computing: what they are, how they work, and what we are (and are not) able to do with them" [192]. However, each course has a different theme, including "Taming Big Data", "Robots Rule", "Game Development", "Can Computers Think", "Creative Computing", and "Programming for Engineers."

Placement Test(s) for Introductory Computing

Several schools have different introductory courses for incoming students, based on the students' prior background. In some cases, a placement test may help guide incoming students into the appropriate course for them.

- Cindy Marling and David Juedes, "CS0 for Computer Science Majors at Ohio University," *SIGCSE '16 Proceedings of the 47th ACM Technical Symposium on Computing Science Education*, March 2016, pp. 138-143 [116]
  At Ohio University, a new CS0 course provides foundational background for students with little or no computing background, whereas students with computing background start with CS1. (Students also take CS1 after successful completion of CS0.) This paper describes the CS0 and CS1 courses and also reports experiences in developing a placement test to guide incoming students into the proper first course.

Introductory Courses with Application Themes

- Deepak Kumar, *Learning Computing with Robots*, Institute for Personal Robots in Education, http://wiki.roboteducation.org/Introduction_to_Computer_Science_via_Robots, 2011. [102]
  A free, online textbook for the use of Scribbler 2 robots with Python in an introductory computing course.
- Mark J. Guzdial and Barbara Ericson, *Introduction to Computing and Programming in Python: A Multimedia Approach, Fourth Edition*, Pearson, 2015. [78].
  A well-documented textbook for introductory computer science (often for non-majors), emphasizing image processing with Python programming.
- Jesse M. Heines, Gena R. Greher, S. Alex Ruthmann, and Brendan L. Reilly, "Two Approaches to Interdisciplinary Computing +Music Courses, *IEEE Computer Special Issue on Computers and the Arts*, Volume 44, Number 12, December 2001, pp. 25-32 [84].
  F. Martin, G. R. Greher, J. M. Heines, J. Jeffers, H.-J. Kim and S. Kuhn, K. Roehr, N. Selleck, and L. Silka and H. Yanco, "Joining Computing and the Arts at a Mid-Size University", *Journal of the Consortium for Computing Sciences in Colleges*, Volume 24, Number 6, https://jesseheines.com/ heines/academic/papers/2011ieee/ieee2011paper-v33-forWebsite.pdf, 2009, pp 87–94 [117].
  Two papers describing extensive experiences connecting computing and music within introductory computing.

Introductory Computing and General Education Requirements

At many schools, a course in computing satisfies general education requirements.

- Andrea Tartaro and Christopher Healy and Kevin Treu. "Computer science in general education: beyond quantitative reasoning," *Journal of Computing Sciences in Colleges*, Volume 32, Issue 2, December 2016, pp. 177-184 [183].
  Results of a survey of 76 four-year colleges with regard to whether computing counted toward general education requirements, and if so, what requirement was satisfied. The article also discusses reasons why computing might be used to satisfy several interdisciplinary requirements.
- Jeff Cramer and Bill Toll, "Beyond competency: a context-driven CSO course," *SIGCSE '12 Proceedings of the 43rd ACM technical symposium on Computer Science Education*, 2012, pp. 469–474 [51]
  A consideration of the question, "What should a graduate of a liberal arts university understand about computational technology?", with the answer driving content for a CS0 course [51, p. 469].

Capstone Experiences

Several schools have a long history of developing software for community-based organizations as part of a capstone experience. Two of the most prominent programs are described here.

- Harvey Mudd Computer Science, *The Computer Science Clinic*, URL: https://www.cs.hmc.edu/clinic/, 2017 (accessed July 2, 2017) [82]
  The Harvey Mudd Clinic was first developed in the 1960s by the Engineering Department, and the work was subsequently offered by the Computer Science Department since 1993. Overall this time, student teams have worked on over 1000 projects.
- Sriram Mohan, Stephen Chenoweth, and Shawn Bohner, "Towards a Better Capstone Experience", *SIGCSE '12: Proceedings of the 43rd ACM technical symposium on Computer Science Education,* 2012, pp. 111-116 [131]
  Rose-Hulman Institute of Technology requires Software Engineering majors to complete a two-semester senior project, and students work in teams to implement software for an external client. So that students can be well prepared for this senior-level capstone experience, students take a Junior Project to gain experience in the software engineering process. Overall, the Junior Project and Senior Capstone Project yield a two-semester upper-level sequence, culminating with the capstone project.

An historical perspective regarding capstone courses comes from 1988.

- Clinton P. Fuelling, Anne-Marie Lancaster, Mark C. Kertstetter, R. Waldo Roth, William A. Brown, Richard K. Reidenbach, and Ekawan Wongsawatgul , "Computer science undergraduate capstone course", *SIGCSE '88 Proceedings of the nineteenth SIGCSE technical symposium on Computer Science Education*, 1988, p. 135 [67].
  Panelists from four universities and two companies discuss their experiences with capstone courses in the light of program accreditation by the Computer Science Accreditation Board (CSAB) in 1988.

Distinctions between capstone experiences and student-faculty research may be fuzzy, as described in the following article.

- Michael Jonas, "Capstone Experience — Achieving Success with an Undergraduate Research Group in Speech," *SIGITE '14 Proceedings of the 15th Annual Conference on Information Technology Education*, October 2014, pp. 55-60 [95]
  At the University of New Hampshire at Manchester, a capstone experience studied automatic speech recognition as a field of study to sharpen problem solving skills in information technology. Over four years, student groups on this topic evolved into a research group.

Beyond these specific references, a search of the ACM Digital Library (with the search term "capstone") or a search of the Web (with the search terms "computer science capstone course") yielded dozens of articles and Web sites describing capstone courses at a wide range of schools.

Student Projects

The concept of substantial student projects within the computing curriculum has been well established for at least 40 years, and many schools provide opportunities for students to work on computing-oriented projects.

- Richard Austing, Bruce Barnes, Della Bonnette, Gerald Engel, and Gordon Stokes, "Curriculum '78: Recommendations for the undergraduate program in computer science", *Communications of the ACM*, Volume 22, Number 3, March 1979, pp. 147–166 [17].
  The 1978 ACM Curriculum Committee recommended student projects throughout the undergraduate curriculum, including a term project focused on the local community in CS 9, "Computers and Society," and a full-semester team projects in CS14, "Software Design and Development" [17, p. 158].

- Cooperative Education and Career Development, Northeastern Development, *Grow. Adapt. Thrive.*, https://www.northeastern.edu/coop/, 2017 [48]
  Northeastern University has run a Cooperative Education and Career Development program for over 100 years. "Students alternate classroom studies with full-time work in career-related jobs for six months." [48, "About" Web page]. In this context, traditional student projects are replaced by actual employer-based job experience.

- Dabin Ding, Mahmoud Yousef and Xiaodong Yue, "A Case Study for Teaching Students Agile and Scrum in Capstone Courses", *Journal of Computing Sciences in Colleges*, Volume 32 Issue 5, May 2017, pp. 95-101 [58].
  Many project and capstone courses include the use of agile techniques. This article provides a descriptive overview of this approach for student projects.

International Curricula

Several articles discuss computing standards and curricular development outside North America.

- Ursula Fuller, Arnold Pears, June Amillo, Chris Avram, and Linda Mannila, "A Computing Perspective on the Bologna Process", *ITiCSE-WGR '06 Working group reports on ITiCSE on Innovation and technology in computer science education*, 2006, pp. 115-131 [68].
  The Bologna process seeks to "to facilitate the mobility of people, the transparency and recognition of qualifications, quality and development of a European dimension to higher education, and the attractiveness of European institutions for third country students" [68, p. 115]. One challenge is to construct curricula that will span European countries, allowing students to transfer freely from a school in one country to another school elsewhere.

- British Computing Society, "Academic Accreditation," http://www.bcs.org/category/5844, 2017 (Accessed July 7, 2017) [177].
  The British Computing Society sets standards and reviews computing programs under a royal charter for information technology.

- Accreditation Board for Engineering and Technology (ABET) and its Computing Accreditation Commission (CAC), *Criteria for Accrediting Computing Programs, 2017-2018*, http://www.abet.org/accreditation/accreditation-criteria/criteria-for-accrediting-computing-programs-2017-2018/, 2017 (accessed July 7, 2017) [2].
  Following in the historical process for the accreditation of engineering programs at the university level, ABET's Computing Accreditation Commission accredits computing programs within the United States.

Although some articles consider computing standards and curricular issues at the university level, other articles examine pre-college computing curricula and issues.

- Peter Hubwieser, Michal Armoni, TTorsten Brinda, Valentina Dagiene, Ira Diethelm, Michael N. Giannakos, Maria Knobelsdorf, Johannes Magenheim, Roland Mittermeir, and Sigrid Schubert. "CS/informatics in secondary education," *Proceedings of the 16th Annual Conference Reports on Innovation and Technology in CS Education-Working Group Reports*, 2011, pp. 19?38. [89]
  This Working Group report proposes "a category system (Darmstadt Model)" to allow unified study, research, and action with regard to computing education at the secondary level throughout the world. [89, p. 19]

- Peter Hubwieser and Andreas Zendler, "How Teachers in Different Educational Systems Value Central Concepts of Computer Science", *WiPSCE '12 Proceedings of the 7th Workshop in Primary and Secondary Computing Education*, 2012, pp.62-69 [90]
  Hubwieser and Zendler discuss "substantial differences regarding the organization as well as the substantial focus of computer science education at their schools" through a survey of teachers in the German states Baden-W urttemberg (BW) and Bavaria (BY). [90, p. 62]

- Craig Marais and Karen Bradshaw, "Towards a Technical Skills Curriculum to Supplement Traditional Computer Science Teaching," *ITiCSE '16 Proceedings of the 2016 ACM Conference on Innovation and Technology in Computer Science Education*, 2016, pp. 338-343 [113]
  This paper describes efforts to meet the needs of "deficient" students entering computing programs in South Africa. Although some secondary schools graduate students who are prepared to start university-level work in STEM fields, other schools do not. This paper proposes an approach to address such problems.

- Raghu Raman, Smrithi Venkatasubramanian, Krishnashree Achuthan, Prema Nedungadi, "Computer Science (CS) Education in Indian Schools: Situation Analysis using Darmstadt Model," *ACM Transactions on Computing Education (TOCE) - Special Issue II on Computer Science Education in K-12 Schools*, Volume 15 Issue 2, May 2015 [155]
  Raman et al. discuss efforts to understand and expand the introduction of computing with secondary schools in India.

IV

Curricular Issues

PERIODICALLY, IN TALKING TO STUDENTS, I quip that identifying a listing of interesting and worthwhile courses is not particularly hard. The challenge is to construct a student schedule that allows on-time graduation (e.g., in four years in the United States). Since time to graduation is constrained, students likely cannot take every interesting course offered, necessitating prioritization and planning. Students, with help from advisors, often must choose which offerings fit best with their long-term goals.

Similar difficulties arise in constructing an overall curriculum. A quick review of college and university catalogs yields an extensive listing of fine titles for courses covering a remarkable range of topics. Identifying possible courses for inclusion in a comprehensive curriculum might be reasonably easy. Further, publishers offer numerous books to support an impressive range of worthwhile courses. With so many options, a computing faculty might contemplate hundreds of possible course offerings, based upon courses and/or books available elsewhere, and the faculty might have additional ideas for new offerings, based on local interests, needs, and opportunities.

Pragmatically, of course, staffing constraints limit the number of overall courses that can be offered, and student demand limits potential enrollment. Just as students must prioritize and plan in choosing the courses they will take, faculty must consider what courses are possible and how those courses will fit together within an overall curriculum.

Developing and refining a curriculum can be further complicated by several high-level themes that may transcend specific topics within individual courses. For example,

- *Computational thinking:* Students should develop problem-solving skills and gain experience working on large-scale projects.
- *Preparation for lifelong learning:* Students should learn how to read materials, such as reference manuals, books, and articles, as they prepare for careers in the rapidly-changing field of computing.
- *Ability and skill to work in teams:* Students should gain experience working within teams and learn skills and approaches for collaborative project development.
- *Communication skills:* Students should develop strong communication skills, including writing and making oral presentations to communicate with team members and clients.

Many of these themes are sometimes included within such general categories as "soft skills" or "professional practice", but typically, these themes are not associated with an individual course. For example, many computing courses help students gain experience with computational thinking and reading technical material; and multiple courses may involve teamwork, writing, and oral presentations.

With these considerations, an overall curriculum requires attention to at least four fundamental elements:

- Content (e.g., Part II of this book)
- Context (e.g., Part III of this book)
- Connections among courses and the overall curriculum (e.g., the "big picture")
- "Soft skills" and "professional practice"

This Part (Part IV) explores the last two of these fundamental elements: "the big picture", "soft skills," and "professional practice." Thus, Part IV complements Parts II and III in reviewing vital components of any computing curriculum. (Looking ahead, Part V will review yet another of these components: computing and mathematics.)

CHAPTER 13

Staying connected with the big picture

This column first appeared in *SIGCSE Bulletin*, Volume 40, Number 4, December 2008, pages 16–17[227] Subsequent reflections on this subject appear later in this chapter.

13.1 ORIGINAL COLUMN

ONE OF THE TRUE JOYS of being a computer scientist is the opportunity to work at multiple levels of abstraction when solving a problem. For example, in writing a software application, we consider high-level requirements, a design that includes encapsulation of data elements and options, individual coding pieces (files, classes, or objects), etc. This column observes that teaching computer science entails the same attention to multiple levels of activity.

Unfortunately in teaching, as in the development of programs, it is easy to become caught up in low-level pieces and day-to-day details, and we can miss the big picture. This column considers how low-level details can connect with overall goals of a course and high-level needs of an overall curriculum.

Problem Solving

- When I was writing my first book, Series Editor Gerald Weinberg observed that examples should be chosen, so that they demonstrate an approach or technique that is worthwhile in a broad context—there simply is not enough time in a course or curriculum for students to unlearn examples when they gain more background. Here are two examples:

  - ◇ A bubble sort may illustrate the use of loops and arrays, but it is never the sorting algorithm of choice. An insertion sort illustrates loops and array manipulation just as well, but the insertion sort can actually be useful for almost-ordered data. Thus, one should teach an insertion sort rather than a bubble sort, so students will not have to forget the awfulness of the bubble sort later on.
  - ◇ A simple recursion to compute Fibonacci numbers ($f(n) = f(n-1)+f(n-2)$) is extremely inefficient, whereas recursion for a quicksort, merge sort, or tree traversal greatly simplifies code and is preferred.

- When introducing a low-level technique, data structure, or algorithm, one should indicate a real-world application in which the approach is useful. Details are important for the careful solution of problems, but students often wonder why these details are relevant. Providing a context connects structures and algorithms with a big picture (and if we cannot provide a context, why do we want to bother with the technique?)
- Many students, particularly women and other under-represented groups, are often motivated by a desire to contribute to society. To build on this interest, it is helpful to identify applications that address human needs and improve the quality of life.
    - ◇ Games and gambling motivate some, but these areas turn off others; finding different examples and applications may connect with a wide range of students and thus are relatively safe.
    - ◇ Example: Some years ago, when introducing loops in CS1, I asked students to simulate the tossing of a coin. In 20 experiments, students were to determine how many tosses were needed to get at least one head and one tail.

      In more recent years, I have found a recasting of this problem connects to many more students: A couple decides to have children until they have at least one boy and one girl (and then they will stop having children). Assume that the likelihood of a boy and a girl is equal and that the gender of one child is independent of any previous children. Write a simulation to determine the family size of 20 couples.

      Of course, the coin toss and couple simulations involve identical programming (only the descriptive variable names and output labeling changes), but my experience suggests the family-size simulation connects with a much wider group of students.

Communication Skills: Reading, Writing, and Teamwork

Communication skills are commonly identified as important skills for graduating computing students, but these skills may be forgotten within courses whose focus involves many low-level details.

- Graduates should be able to read documentation and manuals, but where do students practice such reading? Teachers should not always spoon feed students by lecturing about all topics, if graduates are supposed to be able to search for sources and read articles.
- Graduates are supposed to be able to write well-structured, clear, concise, and nicely phrased descriptions of their ideas and work. In order to gain such skill, what do students actually write about when?
- Many studies demonstrate that professionals do better work when working in teams, and many projects require working in groups. What courses actually utilize team work and provide students with opportunities to develop their collaborative skills? Further, if such skills are important, how is this importance demonstrated in grading scales?

Ethical Issues

The products of computer science impact society, and some schools offer a course or two on social and ethical issues. However, a separate, isolated course can send the message that one only must consider the consequences of one's work when taking a special course.

- How do students come to appreciate that each software package and project has potential consequences?
- How do students learn to assess the societal impact of their work?
- When do students consider what might go wrong in a software package, what impact such problems might have, and how such issues can be addressed?
- When are students asked how specific projects relate to the ACM Code of Ethics? It may be technically possible to produce a system, but when are students challenged to consider whether or not a system should be developed at all?

News and Current Events

Computing appears regularly in news releases. These stories provide a wonderful opportunity to connect class work with the real-world applications that motivate many students. However, reporting accuracy varies widely. Although some computing faculty may argue that public relations is not their job, one might ask "whose job is it?" If computing folks do not pay attention to public perceptions of their field, then who will?

- To what extent is it possible to connect daily headlines with specific course content?
- How can class work help students separate myths and misconceptions from facts in news stories?
  - ◇ Example: Despite many news reports regarding job outsourcing to countries internationally, the field of computing has remarkable potential for long-term careers. Within the United States, employment in the IT sector increased 17% from 1999 to 2004 — even with all the news of the dot-com difficulties. Also, the U.S. Bureau of Labor Statistics projects that computing is the field with the greatest potential for growth through 2014.
  - ◇ Example: Money Magazine and Salary.com identified "software engineer" at the very top of their listings of the "Best Jobs in America." Further, number 7 on the listing was "computer/IT analysis". In describing the position of software engineer, www.calary.com wrote, "The profession's strong growth prospectives, average pay of $80,500, and potential for creativity put it at the top of the list." www.salary.com
  - ◇ Additional facts regarding employment in computing may be found at http://computingcareers.acm.org/

Exams and Grading

Exams and grading scales send powerful messages to students regarding what is really important in our courses and our discipline.

- If computer science is not the same as programming, do exams have non-programming components?
- What actually gets credit on assignments and projects? (If all credit comes from correct code, the exams convey the clear message that only programming counts—even if we say computer science is a much broader discipline.)
- If communication skills are important,
  - ◇ does writing and oral presentation count in grading?
  - ◇ can students lose points for bad writing?
- A common approach to grading separates content from presentation. If the presentation is muddled, however, how does an instructor know that the underlying thought isn't confused?
  - ◇ For years, I thought some poorly-worded responses on tests came from time constraints and pressure.
  - ◇ Later, in working directly with students in lab, I learned such responses often indicate confusion and misconceptions.
- Should it be possible to get an "A" on an assignment if the ideas suggest fine insight, but the presentation is mediocre?

In summary, courses must cover many topics at multiple levels. Although details can be captivating, class work offers an opportunity to connect those low-level details with high-level themes, principles, ethical considerations, and student motivations.

13.2 SUBSEQUENT REFLECTIONS

In writing the original column, my focus was on curricula: details of courses need to be consistent with the "big picture" of curricula and programs.

Broadening the notion of the "big picture," similar comments apply more generally, including such areas as course scheduling, prerequisite chains, and facility availability.

Course Scheduling

Traditionally in the United States, students are supposed to be able to complete their undergraduate program in four years. Other environments vary, with three years being the expected time for a bachelor's degree at some universities in Europe and elsewhere. With such expectations, a basic question is whether course schedules allow students to take the courses they need within the expected time frame.

This issue of on-time graduation is compounded by at least three factors:

- In times of high student demand, enrollment in specific courses may be limited, due to seating capacities, workstations available in a lab, desired pedagogy, or other factors.
- Students entering an undergraduate program may have an idea about their intended major (at some schools, students declare a major when they first enroll), but college is a time when students explore their interests, and their perspectives often change.

- Some students may wish to spend a semester in off-campus study, such as study in another country.

Clearly, course scheduling cannot address all student interests and concerns, but students have reasonable expectations that they will be able to make steady progress toward their degrees once they start. At one extreme, a student should not expect to take CS1 at the beginning of the senior year and graduate with a CS major at the end of that academic year. On the other hand, at one school, students completing CS1 one semester could not continue their studies in computing the next semester; they did not satisfy the prerequisites for any of the offered courses (e.g., CS2 or similar was not offered).

One common approach for evaluating course schedules is for faculty to review course schedules on a regular basis.

- If students start CS1 in their first, second, or even third semesters, will the courses offered allow them to complete a full computing major and then graduate at the expected time?
- What schedule might students follow if they want to study a semester or two off campus (e.g., in another country)?
  - ◇ If a students want to study abroad, will they need to take one or more computing courses while off campus to ensure they can still graduate on time?
  - ◇ Are some off-campus study programs preferred or recommended to allow students to meet identified graduation requirements?

In responding to such questions, transparency facilitates both advising and student planning. For example, typical student schedules might be posted on program Web sites or in planning booklets. Such schedules might show hypothetical schedules for students starting computing in their first semester, second semester, or perhaps even their third semester. Posted schedules also might illustrate options if students plan to be studying off-campus.

In parallel with checking student schedules, computing faculty should consider posting tentative 2-4 year schedules of what courses are planned to be offered when. Of course, changes in staffing may require adjustments to multi-year course planning, but even with some uncertainty publicizing tentative schedules can be quite helpful for student planning and for advising.

Also, by developing and publicizing both potential student schedules and course offerings over multiple years, faculty can check that courses will be offered at relevant terms, so that students beginning their study of computing at various times will have plausible options for completing their degrees.

Prerequisite Chains

Chapter 2 discusses "Prerequisites: shaping the computing curriculum," describing several purposes for prerequisites, including content prerequisites, maturity prerequisites, filtering, requirement enforcement, and historical requirement(s). Although the chapter was reasonably wide ranging, much discussion focused upon prerequisites for individual courses.

Turning to the curriculum as a whole, a faculty periodically should review prerequisite chains to determine the extent to which students take courses in a reasonable progression (without unreasonable jumps in level). As an example within the field of economics, at some colleges, both micro-economics and macro-economics are offered at the 200 level with only introductory economics as prerequisite. However, students may be strongly advised to take one or two economics courses after the beginning course before taking either micro- or macro-economics. In principle, students can take these courses relatively early in their studies, and occasionally early enrollment in these courses might be appropriate. However, most students may discover that expectations for micro- or macro-economics are substantially higher than what they have encountered previously, and additional background is strongly recommended.

At some schools, similar jumps may be encountered in taking courses in abstract algebra or real analysis. In principle, students may be able to take these courses immediately after calculus (or perhaps linear algebra), but additional background is strongly suggested. Sometimes a mathematics program may offer "bridge courses" that provide important background and experience regarding mathematical reasoning and the writing of proofs, but these courses may or may not be required for upper-level mathematics.

Similarly in computing, computing faculty should consider the extent to which prerequisites are explicit. Will the stated prerequisites provide the appropriate background? For example, if just CS1 or CS2 are listed as prerequisites to a "systems" course, will students be expected to also have background in computer organization and architecture as well? Or, if CS2 is the only prerequisite identified for a course covering the analysis of algorithms, will students also be expected to know fundamentals of discrete mathematics?

Such matters might naturally be considered when faculty review typical student course schedules and tentative multi-year course offerings. If a student follows a recommended sequence of courses over three or four years, faculty can check that students will have the needed background for each semester of their study.

Facility Availability

Much discussion within contemporary education considers various accommodations and infrastructure required for students with a range of disabilities. For example, in the United States, both faculty and institutions are attuned to requirements from the Americans with Disabilities Act, and implications arise for pedagogy and physical facilities.

Although this activity receives much attention, other basic needs may be forgotten. Here are several examples.

- If students need to work in a lab as part of their homework, when are the labs open for their use? If classes meet in a teaching lab through much of the day, students may not be able to start homework assignments until evening. Faculty may complain that students do not plan ahead but rather leave their work until the last minute, but facilities may not be available to support students through the day.

- If software licenses allow only a limited number of concurrent users, to what extent can students use the software for their homework and projects? If license capacities are keyed upon the number of students in one section of a course, then students in each section of the course may have appropriate access during class, but students from multiple sections may not have adequate access when trying to complete homework.

- If students need access to videos or equipment or other materials, how will this access be obtained? If readings or videos or equipment must be checked out of a library (perhaps on 3-hour reserve), then students working off-campus may not be able to check out or return materials during hours when the library is open.

- If students are supposed to have access to textbooks from a central repository (e.g., from a library reserve desk), how much notice will the library need for new courses or changes in books for ongoing courses? Some schools maintain extensive textbook collections, so students do not need to purchase all of their textbooks, and so faculty know students will have access to required readings. However, in some settings, faculty may need to notify the lending library four or more months in advance of textbook requirements—well before new courses are approved and before new textbooks appear.

In summary, schedules, prerequisites, and logistics may work well for certain students. However, faculty need to review how well policies and practices work in practice, to determine their impact on students with different backgrounds and circumstances.

CHAPTER 14

Balancing the forest and the trees in courses

This column first appeared in the *SIGCSE Bulletin*, Volume 32, Number 4, December 2000, pages 17–18[210] Subsequent reflections on this subject appear later in this chapter.

14.1 ORIGINAL COLUMN

THE FOLLOWING QUOTES represent paraphrases of comments I have heard in recent years.

- "Undergraduates are able to learn this material, so it is important to include it in the curriculum."
- "This topic may help some of our graduates, so we need to require it for our major."
- "We need to offer two undergraduate courses on this topic, because so many algorithms can fit into this area."
- "We want our students to have so much practice with this topic that they can solve problems without thinking."
- "My course thoroughly covered the following 53 algorithms: .... "

Such comments have prompted me to think about the need for balance in the CS curriculum in general and in specific courses in particular.

So Much to Do, So Little Time

With the explosion of knowledge within computer science, virtually any sub-discipline contains far more material than can reasonably fit into a single course. Further, four or even five years clearly are inadequate to master all areas within the field, and any undergraduate curriculum represents a balance of conflicting interests and demands. Such considerations were among several factors mentioned by William Wulf in his SIGCSE 1998 keynote address, when he commented that the undergraduate CS major should not be considered as a professional degree. Other professions (e.g., law, medicine, engineering) do not pretend an undergraduate degree pro- vides adequate training for full certification, and the same holds for computer science. In considering implications of the need for balance in CS education, I find it worthwhile to first review some parallels from the "Calculus Reform Movement" of the past decade.

A Summary of the Calculus Reform Movement

A brief summary of the Calculus Reform Movement might stress two points: the need for active learning in the classroom, and the caution that a multitude of details can cloud main ideas. Picking up on the second of these points, calculus reformers observed that many topics had been added to calculus courses over the years, but few topics were deleted. With such a mass of material, students often missed important points and got confused. Students might master specific details, but did not understand concepts, see connections among those ideas, or appreciate how details fit together into an overall whole. Reformists commented that it was not necessarily important what an instructor covered in a course, but rather on what students learned.

Anecdote from the Calculus Reform Movement

Roughly eight to ten years ago, I reviewed a manuscript for introductory calculus, billed as supporting the then-new "lean and lively" calculus approach. The manuscript had several notable characteristics:

- the manuscript contained well over 1000 pages for a one-semester introductory course,
- the text was organized with one section for each class meeting,
- each section first introduced a new formula,
- the bulk of each section consisted of several (> 4) examples, each showing numbers being plugged into the specified formula,
- the end of each section included at least 40 exercises, each of which involved plugging more numbers into the section's formula,
- connections rarely were made between the content of one section and that o f another,
- in the prospectus, the authors commented that students using this text performed well on tests, course grades were high, and "students liked the course."

A colleague of mine commented that the manuscript should be classified as "fat and fatuous". I speculate that students might perform well in the short term, if tests consisted of more plug-ins. This would naturally lead to high grades, and studies show that course evaluations correlate positively with grades. The author's charisma in class might further affect evaluations, through the Hawthorne effect.

On the other hand, with new, unmotivated, disjointed formulae presented without context in each class, one wonders whether students remembered anything six months later. We could question whether students could apply calculus to new problems, whether the course promoted problem solving of any kind, and whether the course reinforced the erroneous impression that mathematics consists of arbitrary rules without underlying ideas. Thus, the proposed manuscript seemed to contradict everything the Calculus Reform Movement represented.

Demands on the Undergraduate CS Major

An undergraduate CS major typically seeks to prepare students for jobs in industry or government, further study in graduate school, careers in fields which use computing, or teaching at various levels. To serve the needs of such career paths, undergraduate programs typically react to demands for introducing specific tools, methodologies, theories, algorithms, concepts, communication skills, ability to work in teams, social and ethical issues, and many more topics. Such matters range from the practical to the abstract. Some argue for an emphasis on specific software tools used by local companies; others focus on foundational principles that apply generally, but which may require additional training for use in a particular' development environment.

Demands on Courses

On a smaller scale, a course too must strike a balance, and it may be worthwhile to consider whether CS courses sometimes fall into the same traps that motivated calculus reformers. To be more specific, faculty might consider the following questions as responses to the quotes beginning this column:

- While undergraduates can learn many topics, how does a topic support curricular goals or prepare students for careers and lifelong learning?
- Although a topic may be vital for some and possibly helpful for all, how does that topic fit with the general goals for all students?
- Will a proliferation of courses and specific topics within those courses provide vital, long-term insights, or will students be so overwhelmed with details that they lose the main ideas?
- Since computers are particularly effective at rote operations while people flourish with complex problem solving and integration of concepts, when is practice with a topic needed to provide insight, and when is practice excessive as duplicating machinery?
- At what point does a course contain so many details that students miss important connections and unifying themes?

As CS faculty revise courses and the overall curriculum to incorporate new topics and understandings in the discipline, CS faculty also need to identify corresponding topics to remove from those courses and the curriculum.

14.2 SUBSEQUENT REFLECTIONS

As the discipline of computing evolves over time, new topics emerge as being important, while some other topics may diminish in importance, some topics may require substantial updating, and yet other topics may become outdated. Often course planning may highlight new or updated material, but dropping topics sometimes can seem more difficult. For example, some traditional topics may seem truly lovely, examples may be well developed, and course materials may be nicely refined. If coverage of a topic went well a few years previously, an instructor may be reluctant to remove that topic from an ongoing course.

With such tendencies, if a past course seemed challenging, but manageable for students, a revised course may become overloaded. The following comments may help an instructor address the possible overloading of a course.

- If only one course can be offered for a topic (e.g., due to staffing constraints or competing curricular demands), an instructor should resist the urge to insert all possibly relevant material into the one course. In principle, a course might cover a vast array of subjects, but students can only absorb a limited amount of material.

- The prerequisites and level of a course (e.g., 100-level or 200-level or 300-level) should indicate the background and sophistication of students enrolled. Introductory courses with little or no prerequisites, for example, likely cannot expect the same insights, cover the same amount of material, or utilize the same assumptions as a course for students nearing graduation.

  To illustrate, one school was able to offer a single, 200-level course on operating systems, but the coverage of material included sufficient scope to fill a graduate-level seminar. In practice, a few students completed the course by neglecting their other academic work, but most students found the course overwhelming, and many switched their majors to other disciplines; the course was simply unreasonable for students with the prescribed prerequisites and background.

  As courses are updated and refined, an instructor should consider whether class work, reading, labs, and assignments are at a level appropriate for the expected students.

- In my experience, interesting potential course projects and activities can easily expand to consume 2+ semesters of work. Also, since students may have little sense of the time required for a project, student proposals for projects often may be over-ambitious. Overall, an instructor may need to review any proposed course projects to determine if the likely work seems appropriate for the time and resources available.

Overall, one common goal of education is to help students move from one level of understanding and mastery to the next, and challenging students can help students develop. However, over-ambitious projects can overwhelm and frustrate students rather than encourage learning. Finding an appropriate selection and balance of topics and course activities requires ongoing faculty review and adjustment.

CHAPTER 15

Guided reading and seminar issues

This column first appeared in the *SIGCSE Bulletin*, Volume 31, Number 4, December 1999, pages 27–28[209] Subsequent reflections on this subject appear later in this chapter.

15.1 ORIGINAL COLUMN

OVER THE YEARS, students frequently have asked me to work with them on guided-reading or independent-study topics. While I greatly enjoy working with students and want to be accommodating, I must be realistic about what activities I can do. As at many (most? all?) schools, student demands and interests are extensive, but faculty time is not. Thus, in such circumstances, I often require students to take initiatives beyond what I require in regular courses. This column reviews some general mechanisms to encourage students to actively engage in a subject. In some cases, I have tried the approaches personally; other ideas have evolved from thoughts heard elsewhere. Most scenarios require the instructor to provide structure, insight, coaching, feedback, and management. Students have varying responsibilities.

Student Lectures

My most successful group-guided-reading course occurred some years ago, when seven students wanted to study a common topic. All were intelligent and motivated; what they needed was guidance and structure. Thus, in consultation with the group and before the semester, I selected two standard texts and developed a detailed, day-by-day schedule for a course that would meet three days a week. The first day, I provided a crash course on mechanics, organizing lectures, public speaking, handout preparation, and other related topics. Each subsequent day, lecturing rotated around the group on approximately a two-week cycle. Each speaker was responsible for synthesizing and presenting the designated material for the full class period. While students could consult me for assistance, they understood they largely were on their own to make up any deficiencies in one day's presentation. As you might expect, this added considerable incentive for presenters to be well prepared.

Overall, the course was remarkably successful. Early on, students learned the need for extensive preparation few were satisfied with their first lectures, but each improved quickly. Also, as the semester progressed, a friendly competition developed among the students, as they strove to outshine previous speakers. By the semester's end, lectures were consistently highly polished. Extensive, well-organized handouts became common; student presenters even came to class in business suits for an added sense of professionalism!

And, grading was particularly easy. With material building though the semester, everyone in the classroom knew immediately when a presenter was unclear about any topic. By the semester's end, each participant (including two sophomores) had presented material typically covered in graduate-level courses; impressive mastery was the norm.

Some Variations on Lectures

In the course just described, the group was small enough to allow informal feedback. In other contexts, students might be asked to complete evaluation forms for each student presentation. As these can be quite harsh at times, the instructor may prefer to collect these forms and then summarize the comments for the speaker using a consistently constructive tone.

As another variation, each presenter may assign 1 to 3 problems as part of the normal class assignment. Note, however, such student assignments often are very hard, and instructor monitoring may be needed. Another variation requires each student to bring a one-page summary of the reading to class. The designated presenter then adds depth and perspective. (One instructor I know actually banishes from the classroom students lacking the summary. While this may seem harsh, this instructor reports that students are never unprepared a second time.)

As a variation for a guided-reading course with one student, the student and I developed an initial schedule of readings and exercises. Through the semester, the student then began our weekly meetings with a 30-minute presentation of the main points, ideas, and experiences of the past week. My role was to listen, raise questions, point out alternatives, and answer questions.

The Moore Method

R. L. Moore developed another variation on this theme for mathematics courses, emphasizing the use of discovery and inquiry-based methods. In this approach, now commonly called the Moore Method, the instructor provides handouts with definitions of terms and statements of results, but virtually no proofs. Thus, handouts provide a reasonable road map to the subject, with results broken down into challenging, but manageable, parts. [112]

Class periods challenge students to explore the subject and develop their own proofs. During class, students present proofs they have developed; the instructor does not fill in the gaps.

If no student figures out the next required proof, the class may brainstorm possible approaches and revisit the problem in the next class. During the semester, each student must present a specified minimum number of results, and more participation is encouraged. Since the results typically vary in difficulty, this approach provides opportunities for students of varying abilities. In mathematics, this approach has a strong reputation as providing excellent preparation for later independent study and research. Additional information is available from http://www.discovery.utexas.edu/dlp/dlp.html.

Student-Based Discussion

Several approaches to active student participation begin with students preparing questions on assigned readings, highlighting topics to be addressed during the coming class. Variations on this approach involve the distribution and use of these questions. The instructor may collect the questions as an aid for lecture preparation or may distribute a question summary;

or questions may be circulated to the full class either with or without attribution. In my experience, students initially need help in framing questions and identifying issues for discussion. When these questions go only to the instructor, students may feel less risk in posing interesting questions. The instructor then may use selected questions anonymously to begin discussion. As the semester progresses, students generally write better questions, which then may be distributed to the class, first anonymously and then with attribution.

Naturally, instructors need not lead all class sessions. Students may lead discussions perhaps based on their own questions or prepared opening statements. While leadership responsibilities commonly rotate following a fixed pattern, this may encourage students to prepare only the parts they will lead. Alternatively, the leader may be selected randomly. Thus, one colleague of mine puts student names on a wheel, which is spun to begin discussions. Another literally draws names from a hat. Alternatively, one can run a program to randomly select a name, displaying the output on classroom monitors. While any random draw is arbitrary, students respond well to a perceived sense of fairness.

Other Ideas

Of course, this column only begins a discussion of ways to encourage students to actively engage with course material and to make them responsible for classroom activities. Thus, I especially welcome readers' ideas and comments on this or other classroom Issues.

15.2 SUBSEQUENT REFLECTIONS

In my experience, upper-level guided reading and seminar courses typically explore a selected topic in considerable depth. Sometimes, study begins with one or a few books for general background, but students progress to locating and investigating relevant articles. Also, guided reading courses often involve few students (e.g., 1-4 students), while seminars may be a little larger, but often under a dozen. In these traditional courses, a faculty member may choose initial readings and develop a basic organization, and the instructor and/or students may identify later readings, projects, or other work. With this format, students work largely on their own to explore a topic in depth. Unfortunately, these courses often require much faculty time—for on-going preparation, class meetings, and student feedback.

Although this course format, as described, may be common, this type of course requires an instructor to devote considerable time and effort for relatively few students. Also, at some schools (and often at small colleges), the supervision of guided reading courses and seminars may not be included in computations of a faculty member's teaching load: that is, such courses may be supported as teaching overloads. Altogether, today's high enrollments in computing courses and common constraints on faculty sizes may motivate exploration of creative alternative approaches for reading and seminar courses. i

The following notes may encourage brainstorming of possible new formats. i

Tutorials at Williams College

Over the past several months, I have been inspired by the notion of a *tutorial*, as offered by Williams College—often at the junior or senior levels. As described by Williams' literature, students within an overall course are divided into pairs. Then, "Every week, the two students take turns developing independent work—an essay, a report on lab results, a piece of art—and critiquing it. With the support and guidance of their professor, they sharpen their critical thinking, improve their writing, develop ideas, and defend positions." [287]

Williams' tutorials typically are limited to 10 students (5 pairs) and thus represent a significant faculty commitment for relatively few students. Altogether such a limit may not scale well to other college settings, but allocating one instructor for 10 students seems more effective than having one faculty member support just 1-3 students as an overload.

An Upper-level Course Combining Technical Depth and Social/Ethical Issues

Personally, I am considering how alternatives to a Williams-style tutorial might be organized within the discipline of computer science and also how a course might capture this general approach—but with somewhat higher enrollments.

Although my thoughts for a different type of reading or seminar course continue to evolve, some ideas for exploration are illustrated in the following hypothetical example. In any implementation, I would expect substantial adjustment in details, but perhaps the style of this course will spark ideas for faculty at a range of schools.

Hypothetical Example:

Consider an upper-level course, perhaps enrolling 20 students, that seeks to explore several technical subjects in some depth while also examining social and ethical issues.

Overall, the course might be divided into 5 segments, each extending about three weeks. Since the course aims to consider social and ethical issues for computing, and since few computing students may have background in this field, the first course segment might provide a framework for exploring social and ethical issues. Two possible sources for this segment might include Michael J. Quinn, *Ethics for the Information Age, Seventh Edition* [152] or Sara Baase and Timorhy M. Henry, *A Gift of Fire: Social, Legal, and Ethical Issues for Computing Technology, Fifth Edition* [18] Subsequent 3-week segments would explore a separate subject, drawn from current active areas of computer applications. For each segment,

- Week 1: Provide general background, perhaps utilizing a book or a sequence of articles.
- Week 2: Delve reasonably deeply into technical issues, perhaps asking students to identify and synthesize additional articles.
- Week 3: Consider social and ethical dimensions of the application. If students were divided into 6 groups of 3-4 students to develop a poster or give 15-20 minute presentation, 3 groups might report at each of two sessions and the others provide feedback.

In this format, subjects for each 3-week segment would depend upon instructor and student preferences. Four possible segment themes, with possible supporting books, follow:

- **Big data (opportunities and risks):** Cathy O'Neil, *Weapons of Math Destruction: How Big Data Increases Increases Inequality and Threatens Democracy* [144]
- **Bitcoins (uses and potential abuses (e.g., in ransomware)):** Arvind Narayanan, Joseph Bonneau, Edward Felten, Andrew Miller. and Steven Goldfeder, *Bitcoin and Cryptocurrency Technologies: A Comprehensive Introduction* [135]
- **Search engines (algorithms and social impact):** Amy N. Langville and Carl D. Meyer, *Google's PageRank and Beyond: The Science of Search Engine Rankings* [105]
- **Impact of computational complexity on widespread applications:** Lance Fortnow, *The Golden Ticket: P, NP, and the Search for the Impossible* [66]

CHAPTER 16

Writing with the computer science curriculum

This column first appeared in the *SIGCSE Bulletin*, Volume 30, Number 2, June 1998, pages 24–25[206] Subsequent reflections on this subject appear later in this chapter.

16.1 ORIGINAL COLUMN

CONTRIBUTIONS ARE SOLICITED for this column which addresses assignments, class projects, research ideas, pedagogical approaches, and other issues related to effective classroom teaching. Send ideas, full articles, and comments to the author.

Introduction

Over the years, several comments concerning the CS curriculum in general and the role of writing in particular have had a substantial impact on this writer.

- If computer science is not programming, why are there so few non-programming assignments in computer science courses?
- While CS faculty usually describe their courses as emphasizing problem solving and other concepts, why do students often talk about "the first C++ course", "the Pascal course", "the Scheme course", or the introductory or advanced "Ada course" ?
- If CS teachers are serious about the importance of communication skills [8, p. 21], why don't teachers take points off for writing errors in student work?
- While employers often highlight the need for their personnel to have strong communication and teamwork skills, why do many CS job descriptions and advertisements omit these points?

This column explores some thoughts arising from such comments.

The Role of Writing

While CS students clearly must become fluent programmers, many sources also emphasize the need for strong communication skills for CS graduates. Computing Curricula 1991 states, "Students should be encouraged to develop strong communication skills, both oral and written" [8, p. 21]. In its description of criteria for accreditation, CSAC states, "The communication skills of the student, both oral and written, must be developed and applied in the program" [45].

Of course, English departments and writing laboratories have important roles to play in the development of student writing. However, several thought-provoking arguments suggest that computer science departments also should work actively to develop student writing. A few such arguments follow.

- When writing is part of a computer science course, students understand that writing is important – not peripheral – to CS.
- When computer science faculty review writing, students learn that principles taught in English apply in technical areas as well.
- When programming is only part of a course, students understand that computer science extends beyond coding.

In addition, it is worthwhile to note that inclusion of writing exercises can be combined effectively with collaborative learning and other team-based approaches to promote active learning. While it is interesting to explore this broader context, such points are beyond the scope of the present column.

Writing in Introductory CS Courses

Programming naturally will be an important part of many introductory courses. However, once an instructor decides to include writing-based assignments as well, it is easy to identify many activities that might have a writing component.

- Explain why something happens.
- Describe what you observe
- Compare two approaches.
- Discuss advantages of one approach over another. o Justify your answer.
- Argue why your testing supports your claim that your program is correct.
- Discuss the purpose of a procedure.
- Carefully state the requirements or describe the design of a problem.
- Justify your selection of data structures or algorithm.

In addition to providing students with practice in writing, this author has noted that reading student prose has the side effect of providing worthwhile (but sometimes frightening) insights into what students are actually thinking. While fuzzy writing may reflect a student's weakness in communication, imprecision or confusion in writing stems from inaccurate or muddled thinking in many cases.

Writing in Upper-Level CS Courses

Computing Curricula 1991 suggests "Written and oral communication skills are also developed ... in an independent student or undergraduate research project under the personal tutelage of a faculty member" [8, p. 21]. Similarly, the Liberal Arts Computer Science Consortium recommends a senior project which includes "The writing of a significant scientific paper or substantial technical document to give the student experience in writing for a scientific audience" [283, p. 93]. While a capstone experience provides one important opportunity for writing, other upper-level courses may utilize writing effectively as well.

For example, *Computing Curricula 1991* identifies abstraction (the scientific method) as a working methodology that should "appear prominently and indispensably" in the discussion of all areas of computing [8, p. 11]. In such a process, computer scientists collect data, form hypotheses, create models, make predictions, design experiments, and analyze results. While programming may help this work somewhat, the complete task requires considerable description and careful analysis. Michael Jipping and others, however, have noted that students often are quite weak in their analysis and experimental design, and considerable effort is needed to teach students to be effective with this methodology [94]. Writing can play an important role in this learning. Additional comments related to effective laboratory pedagogy will appear in a later column.

More generally, design and analysis are vital to many areas within computer science, and writing can help clarify students' thinking. For example, a first step in developing an object-oriented design might involve writing an explanation of relevant classes, the relationship among such classes, and the flow of messages required for various transactions. In such assignments, writing can help focus attention on concepts, principles, and analysis—all of which are important parts of upper-level CS courses.

Some Grading Practices

While instructors often talk about the importance of clear and accurate writing, students may pay more attention when the quality of writing has a direct impact on grading. The following list indicates some ways that instructors can emphasize the importance of written communication in tangible ways.

- Require comments in programs (e.g., in program, procedure, and function headers). Programs without comments might be returned ungraded.
- Require comments in programs before students can obtain help from an instructor. "If you cannot describe what a procedure is to do in English, how can I help you get the procedure right?"
- Take off for grammatical errors and unclear English. Inaccurate writing often indicates confused thinking.
- Grade on the basis of what the paper says, not on the basis of what the student might have meant. "If that is not what you meant, why did you say it?"

As a simple example, when students in a CS1 lab were to compare two algorithms, one student wrote:

> Solution one is longer, and solution 2 is more concise. It will take longer to compile, but from a readers point of view, it is very easy to read and understand.

Solution 2, on the other hand....

Here, the first pronoun "it" refer to solution 2—an incorrect reference. While one might hope the student meant solution 1, this could be a clever ploy by the student to avoid answering the question. Deducting points for the incorrect reference instructs the student regardless of his/her intent. If the student really knew the answer, this deduction demonstrates that pronoun antecedents matter. If the student meant what is written, losing points is justified on content grounds.

Concluding Observations

As CS faculty consider the appropriate use of writing in the CS curriculum, some insights may come from experiences of the calculus reform movement which often has expanded the role of writing in mathematics:

- Writing assignments in mathematics help raise conceptual issues and highlight relationships in ways that traditional assignments do not.
- Writing assignments require students to articulate their ideas – not just base work on intuition or guesswork.

but

- Grading writing assignments often takes more time than grading traditional ones.
- Undergraduate graders and teaching assistants often are ill-prepared to help with the grading of written assignments.

16.2 SUBSEQUENT REFLECTIONS

As noted in the original column, *Computing Curricula 1991* states, "Students should be encouraged to develop strong communication skills, both oral and written" [8, p. 21]. Similarly, *Computing Curricula 2013* identifies "verbal and written communication" as part of "soft skills" which "play a critical role in the workplace" [10, p. 15]. However, the incorporation of writing within *Computing Curricula 2013* curricula seems somewhat fuzzy: several course exemplars in these recommendations utilize writing as part of assessment, but the actual recommendations specify only one core hour for "Writing effective technical documentation and materials", under "Professional Communication" within "the Social Issues and Professional Practice (SP) Knowledge Area" [10, p. 99].

Ultimately, faculty must decide how technical writing fits within an undergraduate computing curriculum. Several course exemplars in *Computing Curricula 2013* provide some suggestions, and the following notes provide further ideas for writing assignments for either non-majors or majors and at levels ranging from introductory courses to advanced seminars.

Writing assignments for both non-majors and majors courses

At a basic level, effective writing typically requires at least two fundamental qualities.

- The writing must have content to convey, and
- The writing must present the content well (e.g., in a well structured, coherent, clear, and concise manner).

In my experience, many incoming students are poorly prepared for writing in both of these areas. Often incoming students are not accustomed to college-level expectations for exploring a topic in sufficient depth to bring new perspectives or structure to a subject. Rather many beginning students collect some ideas from a few sources and paste the ideas together (sometimes as quotations or paraphrases and with or without citations).

Similarly, although students may be able to write grammatically-correct sentences, these may or may not be organized into coherent paragraphs. Also, an introductory paragraph may or may not identify the topic and provide a context, but in my experience, the opening rarely indicates how the rest of a paper is organized.

Altogether, incoming students have received foundational instruction in writing (often by beleaguered teachers striving to provide feedback in several large classes), but college-level work raises expectations to a higher level.

Research Exercises

When exploring a subject, students (particularly at the introductory level) often lack experience in identifying sources and analyzing resources. Students may believe they are proficient in using search engines, but they do not consider issues of correctness and bias. (Recently, I heard of a student who used Facebook as an authoritative, primary source, for example.)

Similarly, incoming students often lack an understanding of basic rules and practices regarding paraphrase, quotation, and citation.

Since students believe they are capable in locating and utilizing appropriate sources, lecturing or routine exercises typically have little impact. Instead, in computing terms, an "interrupt" is needed.

One approach, described in [279], utilizes a series of exercises, in which students are asked to answer apparently simple sounding questions that have subtle or surprising answers. Students also write a paragraph explaining the process they used in finding their answer. At first, students may utilize their familiar search strategies, but when they report in class, they discover deficiencies in their answers. In my classes, a librarian attends each reporting session and subsequently comments on how students might be more effective and efficient.

These exercises also provide a vehicle to review matters of quotation, citation, and bibliographies. Providing detailed feedback on student answers takes time, but my students often remark that they apply the lessons learned throughout their college careers.

Risks of computing

Another type of writing assignment provides considerable guidance, while allowing latitude and creativity for the student for a relatively short paper on computing and its impact:

> Much of computing has an impact on society; computers are used in many applications, because they help people solve problems. Computing tools support tasks and activities. However, technology also has the potential to go wrong, and errors can have a significant impact. The occurence of errors also can highlight issues related to technology, management, and ethics.
>
> The Risks Digest [4], moderated by Peter Neumann for the ACM Committee on Computers and Public Policy, documents many examples of practical difficulties that have arisen in computer hardware and software. This paper asks you to review some of this material to identify two or three major types of difficulties that can arise from the use of computers. In particular,

> - Read several issues of The Risks Digest [4],
> - Identify two or three themes that are common to several of the examples,
> - Use a few examples to illustrate these themes,
> - Indicate possible consequences of such problems, and
> - If appropriate, suggest some conclusions about the use of computers, or propose some guidelines that might help minimize the difficulties.
>
> In this paper, you will need to define your topic clearly. Thus, you will need to make a special effort to focus your paper clearly and put your topic in context. Your paper should also discuss your topic in some depth (within the constraints of a 2-3 page paper). [272]

Additional instructions may suggest the target audience, length, citation style, and elements of writing to be highlighted for this paper.

Mixing Research and Social Impact

The "Subsequent Reflections" section of Chapter 15 provides additional possibilities for mixing research into a computing topic with an exploration of its social and ethical impact.

Writing Assignments for a Non-majors Course

Since non-majors typically have little technical background, paper assignments can encourage students to investigate a topic in moderate depth while also considering its implications.

Social Impact of a Computing Application

Some courses for non-majors (and some for computing majors) include initial readings that provide a framework for analyzing the social and ethical impact of a computing application. The following paper assignment illustrates a type of activity that may be given shortly after those readings, with the intention of bringing abstract concepts into a practical focus.

> The beginning of this course has described several ethical viewpoints. This assignment asks you to consider a practical use of technology from one or more of these perspectives. More specifically, you should follow these steps.
>
> 1. Explore sources in the library to study the ethical dimensions of a computing example or case study. For example, you might consider how one or more ethical perspectives apply in specific circumstances, such as
>   - sending spam,
>   - posting assignment answers on the World Wide Web (by an instructor after the assignment has been collected),
>   - collecting information on customers of an e-commerce site,
>   - gaining access to the Web site of a political candidate and editing the content without authorization,
>   - facilitating stalking by gathering telephone numbers and addresses from all telephone directories in a state, country, or region,

- demonstrating security weaknesses at a Web site by mounting a denial-of-service attack.

Whatever example or case study you choose, the discussion should focus on a concrete and specific situation rather than an abstract possibility.

2. Write a 2-3 page paper that describes your specific example or case study. Your discussion should include:

    - a description of the specific example or the circumstances of the case study,
    - an identification of the ethical issue(s) raised by the example or case study, and
    - lessons (positive or negative) that may be drawn from your example.

> You should write this paper at a level appropriate for others in the class. Thus, you can assume your reader has understood material discussed in class and in readings, but you should not assume any ideas specific to your selected example.
> As in all papers, the paper's introduction should indicate how your material is organized, as well as identify the main topic of the paper. [226]

Research Paper

A more open-ended research paper for non-majors might focus both on writing and a computing-related topic of each student's interest. In the following example (heavily edited from [273]), students initially bring a draft of their papers to class to obtain feedback on writing from other students. Later, since students have spent time and effort researching a topic, they share their findings with others in the class through oral presentations.

> This course has discussed many aspects of computing, including ethical issues, hardware, data representation, the nature of operating systems, software packages, software development, applications, networking, file sharing, privacy and security, the Internet, and e-voting. This paper assignment asks you to explore your interests by choosing a topic that you have not written about previously (at least in this course). To get started, you will need to do library research covering at least five sources (with at least three from published sources obtained from either the library or electronic databases), and you will need to organize your ideas and findings in a 4-6 page paper.
>
> Since you can learn a great deal about writing and about subject matter by reading other papers, the writing process for this paper is divided into several steps as follows:
>
> - **Report Topic to Instructor:** *3 weeks from the end of the term*
> Your report should identify your topic and also present a preliminary listing of source materials. Overall, reporting your topic can help you get started.
> - **Draft Due:** *2 weeks from the end of the term*
> Bring to class 4 copies of a draft of your paper. During the first part of this class session, students will break into groups to exchange papers and make some introductory comments about their papers. A copy of the draft will also be handed to the instructor. (This copy will not be graded, but the instructor will check that the paper is in a form appropriate for review by others.)

- **Feedback Due:** *class session after drafts due*
  In preparation for this class, read carefully each paper by the others in your group and write constructive comments. For example, comments might report your overall impression, the paper's good points, and one or two suggestions for improvement. Then bring TWO copies of your comments to class—one for the student author and one for the instructor. During the first part of this class, groups will meet again to return the annotated papers and make final comments.

- **Presentations Due:** *last week of term*
  Since you have explored a range of topics, you can share your research in an 8-10 minute presentation during class time.

- **Final Paper Due:** *last day of classes*
  In preparing the final paper, you may draw upon the feedback you received on your draft, rewriting the entire paper if you wish, or you may choose to ignore the feedback as being not particularly helpful.

Since the success of this assignment requires all class members to give and receive feedback and to report findings to the entire group, *failure to meet any of the above deadlines will automatically drop your semester grade by two-thirds of a full letter grade:* e.g., from A− to B or from B+ to B−. Only a verified medical excuse or a statement of extraordinary circumstances from Student Affairs will be considered in changing this penalty.

Writing Assignments for Majors Courses

Within courses for computing majors, a common writing activity involves the documentation of a software package, perhaps including specifications, design, a description of algorithms, testing protocols, and user documents. Although such writing serves many purposes, other types of activities may also be incorporated into courses—usually at the upper-level.

- Students might create posters to document their projects—either software packages or other research activities. At some schools, these posters are publicly displayed in hallways to highlight ongoing activities, create an atmosphere of excitement within the computing program, and help recruit both prospective and enrolled students. In some cases, software development courses or other scholarship-based courses may hold poster sessions for which the entire campus is invited. Poster sessions also can be popular during Parents' Weekends or events sponsored by the Admission Office.

- At some schools, writing activities may arise in two parts when an undergraduate engages in student-faculty research.

  ◇ *Initial proposal:* A description of background for the proposed work, such as a literature review, the resources available, and the work to be done—not unlike a proposal at the beginning of work for a master's or doctoral project.

  ◇ *Technical article:* A paper, in the style of a journal or conference submission. The initial proposal might identify venues for the article's possible submission and presentation, with the decision to submit deferred until the project's end.

Yet another approach for integrating writing with upper-level explorations, technical analysis, and social/ethical issues may be found in Chapter 30, "Combining technical depth, social/ethical issues, and active student involvement."

CHAPTER 17

College courses of varying credit

This column first appeared in ACM *Inroads*, Volume 5, Number 2, June 2014, pages 26-28[251]

17.1 ORIGINAL COLUMN

CHALLENGE

Develop a structure for a curriculum, so that

- needed topics (e.g., from the newly-released CS2013 recommendations from ACM/IEEECS [10]) are packaged nicely into courses:
  - ◇ Coverage of all Tier 1 topics is a priority for all students, while
  - ◇ Flexibility allows tailoring Tier 2 topics to meet 80% of identified topics.
- course offerings meet demands of a diverse student body (with different interests and priorities)
- credits earned by students reflect each course's rigor and scope
- time required for faculty reflects the teaching credit earned

This column identifies a traditional approach to meet this challenge and then outlines a possible alternative.

A Typical Curricular Approach

Many colleges and universities package topics into courses of a standardized scope, with topics divided among required core courses and electives. Often major requirements specify core courses (with Tier 1 topics) plus elective options (for Tier 2 topics). (See CS2013 [10] for details regarding Tier 1 and Tier 2 topics.)

To be more specific, the definition of a "course" may differ among schools and governmental jurisdictions. Common approaches often utilize the notion that "one college credit" corresponds to a class meeting one hour per week over a period of about 14 weeks:

- in some definitions of a college credit, "1 hour" refers to classroom instruction
- students may be expected to complete 2 hours of homework for each hour within the classroom
- about 2 hours of structured laboratory activity may be considered as equivalent to 1 hour of regular classroom instruction
- a final examination may be administered in the 15th week of a term

In this common accounting, 14 hours of classroom instruction may correspond to one credit.

Although 1 credit may be a basic unit, my experience suggests that most courses involve several credits. For example, at many colleges and universities, a typical course might earn 3 or 4 credits. In some cases, however, "credits" may not be discussed at all, but a "course" might be a standard, but possibly undefined, unit of measurement. In this context, a standard faculty teaching load might be 4, 5, or 6 courses per year. These courses may meet 3 or 4 times per week for a semester (or quarter), with some classroom instruction possibly replaced by 2-4 hours of structured laboratory activities.

Regardless of how credits or courses are formally defined on a campus, my experience suggests that most regular courses meet for 3-4 hours per week of classroom instruction, structured labs may replace some classroom hours, and expected homework aims at 6-8 hours per week.

Use of Traditional 3-credit or 4-credit Courses

As curricula have evolved over the years for common courses (e.g., CS1, CS2, basic courses in architecture and operating systems, upper-level courses in algorithms, etc.), many traditional courses fit within a standard format that meets 3 or 4 times per week. Early curricula from ACM/IEEE-CS (e.g., Curriculum 1968 [5] and Curriculum 1978 [17]) discussed content in terms of these specific courses. (For example, from time to time I still hear educators and publishers referring to a course on "Data Structures and Algorithm Analysis" as "CS7", using a designation CS7 from Curriculum 1978 [17].) More recent curricular discussions from the Liberal Arts Computer Science Consortium [109] also outline content for courses that typically would receive 3 or 4 credits.

Altogether, this common understanding of a course has served well for clearly-focused, coherent courses that have evolved over the years. When current courses cover content in the well-defined Tier 1 core of CS2013, it seems to me that such courses still may serve a useful purpose. CS2013 recommendations indicate that all students should cover all Tier 1 core topics, and 3-credit or 4-credit courses may provide an efficient mechanism to package that material [10].

Advantages of 1-credit and 2-credit Courses

However, in several settings, 1-credit and 2-credit courses may be natural. Sometimes, a 4-credit course covering Tier 1 topics may naturally break into pieces. For example, some years ago, when I was a visitor at another campus, I taught a math-for-computer scientists course that contained standard topics in discrete mathematics as well as substantial coverage of formal program verification. For this course, the topics of discrete mathematics had little to do with program verification, but the single course provided a convenient mechanism to treat both topics. Thus, in teaching the course five days a week one summer, I covered discrete mathematics on Monday, Wednesday, Friday (homework due on Monday), and program verification on Tuesday, Thursday (homework due on Thursday). Effectively the 4-credit course consisted of two 2-credit courses that had common tests and a single, unified semester grade.

Turning to Tier 2 topics, it sometimes may be appropriate to consider courses of modest scope. In particular, CS2013 suggests students cover only 80% of Tier 2 topics, allowing some flexibility for curricula to address particular interests and priorities of students and/or faculty.

I think of 1-credit and 2-credit courses as being "lean and lively" (using the jargon of the Calculus Reform Movement of the 1990s). Such offerings may have several potential advantages over larger, 4-credit courses.

- By packaging topics into relatively modest units, a school may be able to offer several options rather than a one-size-fits-all offering.
- Although courses for modest credit should not attempt to cover the full range of content for 4-credit courses, modest courses can provide students with a moderate level of background.
- Students may be able to enroll in 1 or 2 extra credits to enhance their background without overloading their schedules with a full sized course.
- Faculty may be able to develop or revise 1-credit or 2-credit courses relatively easily, allowing these small courses to connect well with new advances in the field.

Some Cautions

Although 1-credit and 2-credit courses provide interesting and practical opportunities to enhance a curriculum, care must be taken to avoid potential difficulties.

- When students and faculty are accustomed to 3-credit or 4-credit courses, expectations for a 1-credit or 2-credit course may escalate. Even if the planning for a course may begin with a modest scope, content may expand to rival courses carrying substantially more credit.
- Escalating expectations can lead to substantial overloads of faculty time, if contact hours expand to rival courses with more credit.
- Homework expectations for students may reach unreasonable levels for the number of credits earned.

Altogether, concern sometimes is expressed that two 2-credit courses or four 1-credit courses may require substantially more work (e.g., for faculty preparation or for student homework) than a 4-credit course. From my perspective, I can see the potential for trouble, but I do not think that such difficulties are inevitable.

In particular, the key to these practical issues seems to be realistic planning and review. The scope of a 1-credit or 2-credit course must reflect that the course meets just one or two hours per week. The list of topics should not rival the range of content found in courses carrying more credit.

Beyond matters of work load, scheduling can be challenging. If the modest courses are offered at numerous times throughout a week, students may find extensive scheduling conflicts with 3-credit and 4-credit core courses. One approach to scheduling may be to identify an established time for a 4-credit course (e.g., Tuesday, Thursday 10:00-noon) and schedule several short courses during that time (e.g., a 1-credit course Tuesday 10:00-11:00, another 1-credit course Tuesday 11:00-noon, and a 2-credit course on Thursday between 10:00 and noon).

Acknowledgments

I am deeply indebted to my colleagues in Grinnell's CS Department for helping me sort out the many complexities of planning and scheduling 2-credit courses within a curriculum. Further thanks to Grinnell's Curriculum Committee and Dean Lopatto for encouraging me to sharpen my thinking and articulating perspectives on this topic. Finally, many thanks to Marge Coahran for her suggestions regarding this column.

Conclusions

Although 3-credit and 4-credit courses may be the norm at many colleges and universities, the time may have arrived for introduction of several 1-credit and 2-credit courses—particularly for a range of Tier 2 topics in CS2013. Modest courses may not provide the full depth of coverage seen in courses for more credit, but these courses may help add diversity to course offerings, provide a mechanism to address a range of student priorities and interests, and give a useful framework for course experimentation and updating—all while allowing faculty to maintain appropriate work loads and students to handle manageable homework expectations.

V

Computing and Mathematics

FROM THE FIRST ACM CURRICULUM IN 1968[5], computing faculty have discussed what mathematics might be appropriate for some or all computer science students. *Curriculum 1968,* for example, specifies these mathematics courses as part of a bachelor's-level computer science degree:

Bachelor's level: at least 30 semester hours including

- Course B3. Introduction to Discrete Structures
- Course B4. Numerical Calculus (including some calculus, linear algebra, and numerical solutions of differential equations)
- Two advanced courses, selected from four alternatives that included
  - ◇ Course I8: Numerical Analysis I
  - ◇ Course I9: Numerical Analysis II
- Supporting mathematics courses: at lest 18 semester hours that included
  - ◇ Course M1: Introductory Calculus
  - ◇ Course M2: Mathematical Analysis I
  - ◇ Course M2P: Probability
  - ◇ Course M3: Linear Algebra
  - ◇ At least two additional mathematics courses, selected from:
    - ∗ Course M4: Mathematical Analysis II
    - ∗ Course M5: Advanced Multivariate Calculus
    - ∗ Course M6: Algebraic Structures
    - ∗ Course M7: Probability and Statistics (highly recommended)

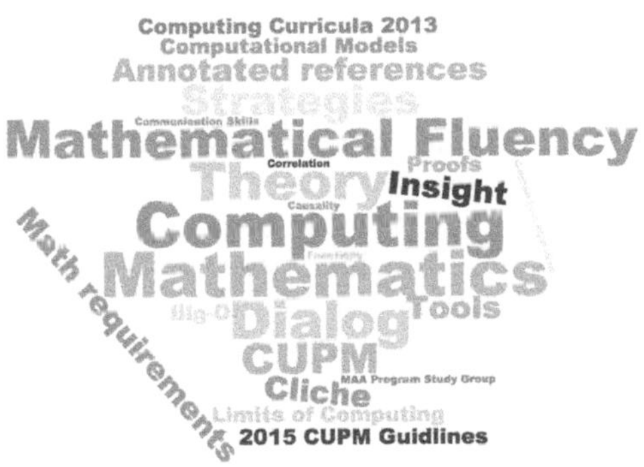

FIGURE 17.1 Computing and Mathematics Word Cloud

In this listing, a "basic-level" computer science course has a course number beginning with "B", an "intermediate-level" computer science course number begins with "I", and a supporting mathematics course number begins with "M".

Overall, an undergraduate computer science degree in *Curriculum 68* includes 8–10 mathematics courses and 6-8 required computer science courses. In addition, *Curriculum 68* indicated some additional computer science electives, and these were organized into several categories for a degree with a specialization of one of several prescribed areas. Together, the number of computer science courses required in a major was not significantly different from the number of required mathematics courses. [5, Section 4, Undergraduate Programs, pp. 161-162]

In the approximately 50 years since *Curriculum 68* appeared, the disciplines of computing and mathematics have diverged dramatically, and specific requirements for mathematics within an undergraduate computer science major have decreased substantially. Today, discrete structures continues as a widespread requirement (e.g., in *Computing Curricula 2013* [10]), and mathematical analysis, proof, and theory continue to provide vital insights within the discipline of computing.

Ongoing questions for an undergraduate computer science major include:

- How many mathematics courses should be required for all computing students?
- Exactly what mathematics topics should be covered in the required courses?
- How much additional mathematics should be specified for various computing subdisciplines?

This part considers these questions in the context of national curricular recommendations, but also by considering general principles, relationships between computing and other disciplines, high-level goals, and the role of theory.

CHAPTER 18

Mathematics and compuitng topics in the classroom

This column first appeared in the *SIGCSE Bulletin*, Volume 37, Number 2, June 2005, pages 15–17 with the title, "Mathematics and CS in the classroom" [218]. The modified title used here reflects the broad usage of the word "computing" world-wide. In North America, "computer science" is often used for some areas of study, but other words and phrases are used for other areas. Internationally, "computing" seems a rather inclusive term. Subsequent reflections on this subject appear later in this chapter.

18.1 ORIGINAL COLUMN

TONY RALSON AND PETER HENDERSON have written columns for this issue discussing mathematics for computer science . (*Editorial note:* See [153] and [85], respectively.) This column considers possible implications for the classroom. While the ideas in this column are not particularly new, perhaps they will stimulate discussion.

Goals for Mathematics

First, to complement the goals and content discussed by Tony and Peter, it seems worthwhile to remind readers of additional goals for mathematics that were identified by the Pedagogy Focus Group on Supporting Courses in the development of Computing Curricula 2001:

> Students of computing should:
>
> - attain sufficient mathematical maturity and insight to work comfortably at an abstract, logical level with computing concepts,
> - attain knowledge of specific topics which support fundamental areas of computer science,
> - understand topics in mathematics, not just by themselves, but as integral parts of numerous areas within computing, and
> - utilize mathematical methods and insights to support and model algorithmic thinking as a means to solve various types of problems. [140]

"Exciting" and "Interesting" Topics

Second, in his editorial, Tony writes, "Calculus is one of the greatest inventions of humankind and educated people should know something about it. The trouble is that too many subjects exist about which we could say the same so that only a polymath knows something about all of them."

My second observation is that Tony's comment likely applies quite broadly. Here are three examples.

1. In the 1980s, many mathematicians observed that the list of topics in calculus courses had gradually expanded over the years, so the 1980s courses often had a new topic each day. Students struggled to learn one formula after another, and there was little time to explicitly link specifics to general themes (e.g., rates of change, areas through sums). Pragmatically, faculty realized that few students would remember the myriad formulae past the end of the semester, so the real focus should be on the use of a few basics within the framework of problem solving. Using the catch-phrase, "lean and lively calculus", courses therefore were redesigned to emphasize main topics (perhaps half a dozen big ideas). Selected formulae illustrated these main ideas, but many favorite topics were dropped. In their place, students were challenged with numerous problems requiring multi-step solutions, and course time was devoted to making explicit connections among themes. For example, my colleagues at Grinnell assign both routine problems ("book exercises") and integrative problems (the "real" problems). Interestingly, with this change in focus, faculty in other departments often took notice. For example, economics faculty comment that today's students seem much more comfortable talking about rates of change and derivatives.

2. In recent years, many discussions have identified new topics that should be added to CS1 or CS2 – without removing other topics. At many institutions, the list of introductory topics likely extends for pages, and faculty report difficulty shoehorning all this material within two semesters. Often solutions to this overcrowding seem to focus either on how to cram more into two courses or how to extend the introductory sequence to three semesters. At individual institutions, such solutions may indeed make sense. However, I wonder if introductory computer science today is in a similar state to introductory calculus of the 1980s. Are students able to keep track of so many details, while they connect these details to the important main themes? If not, should we be striving for "lean and lively CS1 and CS2", so students focus more on computing in support of problem solving and fundamental principles?

3. Some years ago, I participated in a conference focusing on an interesting subdiscipline of computer science and its role within the undergraduate curriculum. Several speakers described fascinating subjects – often research-level topics. Presentations frequently described an advanced undergraduate course on the subject – often with the rationale that "undergraduates can learn [this topic]." In isolation, each course might provide an outstanding experience for undergraduates. However, the prerequisites for these advanced courses, coupled with the core courses for Computing Curricula 2001, would likely require far more than four (or even five) years of undergraduates study.

This leads to my next observation that "undergraduates can learn [this topic]" may not be a strong rationale for inclusion of a topic or course within a curriculum. A more compelling rationale would consider whether "undergraduates should/must learn [this topic] instead of [that topic]".

In summary, it seems easy to identify a wealth of topics that might be interesting at the undergraduate level. However, coverage of all those topics might easily require 6 or 8 or more years – far beyond what is considered appropriate for an undergraduate program. Further, with research producing new material at a rapid rate, there seems little hope that any undergraduate program could realistically cover everything that every employer or graduate school might dream about. Instead, curricula represent a compromise of topics within a reasonable time. Sometimes a reduction in material may even increase students' ability to synthesize ideas and improve their problem solving and communication skills.

Explicitly Making Connections

Many faculty comments reinforce my own observations that students often do not connect different ideas within a single course or related ideas in distinct courses. Regardless of how obviously faculty may believe different ideas connect to each other, these relationships must be made early and often before many students recognize them. This suggests that both class time and homework exercises must help student integrate ideas.

In particular, there is a significant danger that students will view separate mathematics courses as being unrelated to computing, unless these mathematical concepts and techniques are used repeatedly in computer science courses. Separate math courses may create a "mathematics ghetto" that students take and forget, if ties are not explicit. On the other hand, I expect my experiences mirror those of many other faculty that student eyes often light up when they see mathematics collaborating with computer science in answering questions.

Course Prerequisites

Prerequisites for a course typically have two purposes:

A. identify subject matter from one course that is utilized in the second

B. assure a general level of sophistication, maturity, or experience

Content background (purpose A) allows a later course to proceed quickly and efficiently, and these prerequisites may serve well. On the other hand, my earlier observations on making connections suggest that students may not understand prerequisites unless the later courses explicitly utilize content from previous ones. In some cases – particularly after a course structure has been in place for several years, it is easy to forget the reasons for prerequisites, and faculty may forget to connect new ideas with content from previous courses.

Of course, from time to time, maturity prerequisites (purpose B) may make sense, but I wonder if this purpose is overused – at least on some campuses.

Prerequisites can provide structure for a curriculum, but they also decrease flexibility in student scheduling. When material from one course is actually used later, and when ties among topics are explicit, then prerequisites may strengthen a program substantially. However, when one course does not really build on another, then a prerequisite may serve little real purpose.

Building on the comments of Tony and Peter, mathematics can provide valuable assistance when it is tied to topics in computing. Further, issues of topic selection, connections among topics, and prerequisites certainly apply to the selection of mathematics within the computer science curriculum. However, the same questions likely apply much more broadly as well.

18.2 SUBSEQUENT REFLECTIONS

Through the years, the overall discipline of computing has expanded greatly. For example, the Association for Computing Machinery, ACM, with its sister societies, presents recommendations for separate undergraduate programs in computer engineering, computer science, information systems, information technology, and software engineering. In addition to these subdisciplines, the *2015 Curriculum Guide* of the Committee for the Undergraduate Program in Mathematics (CUPM) of the Mathematical Association of America (MAA) discusses the related fields of computational science and big data. [44]

Within computer science, the 2013 ACM/IEEE-CS recommendations identify Discrete Structures as a Knowledge Area that is important for all undergraduate computing majors [10]. However, beyond this base material, each subdiscipline has its own needs and interests. A few examples follow:

- Human-computer interactions (HCI) likely draws upon probability and statistics as data on user experiences are gathered and analyzed.
- Graphics utilizes linear algebra in transforming images.
- Cryptography and cybersecurity rely upon concepts of number theory and abstract algebra (e.g., groups, rings, and fields).
- Artificial intelligence, once largely qualitative, now draws heavily on multi-variable calculus in computing gradient vectors to adjust weights in neural networks.
- Big data relies upon mathematical modeling, probability and statistics, and multi-variable calculus.

The *Summary Report of the MAA Program Study Group on Computing and Computational Science* provides a more detailed description of the types of mathematics that may be useful in various subdisciplines related to computing. [280]

This range of mathematical material that supports one or another part of computing presents a challenge for undergraduate curricula: what mathematics background should be required for computing students?

- Together, the mathematics needed for various computing subdisciplines likely covers an entire mathematics major, perhaps more. *From this perspective, every computing major should also double major in mathematics!*
- Pragmatically, however, beyond Discrete Structures, specific areas of mathematics may be needed only for certain computing subdisciplines. For example, a person interested in HCI may not need background in linear algebra or multi-variable calculus. Similarly, a person interested in graphics may not need to study abstract algebra.

Within an undergraduate program, therefore, recommendations for mathematical background may need to be tied to specific subdisciplines. For example, at some schools, a mathematics requirement might entail a course in Discrete Structures plus one or two additional mathematics courses, with specifics determined through student-advisor discussions.

Computing students will need to take mathematics to support their interests, but few students will have the time and interest to take all the mathematics that might be useful in some area of computing. In the coming years, structuring both requirements and advising for computing students likely will require considerable brainstorming and creativity. Using the Summary and Supplemental Reports of the MAA Program Study Group on Computing and Computational Science [280, 281] might help in initial student-faculty discussions.

CHAPTER 19

An opportunity for computing-mathematics dialog

This column first appeared in ACM *Inroads*, Volume 4, Number 2, June 2013, pages 32–34[249] Subsequent reflections on this subject appear later in this chapter.

19.1 ORIGINAL COLUMN

MOST READERS OF THIS COLUMN know that the ACM with IEEE or AIS regularly develop new undergraduate curricular guidelines in five programmatic areas: computer engineering (CE), computer science (CS), information systems (IS), information technology (IT), and software engineering (SE). For example, an initial draft of the new ACM/IEEE recommendations for undergraduate computer science appeared in 2012, and a revised draft appeared in January 2013. Similarly, within mathematics, the Committee on the Undergraduate Program (CUPM) of the Mathematical Association of America (MAA) publishes recommendations for undergraduate mathematics programs, with the next complete report due in 2015.

Study Groups to Inform the 2015 CUPM Undergraduate Mathematics Recommendations

In anticipation of the 2015 recommendations the MAA's CUPM created about 20 Content Area Study Groups to examine various subject areas within mathematics and also about 9 Program Study Groups to examine possible connections between mathematics and other disciplines. For example, Program Study Groups were to consider how mathematics programs might collaborate with degree programs in other disciplines, such as majors, minors, concentrations, and interdisciplinary programs.

One Program Study Group dealt with computing and computational science. Appointed in Summer 2012, I was asked to serve as Chair, together with Douglas Baldwin (SUNY Geneseo) and Daniel Kaplan (Macalester College). The Group's charge involved two main parts.

- *Study:* Investigate the needs for mathematics for majors in various areas of computing and computational science, double majors, interdisciplinary majors, and several combinations of majors and minors.
- *Reporting:* Prepare a summary report (5 pages maximum for written publication) and a supplementary report (length as needed for Web posting) by December 1, 2012.

Process for the Program Study Group on Computing and Computational Science

Over the fall, the group talked with a range of people from both academia and industry, examined national curricular recommendations by ACM and its collaborating societies, prepared draft materials. presented ideas at several regional meetings, and solicited feedback from several groups of faculty. Time constraints prevented widespread circulation of drafts, but moderate distribution and feedback was possible.

Some Important Themes for Computing and Computational Science

As this work continued, several important themes emerged:

- Many members of the mathematics community consider computing and computational science as being reasonably monolithic. In contrast, the computing community may take for granted differences in curricular needs for programs in CE, CS, IS, IT, SE, Computational Science, and the emerging field of Big Data.
- Various recommendations for undergraduate computing programs vary widely regarding expectations and requirements for mathematical preparation. For example, CE generally requires more mathematics than CS or SE, which in turn require more mathematics than IS and IT.
- Subdisciplines within computing and computational science require varying amounts of discrete mathematics, continuous mathematics (e.g., calculus and analysis), probability, statistics, and other areas of mathematics. For example, requirements for computer science generally focus upon discrete mathematics. The ACM/IEEE requirements for CE specify considerable discrete mathematics, logic, 1-variable calculus, matrices within linear algebra, several topics involving probability, and hypothesis testing and sampling within statistics. Also, the relatively new fields of computational science and big data may require more continuous mathematics and statistics, but relatively less discrete mathematics, than the traditional computing fields.

Current Reports of the MAA Program Study Group on Computing and Computational Science

Currently, full Summary and Supplemental Reports have been presented to CUPM for their review. In time, CUPM likely will incorporate this material into their own reports and Web pages. In the meantime, these materials are available on my personal Web pages within the directory

http://www.cs.grinnell.edu/~walker/maa/program-study-group/

Future Development and Refinements

Since the Program Study Group on Computing and Computational Science serves at the pleasure of the MAA's CUPM, future refinements of these reports will depend upon directions from CUPM. However, preliminary feedback to all Program Study Groups suggests that CUPM encourages interested people in all disciplines to provide comments to the relevant Program Study Group Chairs.

With this in mind, I invite readers of this column to email me at walker@cs.grinnell.edu regarding your thoughts regarding these [draft] reports. Please include "Feedback: PSG on Computing/Computational Science" in your Subject line. At this writing, timing is unclear, but comments within a month or two would be most likely to fit with directives that might be received from CUPM.

Use of this Material for Computing-Mathematics Dialog

Regardless of the possibilities of revision, these reports provide a natural opening for expanded conversations between computing faculty and mathematicians/statisticians. These reports highlight different needs for mathematics within various computing-related fields, and this raises natural questions of what courses would work best in a particular institution. ACM and its sister organizations have published many recommendations, and the forthcoming revised recommendations from MAA's CUPM will provide similar guidance from the perspective of mathematics and statistics.

Let the interdisciplinary conversations begin [or continue]!

Acknowledgments

Many thanks to group members Douglas Baldwin, SUNY Geneseo, and Daniel Kaplan, Macalester College, for their many contributions to this project. Their time and effort in brainstorming, investigation, analysis, and writing were instrumental in guiding the work of this Program Study Group and creating the reports.

Special thanks to Douglas Baldwin and to Marge Coahran, University of Toronto, for their suggestions on clarifying and refining this column.

19.2 SUBSEQUENT REFLECTIONS

The original column suggested that the appearance of CS 2013 [10] and CUPM 2015 [44] provided a wonderful motivation and/or excuse for discussions between computing faculty and faculty from client departments. Revised national recommendations for computer science and for mathematics had recently emerged, so faculty of both departments would naturally be reviewing their curricula and course offerings.

Arguably, however, almost any time may be good for fruitful dialog. If one or both parties are interested, and if the other party is not openly hostile, then discussion and collaboration may be productive. For example, here are just a few possible motivations for starting or continuing discussions,

- One department may be engaged in a periodic external review.
- The computing faculty may be considering changes in its curriculum or courses.
    - ◇ Mathematics students may enroll in computing courses (some computing courses may even be required for a math major), so both computing and mathematics faculty need to know the impact of possible changes.

  - ◇ Mathematics faculty may be able to offer feedback on how well current computing courses are meeting the needs of their students; will new or revised courses be helpful, counter-productive, or neutral?

- With enrollments large and increasing in computing, the computing faculty may want some mathematics faculty to help teach courses related to computing (e.g., discrete structures, computational science, big data).

- Similarly, mathematics faculty may be interested in new or revised computing courses to support their programs, such as applied mathematics.

In short, both computing and mathematics departments may consider each other as client departments, so each have a clear interest in meeting needs of the other's students.

More generally, computing faculty may want to initiate discussions with a range of client departments, such as STEM fields, social science departments, etc. Students throughout a college or university may need background in computing-related fields as they consider possibilities for scholarship and professional work. Through dialog, computing faculty can determine how well those needs are being met—what goals and interests are well supported, and what gaps and deficiencies remain. A computing curriculum likely cannot do everything for all possible clients, but discussion can help determine the extent to which current courses and programs are meeting intended goals.

And, as a side effect of dialog with faculty throughout a school, computing faculty may be able to identify what possibilities might be available for computing students to help with computing-related needs with other faculty. With high computing enrollments, computing faculty may not be able to support all requests for independent projects and student-faculty research, but it might be possible for faculty in other departments to supervise meaningful projects in which computing students might make worthwhile contributions.

Beyond the cliche: mathematical fluency in the computing curriculum

This column first appeared in ACM *Inroads*, Volume 6, Number 4, December 2015, pages 24–26[256] Subsequent reflections on this subject appear later in this chapter.

20.1 ORIGINAL COLUMN

COMPUTING FACULTY LARGELY AGREE upon the importance of discrete mathematics within the undergraduate curriculum. For example, CS2013 states, "We recognize that general facility with mathematics is an important requirement for all CS students." [p. 49] Curr2013, and the CS2013 Body of Knowledge recommends 37 Core-Tier1 hours and 4 Core-Tier 2 hours [10, pp. 76-81]. Further, many [most?] schools require at least one course in discrete mathematics as part of a major program in computing. Beyond specific topics, however, computing practitioners require mathematical insight and intuition. For example, what mathematical statements make sense within a specific context, when do mathematically-sounding statements utilize terms and ideas precisely and correctly, and how do mathematical concepts apply in practice? This column explores these questions at considerable length, giving several examples and providing follow-up questions for a computing curriculum. As an anonymous reviewer noted, this 'column is looking in depth at what the cliche of "mathematical maturity" means for computing education.'

This column discusses three examples that illustrate the need for mathematical insight and maturity. Readers are challenged to consider how a computing curriculum might help students attain an appropriately mature level of mathematical understanding.

Example 1: Digits of Accuracy

In some respects, computers make arithmetic computations easy. A user enters numbers, the computer translates the data to a binary form (e.g., IEEE 32-bit or 64-bit format), the CPU performs arithmetic, the machine translates the results back to decimal, and the results are printed. Of course, in the translation to binary, the computer has no way to know the accuracy of the input, and the input typically is treated as exact (at least up to the constraints of the data representation). At the end, the translation to decimal utilizes the full accuracy of the stored binary representation. With such an easy process, programmers may forget to ask, "how many digits of accuracy should be printed?"

As a simple illustration, a handbook typically declares that 1 quart = 0.946333 liters. This formula, of course, assumes the measurement of 1 quart is exact. However, if only one digit of accuracy is reported, 1. quart might represent any size between 0.5 and 1.5. Using the same conversion factor, 0.5 quarts yields 0.47317 liters and 1.5 quarts yields 1.4195 liters. From this perspective, the middle value 0.9 liters may serve as a reasonable estimate for 1. quart; further accuracy seems unrealistic and possibly misleading.

Questions Impacting Teaching and the Curriculum:

In considering the conversion of quarts to liters (or similar computations),

- How should one report the number of liters for 1. quart, 1.0 quarts, 1.00 quarts, etc.?
- What guidance should programmers receive in specifying the number of digits to be printed for liters?
- What is the appropriate response if a program, written for a course, always prints 6 digits of accuracy?

As a more sophisticated illustration, consider grading a group project. Suppose groups have 10 or fewer students, and suppose each student must rate the contributions of each other student in the group on a scale of 0 to 10. The ratings are to be averaged and a resulting score reported. How many digits of accuracy should be reported?

In this context, the computation of an average is easy, but what about the accuracy of the input? Typically, a rating by a student will be subjective—what does a score of 7 mean and how is that different from a score of 8? Would a score to two decimal places (e.g., 7.39) have meaning beyond a score rounded to 1 or even 0 decimal places (e.g., 7.4 or 7)? Can students assign meaningful and consistent scores as integers on this scale, to one decimal place, etc.?

Although such questions seem natural, I am aware of one class environment in which class management software reported student averages of this type to 6 decimal places—faculty argued they were just using the default printing from their course software.

Questions Impacting Teaching and the Curriculum:

Where might questions of numeric accuracy be discussed in a computing curriculum—perhaps in the context of the printing of output?

- How should students respond to the above system which presents averages to 6 decimal places?
- Does such a reporting system encourage an inappropriate use of mathematics?
- How should a department chair or dean respond after hearing of such a reporting system?

Example 2: Interpreting English Descriptions of a Functions

A common activity within computing involves measuring, estimating, or analyzing quantities. For example, we may run experiments to time algorithmic performance, estimate calls for dynamic memory allocation, and analyze the secondary memory required for an algorithm. Sometimes the description of a size function may be clear: how many packets of network traffic does an algorithm generate? However, sometimes a description of a quantity may be sloppy: in timing an algorithm, does a measurement include CPU time only, delays for moving data between a register and memory, operating system work for swapping data to secondary storage, time for other processes scheduled on the same processor, etc.?

Questions Impacting Teaching and the Curriculum:

- When discussing algorithmic performance, to what extent do students gather actual experimental data?
- Timing of algorithms for different data sets may yield expected patterns in most cases, but a few data points sometimes lie far from an expected graph. How should students respond to such messy data, and how much guidance should they receive?

As a more sophisticated illustration, several weeks ago, I attended a lecture discussing brain activity during various tasks. The primary source [182] focuses upon wearable sensors to measure skin conductance, and various studies report changes in skin conductance are correlated with learning, emotion, and other factors. Mathematically, changes in functions normally indicate derivatives. Thus, if s(t) represents skin conductance, then the derivative s'(t) would be correlated to learning, and a high value for s'(t) might suggest significant learning.

To highlight the versatility of the new sensors, [182] reports measurements attached to a student over a week—during lecture, in-class labs, study, exams, sleep,etc. As a result, Figure 9 in [182] is commonly cited to show the lack of learning during lecture—dramatically less than during sleep or most other activities.

This research seems clear until the term, electrodermal activity (EDA), is introduced. Consistent with numerous sources on the Web, the first paragraph of [182] states, "Changes in skin conductance at the surface, referred to as electrodermal activity (EDA), reflect activity . . . ", and the relevant axis of Figure 9 is labeled EDA in [182], not skin conductance.

All of this raises the question (debated at length after the lecture I attended) about how to interpret Figure 9 in [182]. Does learning relate to changes in the Figure 9 graph (i.e., does EDA = s(t)), or to the height of the Figure 9 graph (i.e., does EDA = s'(t))? Subsequent analysis and consultation with Clark Lindgren, a neurobiologist at Grinnell, indicates researchers use EDA as a synonym for s(t), the graph was measuring skin conductance, and EDA indeed is s(t). Overall, although it seems common in the literature to refer to EDA as changes, researchers actually focus on the changes in EDA.

Altogether, careful exploration suggests a common lack of precision in discussing EDA. Many references identify EDA as changes in skin conductance—apparently s'(t) as changes in a function, and the term "activity" may further suggest changes. However, in actual usage, EDA is the same as skin conductance, s(t).

Questions Impacting Teaching and the Curriculum:

- How do/should students understand statements, such as "Changes in skin conductance at the surface, referred to as electrodermal activity (EDA)"?
- How might students track down whether EDA was the same as s(t) or s'(t)?
- To what extent could students interpret the graph s(t) or s'(t) to indicate learning?

- What reaction do students have to this question of the definition of a term? Can this example be used as part of a case study somewhere in the curriculum?

Example 3: Significance of Big-O Growth Rates

Big-O analysis provides a high-level, macro analysis of algorithms that can provide insight regarding time or space efficiency. In my experience students often seek shortcuts to possible answers rather than performing careful analysis, and the significance of Big-O may seem elusive. Here are some scenarios.

As a simple illustration regarding the consequences of Big-O analysis, suppose an algorithm processes 40,000 items in 20 seconds. What can one expect for the time required for 80,000 or 160,000 items, assuming the algorithm has $O(1)$, $O(n)$, $O(n \log n)$, and $O(n^2)$?

Questions Impacting Teaching and the Curriculum:

- Are students asked to use formal Big-O results to predict run times?
- What variation in answers reasonably might be observed experimentally?
- Where are students asked to compare predicted times with experimental times?

As a second illustration that asks students to consider the meaning of $O(n)$, $O(n \log n)$, $O(n^2)$, consider the following question that I asked on an in-class test for an algorithms course.

> A programming assignment asked students to code several sorting algorithms and then time the code on several data sets. I received a wide range of timings for various algorithms on data sets of various sizes. Table 20.1 shows a hypothetical table of timings that might have been obtained (based on some [unspecified] time units):

A Consider the timings for Algorithms A, B, and C, and consider timings that might be expected for an insertion sort, a quicksort (with random pivot), a heap sort, and a radix sort. For each of these algorithms, determine whether

- ◇ most timings are consistent with the run time expected of the algorithm,
- ◇ several timings seem consistent with expected run times and several seem inconsistent, or
- ◇ most timings are far from what might be expected for the run time of the algorithm.

Note: One set of timings may be appropriate for 0, 1, or more of these algorithms.

B Considering the timings of Algorithm D, does it seem likely that the program generating this output provides a correct implementation of a merge sort? Explain.

Table 20.1: Algorithm with timings*				
Array Size	A	B	C	D
Ascending Data				
10000:	1	2	1	13
20000:	2	4	1	50
40000:	4	6	0	12
80000:	15	13	0	48
160000:	25	28	1	208
Random Data				
10000:	1	1	29	2
20000:	2	3	115	6
40000:	4	5	465	28
80000:	8	12	1857	120
160000:	26	27	7440	512
Descending Data				
10000:	1	1	58	1
20000:	1	2	232	4
40000:	3	5	929	12
80000:	8	10	3719	56
160000:	50	21	14856	224
* Timings based on unspecified, but consistent, time units				

Curricular Questions:

Beyond formal algorithmic analysis,

- to what extent can students relate specific algorithms to experimental data?
- to what extent can/do students utilize performance data to determine the likelihood that errors exist in code? (Timings consistent with expectations may say little about whether the code is correct, but inconsistent timings may suggest coding errors.)

In future columns, if there is interest, I may suggest additional examples—let me know your thoughts!

Conclusions

Computing programs typically require students to take a course on discrete structures or discrete mathematics, and such courses often provide reasonable coverage of numerous topics in mathematics that are useful within computing. However, a formal understanding may not provide mathematical intuition or a working fluency with material as it applies in day-to-day work. Issues may arise with topics as simple as the presentation of data, the definition of mathematical terms, and the relationships between formal analysis and experimental results.

The examples provided here only begin to describe the practical use (or misuse) of mathematics—sometimes by students and sometimes by professionals. As a curricular matter, I encourage readers to consider where such practical issues of use might be addressed within a computing curriculum.

Acknowledgments

Many, many thanks to Marge Coahran, Research Associate, Toronto Rehabilitation Institute, for her excellent insights and feedback for this column. Thanks also to Clark Lindgren, Professor of Biology, Grinnell College, for helping me understand the research and graphs behind [182], and to the anonymous Inroads reviewer whose suggestions helped me tighten the language for this column.

20.2 SUBSEQUENT REFLECTIONS

The second paragraph of the original article noted that the discussion would highlight "three examples that illustrate the need for mathematical insight and maturity." Through the years following the column's appearance, I have continued to observe misunderstandings and misuse of mathematics within the realm of computing. Here are two more brief examples:

Example 4: Computing Averages

Recently, I became aware of a grading program that computed semester averages, based upon student scores. In testing the program, I observed the following data for a hypothetical, poorly-performing student: reading-homework: (3 of 28), papers (one not submitted, one with a C+ grade), labs (41 of 150), mid-semester test (15 of 108), and final exam (32 of 108). In summary, one paper had a passing grade, and the rest of the scores were below 35%; grades for some categories were below 25% or even below 15%. Yet the grading program generated a semester average of 55%. Naturally, my immediate reaction was that averages should not work that way! Upon investigation, it turned out that not only was the program computing averages, but it also was scaling scores upward.

Originally, faculty had been using the program with little analysis, and the results were being taken for granted. The scaling computation came to light only after it was observed that the averages produced were substantially higher than all but one of the inputs.

Example 5: Correlation and Causality

Several papers I have read recently have observed a correlation between two phenomena and then made statements about causality. Statistically, correlation identifies an association between two collections of numbers, but correlation provides no evidence regarding the causes behind this association.

For example, in the years since World War II, prices have risen; inflation has been widely reported in the United States, and the cost of living is higher now than in the 1940s. Similarly, as a member of the baby boom generation, my age has increased steadily since my birth. Putting these two observations together, there is a positive correlation between my age and the cost of living in the United States. However, rises in the cost of living has not caused my age to increase, and similarly the increase in my age has not caused the cost of living to increase.

In summary, even though correlation and causality are different concepts, some articles continue to tie these two elements together.

Altogether, the uses and misuses of mathematics seem to pervade contemporary society. In formulating appropriate hypotheses, supporting claims, and utilizing computations, computing students, faculty, and professional practitioners need to attain a solid level of mathematical insight and maturity.

CHAPTER 21

Why a required course on theory?

This column first appeared in ACM *Inroads*, Volume 6, Number 2, June 2015, pages 24–46[262]

21.1 ORIGINAL COLUMN

DURING A RECENT TELEPHONE INTERVIEW for a tenure-track position at Grinnell, a candidate asked, "Why would a computing major require a course on the Theory of Computation?" Over the years, I have heard many perspectives on this question, and I was interested that the question continues to be asked. Since the shaping of a course and a curriculum should reflect educational priorities and objectives, a discussion of theory naturally turns to consideration of desired skills and background for our graduates.

Computer Science Curriculum 2013 (CS2013)

CS2013 [10] from the ACM and IEEE-CS provides a starting point. The Algorithms and Complexity Knowledge Area includes three Tier 1 hours, three Tier 2 hours, and identifies other topics as electives [pp. 59-60]:

- Basic Automata Computability and Complexity
  - ◇ Tier 1: Finite state machines, regular expressions, the halting problem (all at modest to intermediate levels of understanding) (3 hours)
  - ◇ Tier 2: Context free grammars, Classes P and NP, and sample P and NP problems (again at modest to intermediate levels of understanding) (3 hours)
- Advanced Computational Complexity
  - ◇ NP Completeness, Cook's Theorem, classic problems, reduction techniques (elective)
- Advanced Automata Theory and Computability
  - ◇ DFAs, NFAs, regular languages, properties, pumping lemma (elective)
  - ◇ Context-free languages, PDA, properties, pumping lemma (elective)

  - ◇ Turing machines (deterministic and non-deterministic)
  - ◇ Chomsky hierarchy, Church-Turing thesis, computability, non-computability (elective)

Additional elective topics might include such material as Syntax Analysis under Programming Languages [10, p. 166].

Altogether, six hours of material is recommended, with other topics identified as elective. Although this coverage touches important topics, students cannot be expected to gain depth or truly appreciate some vital consequences of basic results. Further, deeper and extended coverage can engage students in skills and insights that can have substantial impact both inside and outside computing. A full course on the theory of computation can provide depth, intuition, and skill in the above areas together with extended consideration of the following:

- Models of computation
- Limits of computing
- Responses to limits
- Connections between theory and applications
- Student construction of logical arguments
- Communication skills

Although, in principle, these areas might be covered individually as parts of separate courses, some vital foundations and insights may likely be missed without systematic coverage—often in a cohesive and integrated course. The following sections look at each of these points briefly.

Models of Computation

Periodically, beginning and prospective students ask me which commercial computer can solve the greatest number of problems. I typically begin by indicating that current models can solve the same problems, but substantial differences may be found regarding speed, memory, available software, etc. Throughout the discussion, students have no experience or understanding of the notion of models of computation—how one might formally define a computational model and prove what it can and cannot do. Rather, beginning students seem to believe that each new gadget has revolutionary impact on what problems can be solved. Without a frame of reference, students may hear something about the equivalence of computing models, but they have no experience thinking about or working with these concepts.

A course on the theory of computing, of course, challenges this naive perspective. Students engage contrasting models of computation, determine what problems each model can and cannot solve, consider when models are equivalent or not equivalent, etc. Overall, a unit extending over several weeks provides an opportunity to challenge and rethink perspectives about the nature of computation, as well as presenting the classical topics of automata, pumping lemmas, Turing machines, and the like.

Limits of Computing

With the many successes of computing in contemporary society, common culture often seems to believe that all problems can be solved—given sufficient time, money, resources, and the like. The popular press regularly contains claims, often from publicly-elected officials, about proposed solutions to problems that a computer-literate citizen would know immediately could not be true. Each model of computation can solve some problems, but formal proofs identify problems each model cannot solve. Further, even for a Turing-complete environment, algorithms for certain problems are not solvable in feasible time for large data sets.

In my experience, this notion of the limits of computing takes students a long time to internalize. If one proposed algorithm to solve the Halting problem fails for a given computation, students may want to tweak the algorithm to work in an additional case. The notion that no algorithm can work in all cases seems to contradict many students view of science in general and computing in particular. As a corollary, students initially may be uncomfortable that no algorithm can be developed to show, in general, that a proposed solution to a problem is totally correct. Similarly, until students have to reduce one NP-complete problem to another, my experience suggests that they may not have truly mastered concepts of non-feasibility of computation.

Perhaps most importantly, issues of non-feasibility or non-computability raise vital social and ethical consequences for hardware and software packages, and these issues similarly can impact public policy. As an historical example, the Strategic Defense Initiative (SDI) was proposed in the 1980s as a computer-controlled system to protect the United States against rocket attacks by identifying and intercepting hostile missiles. Since such systems could not be fully tested with real, incoming missiles, automated means would be found to prove the extensive software was totally correct. At the time many computer scientists observed that a complete and formal specification was almost certainly impossible, but even if the specification could be written, a proof of correctness would likely require a solution to the Halting problem. Interestingly, such arguments were largely ignored in the press in favor of proposing more money for SDI.

Although a single course may not revolutionize the world, a course can help raise these fundamental issues for future computing practitioners and leaders .

Responses to Limits

Over the years, I have been impressed with the many innovative approaches that seek to sidestep issues of non-feasibility and/or non-computability. For example, approximation algorithms and probabilistic algorithms do not seek a perfect solution, but rather one that may be workable most of the time. At a basic level, such research is motivated by an understanding of what is and is not possible and feasible—and that foundation comes from a thorough knowledge of computational models.

Connections between Theory and Application

During my mathematical training, I loved theory, but I also noticed a substantial divide often evolved between mathematical theory and its application. In many cases, theory was wonderfully elegant and rich, but pure mathematics and applied mathematics seemed to represent two different worlds.

In contrast, as I moved from mathematics to computing, I found a wonderful interplay between theory and practice. Deterministic automata map to table-driven algorithms and regular expressions—perhaps to recognize tokens in an input stream. Context-free grammars connect with parsers of programming languages. Problem reducibility provide approaches for replacing one problem by another. Altogether, computing theory often informs fundamental and practical tools for the solving of real problems, and insights/generalization from real problems can inspire and inform the development of theory.

Overall, generalization is commonly promoted as a key problem solving technique within the field of computing, and an understanding of the theory of computation provides important depth and insight within the generalization process.

Development of Logical Arguments

Beyond specific topics within a theory course, study and practice with formal proofs helps students formulate logical arguments. Much of a theory course requires analysis of proposed arguments and development of proofs of correctness. Theory highlights axioms and assumptions, and requires students to determine what does and does not follow from basic principles. Careful reasoning can separate wishful thinking or hype from reasoned conclusions, and can help distinguish anecdotes or experiments from logical arguments.

Of course, the algorithmic development and coding require students to think carefully, step-by-step; and students within a computing curriculum gain considerable experience in translating high-level ideas to detailed code. However, it seems curious that the ability to write structured, low-level code often does not translate to the ability to convey a logical argument to peers.

At its basic level, a mathematical proof represents a logical argument, designed to guide the reader from an initial framework to a logically-reasoned conclusion. Starting with initial assumptions, a tight chain of statements and justifications should lead to a stated conclusion, just as a correct algorithm should lead from given initial data to a desired result. Yet, students often have great trouble organizing their thoughts and presenting their arguments.

More generally, much public discussion and debate are supposed to be based on logical argument. Yet, public meetings, political debates, and press releases often contain astonishing and transparent errors in logic. Although it would be foolish to suggest that a single course in the theory of computation will resolve all of society's woes regarding illogical pronouncements, a theory course can help computing students make the transition from the writing of logical code to the statement and analysis of arguments in a general context, including computational theory.

Communication Skills

Not only should computing students and professionals be able to construct logical arguments, computing scientists need to be able to present those arguments clearly, correctly, cohesively, and compellingly.

From my perspective, a proof is a vehicle for the communication of a logical argument. As with any communication, one must consider the audience when writing or speaking (e.g., the amount of detail, the background of the reader, etc.) For example, when I teach a course on the theory of computation, in the first week I ask students to review and critique seven different proofs of the result that a set of N elements has 2N subsets. The different proofs involve different techniques (e.g., a direct proof, proof by induction, proof by contradiction, etc.), but they also represent different levels of detail, different organizational devices, etc. In the resulting class discussion, students identify some proofs that have too much detail, some with too few details, and some that are hard to follow. Some proofs guide the reader clearly from step to step, some seem ponderous, and some provide few transitions. Such discussion helps students consider how to construct logical arguments—how a proof might be organized, what is important to include, what should be stressed, what can be downplayed, etc. Students then can practice and hone their skills over a semester—as is important in the development of any writing.

In addition to writing—and if a class is small enough, students also can develop their skills in presenting arguments and ideas orally. For example, when I teach a theory course, I ask students to work through topics and present their findings to the class. As one exercise, I give the students a list of 10-15 NP complete problems. In groups of three, the students pick a problem, review at least one solution, discover why the problem is in Class NP, learn a reduction of a known NP-complete problem to their chosen one, and present these elements to the class. Not only does such an exercise help solidify student understanding of NP-completeness, but students also gain experience in the organization and presentation of technical talks.

Conclusion

A course on the theory of computation covers central material within computing, but also has great potential for a broad impact. For example, understanding of computational models helps students (and general citizens) consider both what is and what is not solvable and feasible. Students come in direct contact with limitations, and they can consider potential social and ethical impacts and policy implications. Further, computing thrives due to the interplay of theory and practice, so short-changing theory can have a direct impact on practice. Finally, the study of theory can help students relate their careful reasoning for coding to the more general context of logical reasoning and its presentation orally or in writing. Although each of these benefits might be incorporated into different courses, a cohesive and integrated course on the theory of computation can have substantial benefit, not only for students within computing, but for their interactions with the broader community regarding social, ethical, and policy issues

Acknowledgments

Many ideas in the article arise from conversations I have had with numerous computing faculty around the country over many years. For example, discussions with members of the Liberal Arts Computer Science Consortium have helped shape my thinking on matters of curricula in general and about the role of theory in particular. In addition, I have learned much on this and other topics of computing education through extensive, on-going interactions with members of my own department, including Janet Davis, Samuel Rebelsky, John David Stone, and Jerod Weinman. Many, many thanks to all of these people in shaping my thinking!

CHAPTER 22

Some strategies when teaching theory courses

This column first appeared in ACM *Inroads*, Volume 5, Number 3, September 2014, pages 32–34[253]

22.1 ORIGINAL COLUMN

DURING THE RECENT ACADEMIC YEAR, I taught two upper-level theory-oriented courses: *Analysis of Algorithms* and *Automata, Formal Languages, and Computational Complexity* (really *Theory of Computation*), Although various elements of the courses seem reasonably common (e.g., a mix of lecture and discussion, small-group assignments, individual assignments), this column describes three additional elements that may be of some interest to readers of ACM *Inroads*:

- comparing versions of proofs
- oral presentations
- oral final exams

My use of these elements has centered on theory-oriented courses, but I might guess that these same elements might also be adapted for other computing courses.

Comparing Versions of Proofs

In my experience, students often have difficulties understanding what to include in a formal argument and how to present their reasoning. What results can be assumed? What details should be stated? How much justification should be included? How should an argument be structured? Etc.

To address such matters, at the beginning of a semester, I commonly brainstorm with students about common elements of writing, such as the need to identify one?s audience, the role of an introductory paragraph, the logical flow of ideas, etc. My students write many papers for humanities and social studies classes, so they can articulate many elements of good writing. However, students often have difficulty connecting writing for other disciplines to writing formal proofs.

Of course, a main theme is that a proof should convey an argument to a reader. A well-written proof presents the appropriate steps clearly and in order, providing sufficient detail, but not belaboring the obvious. To illustrate the notion of a well-written proof, I consider a result that students have seen in an earlier course on discrete structures:

Theorem: If S is a set with n elements, then S has 2^n subsets.

After stating the result, I distribute six different proofs, three by induction, two presenting a 1-1 correspondence between subsets and binary numbers $0, \ldots 2^n - 1$, and one by contradiction. See [198] for details of the proofs.

Students read the first three proofs for the next class. The subsequent class discussion highlights strengths of each proof, weaknesses, how well the argument is conveyed, the level of detail presented, etc. Then we brainstorm how the proof might be improved. Following the discussion of the first three proofs, students read the next three for the following class, and again class discussion considers the relative merits of each proof.

Although I typically focus on class discussion, other exercises might be considered as well:

- In preparation for a class, students might submit their comments in writing (perhaps electronically or in an on-line forum).
- Discussion could be in small groups rather than a group of the whole.
- Following discussion, students might be required to write their own proof that addresses the points identified during class.

Although this exercise only starts the discussion of writing proofs, it seems to help students consider issues of argument and writing in the context of computing theory.

Oral Presentations

When discussing NP Completeness, it seems common for students to be able to parrot the definitions and high-level approaches, but understanding may seem shallow. To encourage students to delve into the subject, I identify 8-10 NP Complete problems (typically from the textbook or common sources). Students then divide into groups of 2 or 3, pick a problem, and present it to the class taking about 15 minutes (5-7 minutes per student). In practice, this approach has several benefits:

- Students have strong motivation to delve deeply into a problem, learning why the problem is in Class NP and how a known NP Complete problem can be reduced to it.
- Students gain experience organizing and giving oral talks on a technical subject.
- All class members hear about a wide range of NP Complete problems, so they see parallels in approach, presentation, and results.
- Each non-presenting student completes a feedback form for each group talk.

In my experience, students do not want to be embarrassed in making presentations, so they work hard to put together reasonable presentations. Also, drawing upon extensive work [118] by Kent McClelland, a Grinnell College sociologist, I utilize a feedback form that covers both technical content and presentation effectiveness [197]. Each listener reports observations for every talk, and I ask listeners to identify themselves. (I have found that reviews are more thoughtful when signed rather than anonymous.) Although these reports could be given directly to the speakers, the quality and scope of the student reviews often vary. Thus, I collect the student feedback forms, review student perceptions, add my own observations, and produce a composite report for each presenting group.

Often when using this approach, students want to give a second (or third) talk during the course, so they can build on their experience and apply the feedback. When students give several presentations in a semester, the quality of talks often improves substantially. Usually a first talk is adequate, but the second (and third) often are much better. Of course, such presentations make most sense in relatively small classes (e.g., under 20 students), but it might be interesting to try a similar approach in large classes with separate recitation sections.

Oral Final Exams

For several years, in upper-level courses, I have given my students the option of taking an oral final examination, rather than the usual written exam. Interestingly, when given the choice of an oral or written format, my students often chose the oral format. Their reasoning often includes these points:

- After years of written exams, they want to try something new.
- For those expecting to enter the job market after graduation, an oral exam may give practice related to a job interview.
- My students note that they regularly must write about algorithms, theory, etc., but they may not talk about the material as often. The oral component helps them build their oral skills as well as written skills (e.g., coding, writing about algorithms, writing proofs, etc.).
- For some students, an oral final exam may seem intimidating, but others appreciate that the exam is over in 30-45 minutes. A relatively short final exam may be enticing, but curiously none of my students have mentioned the exam length as a reason they wanted to take the oral exam.

Pragmatically, oral exams provide some challenges (at least to me). Here are some ways I try to cope with the logistics:

- Although I expect each oral exam to take about 30 minutes, I schedule 45 minutes to allow students to think through their answers as needed and to reduce stress. (With such a time commitment, I have tried oral exams in classes up to size 20, but this format seems impractical for larger classes.)
- In advance of the exam period, I organize the course into sections, and allocate points to each section (just as I would in designing a written exam). For example, my high-level outline for Theory of Computation for Spring 2014 included:
    - ◇ General/Definitions: 15 points
    - ◇ PDAs/Context-Free Grammars: 15 points
    - ◇ Turing Decidability/Recognizability: 30 points
    - ◇ DFAs/Regular Languages: 20 points
    - ◇ Reducibility: 15 points
    - ◇ RSA Encryption: 5 points

- Within each topic, I identify several possible questions. With an oral format, I can expect students will talk after early oral exams, so I want a range of questions for each section of the exam. For some sections, my outline might indicate I will ask just one question; for other sections, I may ask two or three.
- Prior to the exams themselves, I print out a grading sheet for each student with possible questions, point values, and a place for notes. (The student does not see my notes during an exam, but I can keep track of who says what through the entire exam period.)
- During an oral exam, I allow the conversation to flow according to student answers. If an answer to one question covers another question I had intended, there may be no need to ask the second question. If a partial answer requires clarification, I can ask for details. If one answer naturally leads to another topic in my outline, I may skip to a later section and come back as needed.
- When listening to a student answer, I try not to lead the student one way or another. The goal is to determine what the student has learned, and I do not want to add stress or confusion by taking a student down a path the student did not plan to follow.

Altogether, preparation for oral exams can be relatively modest, but substantial time may be required to listen to each student. However, with the grading sheet, I can add point values as soon as an exam is over, and little grading is needed after students finish.

Overall, oral exams (with relatively small classes) provide a mechanism to help students improve their oral skills, cover a planned scope of material, and allow a structured mechanism to assess students's work within categories.

Acknowledgments

Many thanks to Kent McClelland for developing a practical feedback form [118] for students giving class presentations. The form I use in class builds directly on his insightful work that helped transform my thinking.

Thanks also to Marge Coahran who provided wonderful feedback in the development and polishing of this column.

Conclusions

Each course presents its own challenges and opportunities for teaching and learning. Over the years, much discussion (in print and at conferences) has been devoted to lectures, active learning, etc. The ideas presented here (multiple proofs, oral presentations, oral final exams) seem to have had less discussion, but may be of interest to teachers in a range of courses.

CHAPTER 23

Lessons from the CUPM

This column first appeared in the *SIGCSE Bulletin*, Volume 35, Number 2, June 2003, pages 18–19[215] Subsequent reflections on this subject appear later in this chapter.

23.1 ORIGINAL COLUMN

IN JANUARY, 2003, the Committee on the Undergraduate Program in Mathematics (CUPM) of the Mathematical Association of America (MAA) outlined its forthcoming recommendations [43] for Mathematics Programs at the Joint Mathematics Meetings in Baltimore. At least three elements of this session seem potentially interesting to the SIGCSE audience: faculty contentment with programs, dialogs between the mathematics community and partner disciplines (including computer science), and selected CUPM recommendations.

Faculty Contentment with Programs

As part of the CUPM presentation, a panelist reported conclusions from a previous CUPM report [187] that had examined various successful mathematics programs. Interestingly, faculty members at the best programs were among the least satisfied with the status quo. The current draft CUPM guidelines [43] draw on [187] as follows:

> Site-visits to ten departments and information on a number of others revealed "no single key to a successful undergraduate program in mathematics." However, there were common features. "What was a bit unexpected was the common attitude in effective programs that the faculty are not satisfied with the current program. They are constantly trying innovations and looking for improvement." Also, one of the "states of mind that underline faculty attitudes in effective programs" is "teach for the students one has, not the students one wished on had."

As I think about my interactions with many CS faculty members over the years, it strikes me that the CUPM discussion might be at least as interesting and relevant for the CS community. A commitment to challenge the status quo, experiment, and improve can add vitality and energy to a program. Further, as faculty achieve success in some areas, and as those successes are publicized, now and unanticipated opportunities can arise. Even small and incremental steps can make a significant difference over time.

Having said this, the speed at which technology advances may seem to work against an interest in constant improvement. A CS faculty member may have a keen longing for something to remain stable over time. See if any of the following perspectives seem familiar.

- Since reviewing the whole curriculum (or a subset of courses) requires discussion among several faculty and since we are all too busy, let?s put the review off again this year.
- Perhaps I can use the same course outline this year as last year, or two years ago, or … .
- Perhaps I could use those 5-year-old labs again this year, so I don't have to rewrite them.
- If some content must be updated, then perhaps I need not consider new pedagogy.
- Since aspects of this course are new to me, I will utilize a relatively easy lecture format this time; I do not have time to develop new active-learning activities this time around—but maybe next year.
- I know our lab facilities (or classrooms or offices) are in poor shape, but I can live with them until I have more time.

Certainly, I have encountered many of these feelings, and time constraints may push me to adopt some of these perspectives—at least for some courses I teach in a particular semester. Few, if any, CS faculty can move forward on all fronts of the curriculum and all parts of all courses at the same time.

Having acknowledged such time constraints, the valuable part of the above CUPM comment—at least to me—was the reminder that I must not let myself become complacent. I need to keep looking for areas that can stand improvement, and I need to try new approaches to improve those areas. Not all experiments will succeed, but the MAA comment reminds us that we need to keep trying.

Dialogs Between the Mathematics Community and Partner Disciplines (including CS)

While circumstances vary widely from campus to campus, CS faculty may be interested to learn that the national (U.S.) mathematics community, through the MAA, is actively seeking dialog with faculty in disciplines that make significant use of mathematics. The MAA includes computer science among such areas, called "partner disciplines".

To promote this dialog and to understand needs, between 1999 and 2001, the MAA (with major funding from NSF) held a series of Curriculum Foundation Workshops, in each of which faculty from a partner discipline were invited to talk about their use of mathematics and the mathematical needs of their undergraduate students. Conversations included needed general themes, specific content, processes (e.g., problem solving, the role of proofs), applications, appropriate prerequisites, etc. While a few mathematicians attended each workshop, their role largely was limited to listening and providing background. After each workshop, disciplinary leaders prepared a summary report of the conclusions reached. Following all 11 workshops, representative leaders met in November 2001 to prepare an overall summary document. While details are well beyond the scope of this column, interested readers are invited to read the various reports at [21]. Within this collection, the CS document parallels many ideas in our *Computing Curricula 2001* [12], but also contains special perspectives and ideas.

Beyond the specific reports, SIGCSE members may want to build upon this MAA effort in several ways. First, the Curriculum Foundation Workshops demonstrate a strong interest at the national level for dialog between mathematicians and computer scientists (and others). Second, the interactions at the national level may encourage additional interactions among CS faculty and mathematicians at a local level. While extensive, constructive conversations are common on some campuses, the national activities may provide a focal point to initiate or renew such conversations where they are now dormant. Third, the CS Workshop Report [21] and the subsequent conclusions in the forthcoming CUPM recommendations provide fresh insights to our own *Computing Curricula 2001* regarding mathematics for computer science – particularly discrete mathematics. As a thoughtful and well-presented document, the materials provide interesting food for thought.

Selected CUPM Guidelines

The CUPM recommendations cover the entire undergraduate program in mathematics including remedial mathematics, service courses, courses to support partner disciplines, and advanced courses for mathematics majors. While the full draft recommendations [43] are much too lengthy for inclusion here, the following selected summary recommendations may be of particular interest to the computing community:

1. Recommendations for departments, programs and all courses in the mathematical sciences

  1.2 Develop mathematical thinking and communication skills

  1.3 Communicate the breadth and interconnections of the mathematical sciences

  1.4 Promote interdisciplinary cooperation.

3. Additional recommendations for students majoring in the partner disciplines

  3.3 Critically examine course prerequisites
  Mathematical topics and courses should be offered with as few prerequisites as feasible so that they are accessible to students majoring in other disciplines or who have not chosen majors. This may require modifying existing courses or creating new ones. In particular,

  - Some courses in statistics and discrete mathematics should be offered without a calculus prerequisite; ...
  - Prerequisites other than two or three semesters of calculus should be considered for intermediate and advanced non-calculus-based mathematics courses.

5. Additional recommendations for students majoring in the mathematical sciences

  5.5 Create interdisciplinary majors

It is interesting to note the highlighting of mathematical-thinking skills (e.g., proof and deductive reasoning) at all levels, the emphasis on interdisciplinary cooperation, the need for one or more courses in discrete mathematics that can be taken early (in the first or second year), and the interest in interdisciplinary majors. At the national level, the CUPM recommendations indicate a strong interest in collaborating with computer scientists and in supporting our students. Since the CUPM recommendations play the same role in the mathematics community as *Computing Curricula 2001* does in computing, the new CUPM materials may provide new or expanded opportunities for exchange, collaboration, and support at the local level as well.

23.2 SUBSEQUENT REFLECTIONS

Since the original column appeared, ACM/IEEE-CS published new recommendations for undergraduate computing as *Computing Curricula 2013* (CS 2013) [10], and the Committee on the Undergraduate Program in Mathematics (CUPM) of the Mathematical Association of America (MAA) also published new recommendations in 2015 (CUPM 2015) [44]. As in the past, CS 2013 provides extensive detail, and the recommendations specify topics rather than required courses. CUPM 2015, on the other hand, emphasizes four high-level cognitive recommendations and nine high-level content recommendations, and the mathematics guidelines provide "Course Reports" that provide considerable discussion for 18 commonly-taught courses for mathematics majors.

From my perspective, the CUPM 2015 recommendations are particularly interesting, because many of the statements for mathematics seem appropriate for computing. For example, the "Designing a Major" section begins, "Major programs in the mathematical sciences should present the beauty, fun, and power of mathematics." The last sentence in the same paragraph states, "Each department should create and maintain a community that welcomes and supports all students, including those from groups that have been traditionally underrepresented in mathematics." [44, Overview, section on Designing a Major] As these examples suggest, one might just change the word "mathematics" to "computing" in many CUPM 2015 recommendations to yield appropriate statements for computing.

Similarly, [almost] all of the CUPM 2015 Cognitive Recommendations seem to apply to computing with only minor editing. For example, Cognitive Recommendation 4 states, "Students should develop mathematical independence and experience open-ended inquiry" [44, Overview, section on Cognitive Goals]. Similarly, Content Recommendation 8 states, "Students majoring in the mathematical sciences should work, independently or in a small group, on a substantial mathematical project that involves techniques and concepts beyond the typical content of a single course" [44, Overview, section on Content Goals].

As the above examples may suggest, the CUPM 2015 recommendations maintain a focus upon the big picture of a mathematics curriculum. Although these recommendation contain some modest details, CUPM 2015 has remarkably less detail than recommendations for computing, as represented in *Computing Curricula 2013* [10]. Of course, computing faculty also need to keep the big picture in mind (e.g., See Chapter 13 or [227]), even as the faculty structure and shape details of programs and courses.

Since CUPM 2015 does a particularly nice job in presenting themes, and since the identified themes often translate nicely to computing (at least in my opinion), computing faculty are encouraged to read CUPM 2015 frequently when developing and refining computing curricula and courses. After extended discussions about details among computing faculty, CUPM may be a useful resource to reconnect with high-level themes that can benefit computing as well as the mathematical sciences.

CHAPTER 24

Selected/Annotated references for relationships between computing and mathematics

BASIC REFERENCES INVOLVING MATHEMATICS AND COMPUTING begin with curricular materials and reports from the Committee on the Undergraduate Curriculum in Mathematics (CUPM) of the Mathematical Association of America (MAA). Just as ACM joins with IEEE-CS about every decade to produce curricular guidelines for undergraduate computing programs, the MAA's CUPM releases recommendations for undergraduate mathematics programs about every decade.

Additional references involve ACM/IEEE-CS curricular recommendations, the report of a SIGCSE Committee, and several additional articles.

Recommendations and Reports from MAA's CUPM

- Committee on the Undergraduate Program in Mathematics (CUPM), Carol S. Schumacher and Martha J. Siegel, Co-Chairs, and Paul Zorn, Editor, *C The 2015 CUPM Curriculum Guide to Majors in the Mathematical Sciences*, Mathematical Association of America [44].
  This *CUPM Curriculum Guide*, updated about every 10 years, represents the primary curricular standard for undergraduate programs in the United States, just as the ACM/IEEE-CS *Computer Curricula* recommendations provide guidance for all undergraduate computing programs in the United States.

- Henry M. Walker, Daniel Kaplan, and Douglas Baldwin, *MAA Program Study Group on Computing and Computational Science: Summary Report [280] and Supplemental Report [281].*
  The 2015 CUPM Curriculum Guide to Majors in the Mathematical Sciences includes "Program Reports", describing 14 client disciplines for mathematics departments. The reports on computing and computational science discuss recommendations for computer engineering, computer science, information systems, information technology,

software engineering, computational and data science, and connections among statistics, machine learning, and databases. Sections include the areas of continuous mathematics, discrete mathematics, and statistics.

Recommendations and Reports related to the 2013 ACM/IEEE-CS Curricular Task Force

- ACM / IEEE-CS Joint Task Force on Computing Curricula, *Computer Science Curriculum 2013*, ACM and the IEEE Press [10].
  Although CS 2013 provides recommendations for the entire computer-science major, two sections highlight mathematical needs at the undergraduate level:
  - "Chapter 6: Institutional Challenges" describes "Mathematics Requirements in Computer Science" on pages 49-51.
  - "Appendix A: The Body of Knowledge" includes "Discrete Structures" as one of 18 Knowledge Areas. Pages 76-81 describe recommendations in considerable detail, including 37 Core-Tier 1 hours, 4 Core-Tier 2 hours, and 36 Student Learning Outcomes.
- Douglas Baldwin, Henry M. Walker, and Peter B. Henderson, "The roles of mathematics in computer science", *ACM Inroads*, volume 4, number 4, December 2013, pages 74–80. [20].
  The Acknowledgments of the article states, "The original version of this paper was written at the invitation of the ACM/IEEE CS Curriculum 2013 committee as a position statement on the role of mathematics in undergraduate computer science education. We thank members of the committee for comments on that version. The views stated in this paper are those of the authors, and do not necessarily reflect the position of the committee or its members."

Additional Reports and Articles

- SIGCSE Committee on the Implementation of a Discrete Mathematics Course, Bill Marion and Doug Baldwin Committee Co-Chairs, *On the Implementation of a Discrete Mathematics Course*, April 2007 [168]
  In 2003, with encouragement from the SIGCSE Board, Bill Marion and Doug Baldwin helped create a SIGCSE Committee on the Implementation of a Discrete Mathematics Course. Over the next 3-4 years, dozens of SIGCSE members helped identify "a small number of models for a one-semester discrete mathematics course which meets the needs of computer science majors" [168, Abstract]. The final report reflects an extensive process of investigation, circulation of drafts, and community-wide discussion.
- "Math CountS" is an ongoing column on mathematics within computer science
  - Peter B. Henderson wrote several columns in the SIGCSE *Bulletin* between June 2007 and December 2009.
  - When the regular columns of the SIGCSE *Bulletin* evolved into ACM *Inroads*, Henderson's column continued in ACM *Inroads* in March and September 2010.
  - When Henderson retired as a columnist, John P. Dougherty continued the Math CountS column in ACM *Inroads*, starting in March 2016.

To locate these columns, search in the ACM Digital Library (http://acm.org/dl) under all terms "Math CountS" and either "Peter B. Henderson" or "John P. Dougherty".

VI

In the Classroom: Basics, Lab-based, Active Learning, Flipped Classrooms

WHEN MY OLDER DAUGHTER WAS YOUNG, PERHAPS 3-4 YEARS OLD, my family embarked on a trip from the midwestern United States to New England. During the trip, we passed the time playing a variety of car-based games. At one point, my daughter suggested, "Guess my favorite number!" Playing with a pre-kindergarten child, my wife and I guessed some obvious numbers: "1," "2,", "3," etc., at which point my daughter said, "I'll give you a hint. It's not 143." At that point, my wife and I decided to shift conversation to another game.

In some respects, discussions of classroom pedagogy can be similarly non-helpful. Numerous studies and discussions indicate that pure lecture is remarkably ineffective. Thus, when asked. "what works well for a class format," a simple answer is "I'll give you a hint. It's not lecture." Unfortunately, such a conclusion provides very little guidance regarding what format or pedagogy might be better.

An underlying difficulty with a lecture format is that students can be passive; the instructor actively engages the subject matter, but students need not focus on the material being presented. Students may be distracted, thinking about other matters, or even dozing. While an instructor is talking, showing slides, or presenting a demonstration, students may be thinking about the material being presented and perhaps taking notes, but students also might be exploring the World Wide Web, making plans for activities after class, considering weekend events, texting their friends, reading email, or otherwise contemplating topics unrelated to the instructor's presentation.

Altogether, although lecture has been widely used for centuries, many pedagogies have evolved to actively engage students, and studies consistently have demonstrated that engaged students learn more and perform better than those connected with a passive, lecture-based pedagogy.

Chapters in this Part (Part VI) explore classroom basics and various pedagogies for teaching. Discussion naturally proceeds from some fundamental and general practices to consideration of several specific strategies and approaches.

- The first two chapters, Chapters 25 and 26, provide down-to-earth suggestions for getting started within a classroom setting. Each chapter contains an elementary list of do's and don'ts for beginning teachers. (The list also may serve as a basis for discussion for mid-career teachers, or even experienced teachers, but the target audience is beginning teachers.)

  To avoid overwhelming new teachers, each suggestion is focused with modest commentary. Further, the full list of do's and don'ts spans two chapters to make each collection relatively unintimidating. Extensive lists and suggestions might be considered for experienced teachers, but short, manageable lists seem an appropriate starting place.

- After general suggestions for teaching, the next four chapters focus upon environments and strategies to engage students. Some basic terminology follows:

  ◇ A *closed laboratory* session is a scheduled class meeting within a teaching lab, in which students are guided through a sequence of tasks; an instructor and/or laboratory assistant typically are available to help students by providing guidance, responding to student questions, clarifying instructions, or other supporting role.

  ◇ An *open lab* is a teaching lab with workstations or places for student laptops, in which students work on activities on their own; tutors or laboratory assistants may or may not be present at times to answer questions, but student work is flexible and not prescribed with explicit steps to be followed.

  ◇ A *lab-based course* or a *course utilizing a workshop format*is a course, in which parts of most class sessions involve some type of closed laboratory. Students typically are expected to prepare for each lab through reading or other activity, and students engage in computer-based activities within each lab session. Students also may need to complete activities that were started in class as homework.

  ◇ *Collaboration within a lab-based course* likely means that students are expected to complete labs with one or more partners; pair programming is a common model for completing lab exercises.

  ◇ A *flipped classroom* represents a generalization of a lab-based course or workshop format. Students are expected to read materials, watch videos, or otherwise explore material in preparation for each class; class meetings involve students in active exercises or activities, based on the prepared materials. The term, "flipped classroom," derives from the observation that the presentation of material and the working of problems is opposite between lecture and the flipped classroom. In a lecture format, presentation of material largely occurs in class, during a lecture, whereas that work is considered out-of-class preparation within a flipped classroom. In contrast, student homework in a lecture format often becomes in-class activities within a flipped classroom.

  ◇ *Active learning* describes in-class activities in which students must take an active role. Rather than listening to an instructor, active learning involves individual work, small- or large-group activities, laboratory exercises, student responses to prompts, or other activities that require student initiative or action

Finally, within this Part VI, Chapter 31 provides a brief list of references, with annotations, for readers interested in further study.

CHAPTER 25

Basic do's and don'ts in the classroom: General environment and course suggestions

This column first appeared in ACM *Inroads*, Volume 7, Number 3, September 2016, pages 20–24[263]. Originally conceived as the first of a three-part series, the series continued in the December 2016 and March 2017 issues of ACM *Inroads* (reprinted here as Chapters 32 and 26, respectively). Subsequent reflections on the first and third columns appear in this book at the end of the third column, Chapter 26.

25.1 ORIGINAL COLUMN

IN RECENT YEARS, I have had occasion to talk with new CS faculty members at several schools, each of whom were within a few months of receiving their Ph.D (before or after). Although some have considerable insight and experience, others seem to be true novices—including some who had been Teaching Assistants leading recitation sections. Also, I know some graduate schools have extensive training for assistants who will be grading, teaching, or otherwise helping in the classroom. But other schools provide less support, and even when schools have training programs I know graduate students who have been missed. All of this raises for me questions of where graduate students and new faculty learn basics of teaching, and how these beginning teachers might receive feedback. Apparently, in some environments, schools assume that new teachers are supposed to learn about teaching elsewhere.

From my perspective, effective teaching develops through an evolutionary process that involves observations of what works for an individual instructor in a specific environment, feedback from students, other faculty, and observers, on-going study of teaching and student learning, brainstorming with educators about common problems and possible approaches, etc. Attendance at conferences and discussions with colleagues provides on-going support and professional development, but new faculty need practical suggestions and basic approaches as they get started. This and my next two columns are intended as starting points for the development of effective teaching for new faculty and as points of discussion for all faculty.

- This column lists 23 do's and don'ts related to the classroom environment, student-faculty interactions, and general classroom parameters.
- My next column (*Curricular Syncopations* for December 2016) (Chapter 32 in this book) will consider practical tips and techniques for planning and organizing a course for the first time—particularly for new faculty who have not taught before.
- My *Classroom Vignettes* for March 2017 (Chapter 26 in this book) will present another list of do's and don'ts for the classroom—focusing on combating implicit bias, details of slides, board notes, and similar day-to-day activities.

These ideas will likely not be new or different for many *Inroads* readers, but beginning teachers need a place to start.

Cautions:

A. These columns represent starting points and are not intended as complete discussions of teaching. A full treatment of effective teaching would fill many volumes and likely be overwhelming to faculty who are teaching for the first time.

B. The target audience is primarily new teachers in small- or medium-sized courses, although experienced teachers might find these comments useful for general discussion. (Presenting lectures to several hundred attendees likely requires different considerations, but I expect new faculty would not be given very large lectures as their initial assignments ?and this column focuses on needs for new faculty who will be teaching small- to medium-sized classes.)

C. In my experience, effective teachers often draw upon their personalities and individual traits as they develop teaching styles that work for them.

- For many years, one of my colleagues built upon his quiet, understated manner to craft a intellectually-focused class atmosphere, mixed with a wry humor, to captivate students and promote a close student-faculty rapport that drew out even the most shy and introverted students.
- A second colleague utilizes his extraverted personality to engage and motivate all students in his classrooms. Although many instructors might hesitate to call randomly on students (students might feel intimidated and nervous throughout class), this colleague creates an enthusiastic spirit of collaboration, and students respond positively.
- Another person in my department couples a sense of theater with puns and a [bad?] sense of humor to create an atmosphere of fun while learning and to draw students into discussions.

In reflecting on these three faculty, all are extremely effective (e.g., students master material on assignments and tests, and student end-of-course evaluations are remarkably positive). However, the atmosphere in classes for each instructor is stunningly different, and it seems clear that none of these instructors would be effective utilizing the styles of the others.

Even with these contrasting styles within the classroom, some characteristics seem common for many effective teachers. The following do's and don'ts reflect some of those characteristics that I have observed during my teaching career.

Basic Do's and Don'ts

Connecting Classroom Planning, Pedagogy and Practices to the Big Picture

1. Do base classroom pedagogy and practices on clearly defined course objectives and public statements regarding content (e.g., in a syllabus).

   Commentary: Considerations for each course should begin with clear statements of objectives (what a student should be able to do at the end of the course), course content (what material students should have mastered), and what background will be expected in later courses (which have the current course as prerequisite).

   - Course planning, practice, pedagogy, and assessment should directly support these objectives and this stated content.
   - A reviewer of a draft of this column also noted, "course assessment (content and process) should have a good fit with the objectives in order to guide student choices, to provide a basis for constructive feedback during the course and so promote learning, and to measure the extent to which the learning objectives have been achieved."

2. Do not waste time and energy on tangents, favorite vignettes, or a stream-of-consciousness flow of random ideas that do not relate to course objectives and promised content.

   Commentary: Typically, expectations regarding course objectives and content coverage require efficient utilization of time within the classroom and for homework. Many courses are packed with material and learning goals, leaving little time for topics outside the scope of the course.

   - Digressions may be entertaining, but they can consume large amounts of time with little actual contribution to priorities in a course.
   - When planning a class session, an instructor might write out, in reasonable detail, the various elements to be presented and then review how each topic flows into the next. Without careful planning, much time can be wasted on transitions between topics, whereas a well-planned flow can use one topic to motivate and set up the next.
   - When a presentation devotes extensive time on specific details of an algorithm or data structure, an instructor might review whether another approach might be more effective. When a presentation seems long and complex, my experience suggests an insight or strategy often is missing. Of course, some material is inherently complicated, but over the years I have been astonished to find cleaner and clearer approaches to many topics that initially seem to require intricate logic and numerous cases.

3. Do strive to expand your understanding of how students learn and what pedagogies are effective in various contexts.

Commentary:

- Reading ACM *Inroads* and other teaching-oriented publications provide insights on what approaches are effective within the classroom and how students learn.
- Attending regional, national, and international conferences allow networking and first-hand contact with results from computing-education research.

Creating a Supportive Classroom Environment

4. Do show respect and support—always!

   Commentary: Not only is concern for human dignity an ethical and moral value, but encouraged students often learn to succeed at higher levels than they may have anticipated—at least in my experience.

5. Don't give up on a student.

   Commentary: In my career, I have worked with several students who were quite bright, but entered college with very weak backgrounds. When I first talked with them, I might have guessed they had little chance to succeed—they seemed far behind and were floundering. However, with hard work, they recovered and went on to be among the strongest in the graduating class!

6. Do encourage a sense of collaborative student-faculty effort to help students learn.

   Commentary: Although some students have high self esteem and great confidence in their abilities, many students (particularly from under-represented groups) may worry about their capacity for success. When students perceive an instructor as a coach, mentor, and listener who cares, the instructor may seem less of a threat and more of a resource, and this constructive relationship can be quite encouraging.

7. Don't suggest, either explicitly or implicitly, that an honest question from a student is dumb, illogical, or inappropriate.

   Commentary: Students take an implicit risk whenever they ask a question or suggest an answer—what will their peers or the instructor think? If students perceive this risk is high, they will not ask questions, and they will not volunteer answers. To promote student engagement throughout the class, students must feel supported. (Such issues may be particularly important for students who may feel isolated—perhaps as part of under-represented groups within the classroom.)

Developing One's Own Teaching Style

8. Do pay attention to how the students react to in-class activities.

   Commentary: When students react positively to a classroom experience, do more of that; when they seem bored or confused or non-focused, try something different.

9. Do attend regional and national conferences and talk to colleagues to learn new teaching approaches, to discuss teaching challenges, and to obtain feedback on your own challenges and experiences.

Commentary: Over the years, I have learned much from colleagues at all levels (e.g., K-12, college and university, graduate school, and industry). I regularly discover others encounter situations analogous to my own—details may differ, but many elements are shared. Networking and brainstorming cannot be overestimated as a remarkable resource, and talking makes a difference. (Even if you are shy and retiring, go to a conference, and ask the person sitting next to you in a session, "what do you teach", "how are enrollments and staffing going at your institution", "what was the high point of the conference for you so far"? Such questions are safe and non-stressful, and they can initiate a longer conversation!)

10. Don't be afraid to experiment with different pedagogical techniques and approaches.

    Commentary: Years ago, at a national mathematics conference, a session discussed common characteristics of distinguished math programs around the country. Although outstanding programs have wildly differing characteristics, a common element was that faculty were not satisfied. However good a program seemed, faculty strove to become even better. From my perspective, experimentation is an important part of trying to improve. While talking to colleagues or attending a conference, I hear about results of educational-computing research and about alternative pedagogies for varying contexts. Such discussions encourage me to consider how I might incorporate or modify these techniques in my own courses.

11. Do find ways to communicate your enthusiasm, interest, and appreciation of the material you are teaching.

    Commentary: As an instructor, you have chosen to work in the computing field, because you have found it captivating, challenging, and provocative. Let your inherent excitement carry through in the classroom. Today, there is much discussion of the joy, beauty, and awe of introductory computing. While I am happy to support such sentiments for beginning courses, I believe most well-conceived and well-taught computing courses should convey such themes at any level.

Student-Instructor Interactions

12. Don't ask rhetorical questions in class.

    Commentary: Rhetorical questions seldom promote discussion—at least in my experience. Students quickly understand that a rhetorical question (with an obvious answer) is more a style of speaking than a request for a student response. Rather than promote discussion, rhetorical questions can undermine student involvement.

13. Do be sure the question asked in class is sufficiently focused that students will understand what is actually being asked.

    Commentary:

    - Even if students want to answer a question, they may not have a clear sense of what the instructor intends or what direction might be desired.

- In some cases, such as courses on learning requirements analysis, an important theme might be how to clarify and focus questions. In such situations, it may be helpful to clarify that vague questions are appropriate for addressing course objectives. However, in many courses, vague questions may promote student confusion and frustration rather than learning.

14. Do wait for a response asked in class; eventually a student will crack under the pressure and discussion is started

    Commentary: Waiting just 10-15 seconds is not really long, but it may seem like hours, and some student likely will respond under the strain of silence.

15. Do not single out an individual (particularly from an under-represented group) to serve as a spokesperson for the group.

    Commentary: Each student can speak as an individual, but groups include multiple perspectives—not just one. A reviewer of a draft of this column notes, "We have to be aware of not placing undue burden on people to "wave the flag" for their minority group simply because they are from that minority group. "

16. Do actively engage all students and give everyone an equal voice.

    A. Respond equally to all raised hands.

    Commentary:

    - Depending upon room layout, an instructor may naturally look at one part of a classroom when speaking. Be sure to scan the entire classroom—frequently—to locate raised hands anywhere in the room.
    - Over the years, studies have shown that, when student hands are raised, teachers tend to call upon some groups of students first. (Yes, I know this may seem hard to believe in todays society, but check in your own classroom and in other classrooms around you.)
    - When several hands are raised at once, deciding which student to select can impact a student's self image and sense of worth. Although I certainly want to answer all student questions, I often acknowledge hands first from women and under-represented groups. Encouraging these students can have a positive impact, and I have seen little negative impact on students from dominant groups. Of course, I never indicate I have called on these specific students due to their gender, race, or other characteristic—but I do want students from under-represented groups to feel they are actively valued!

    B. To the extent that class size permits, call on all students rather than relying upon the same students all the time.

    Commentary: In large classes, it may be impractical to actively engage every student on a regular basis. Asking each student a question requiring a 1-minute answer might consume an entire class or even one or more weeks of class time. However, in small classes, instructors may seek to call on each student one or more times each class.

    - Some students may try to show off or take control of a course by trying to answer every question. In such cases, it is fine to encourage others by indicating one person already has answered several questions, and an instructor wants to hear other perspectives.

- Even if one student often is correct, other students need to be engaged in the class and have a chance to contribute. Other students should not be able to let someone else do their thinking for them.
- Some faculty write student names on a deck of cards and then call on students based on cards they draw. Other faculty utilize a random-number generator to pick when calling upon students. Such approaches encourage participation by all students and reduce the risk of bias in student selection.
- When calling on students, students might be expected to either answer the question or ask a clarifying question. Students should not be intimidated with the pressure of having answers to any possible question. However, students might be expected to be following the discussion enough to respond with a question about some part of the problem at hand. (Passing might not be permitted, as that might encourage students to daydream.)
- If at all possible, ask a teaching assistant or a colleague to observe a class from time to time and to record which students are called upon and when. It is not uncommon for an instructor to look at one part of the room more than another or to respond to particularly vocal students or groups. When identified, such tendencies can be easy to address.

Pace and Level of Discussion

17. Do try to maintain a reasonably steady pace in the presentation and discussion of material throughout the semester.

    Commentary:

    - Ideas, techniques, perspectives, etc. require time for students to synthesize and master. If material proceeds slowly during part of a semester, students are unlikely to develop the habit of thinking through material, applying algorithms, experimenting with structures, etc. If material proceeds too quickly, students are likely to become overwhelmed if they have not set sufficient time aside. Overall, a reasonably steady pace helps students manage their time, set expectations, and develop the habits necessary to handle a course's work load.
    - Too slow or too much repetition of known material is boring and inefficient. Too fast or too little repetition can overwhelm students who are just learning new material.

18. Don't rush through material for the sake of completeness.

    Commentary:

    - Remember the goal of classroom work is student learning.
    - It is not particularly important what an instructor covers in class—presumably the instructor already knows the material. Rather, it is important that each student learns the material targeted by course goals and covers material that will be used in courses for which this is a prerequisite.
    - Advanced planning and scheduling may be needed to allow appropriate time for material without rushing at the end!

19. Do pay attention to the level of material covered

    Commentary:

    - Typically, stated prerequisites indicate background that students should be expected to know. Although a little review may be appropriate, extensive discussion of prerequisite material should not be needed. (If substantial repetition of previous courses turns out to be necessary, prerequisite courses might be analyzed for effectiveness.)
    - Instructors should not assume more than the stated prerequisites—how will students learn this additional material, how would they even know that further background is necessary? In practice, hidden prerequisites may discourage students from under-represented groups—particularly if they are feeling marginalized to begin with.
    - Altogether, the level of the material should directly reflect student background and their expected capabilities.

Selecting and Presenting Examples

20. Do find new examples rather than repeating the same ones from the book.

    Commentary:

    - If a textbook presents an example in a reasonable way, students can read through the example in detail at their own pace, and there is no reason to repeat the same material from the same perspective. Much better to provide a fresh perspective and/or a different example.
    - If the textbook does not present an example well, why is the textbook being used?

21. Don't give examples that serve only in the short-term, but will have to be unlearned later in a more advanced course.

    Commentary: Early in my writing career, series editor, Gerald Weinberg, observed that students have the overwhelming task of learning a huge number of concepts and techniques. Rather than having to unlearn topics later on, each example should provide a legitimate mechanism for solving a problem—at least in some circumstances. (Of course, examples that explicitly illustrate inefficiency or poor practice may be appropriate if that is the point of the example, but examples should not be put forward as constructive at one point, only to be shown inappropriate in all contexts later on.)

22. Do ensure that examples are time-efficient and clearly illustrate the intended point(s).

    Commentary: Long, detailed examples can consume substantial class time. Sometimes working through all aspects of such examples can provide considerable insight. However, sometimes main ideas can become lost in a sea of detail.

    - To save class time and maintain focus, a handout might contain the numerous, step-by-step details. Class time then can highlight selected steps that illustrate the main parts of an algorithm or structure.

- When a long and complicated example seems necessary, one approach involves developing the example as a running illustration that extends over several concepts and provides multiple insights.

23. Don't select examples that place a disproportionate cognitive burden on the students.

    Commentary: Some computing applications may require extensive background in biology or economics or another field. If students in a class have diverse interests and backgrounds, presenting such applications may force students to spend substantial time on topics that are separate from the primary topics under study. Student effort generally seems a positive trait to be fostered, but effort spent on material outside the goals of the course may create more frustration than relevant insight.

Acknowledgments

Many thanks to Charlie Curtsinger, Grinnell College, and Titus Klinge, Iowa State University and Grinnell College, for their insightful feedback and suggestions on various details of this column. Thanks also to the reviewers of this article for their suggestions.

CHAPTER 26

Basic do's and don'ts in the classroom: Combating bias, making presentations, and developing slides

This column first appeared in ACM *Inroads*, Volume 8, Number 1, March 2017, pages 12–15[268] Subsequent reflections on this subject appear later in this chapter.

26.1 ORIGINAL COLUMN

THIS COLUMN IS THE THIRD AND FINAL INSTALLMENT in a series designed to help new and inexperienced faculty get started in their teaching.

- The *Classroom Vignettes* column for September 2016 (Chapter 25 in this book) listed 21 do's and don'ts related to the classroom environment, student-faculty interactions, and general classroom parameters.
- My *Curricular Syncopations* for December 2016 (Chapter 32 in this book) presents practical tips and techniques for planning and organizing a course for the first time.
- This column presents another list of do's and don'ts for the classroom—focusing on combating implicit bias, details of slides, board notes, and similar day-to-day activities.

As with the previous columns, the reader is warned that what follows is designed as a starting point rather than a definitive or complete statement regarding effective teaching. Although some elements of effective teaching may vary according to faculty personalities and perspectives (as noted in my September 2016 column), the suggestions here are widely observed as practices and approaches that apply in many classroom settings with small and mid-sized enrollments.

FIGURE 26.1 Two lab-based sections of CS2: celebrate diversity

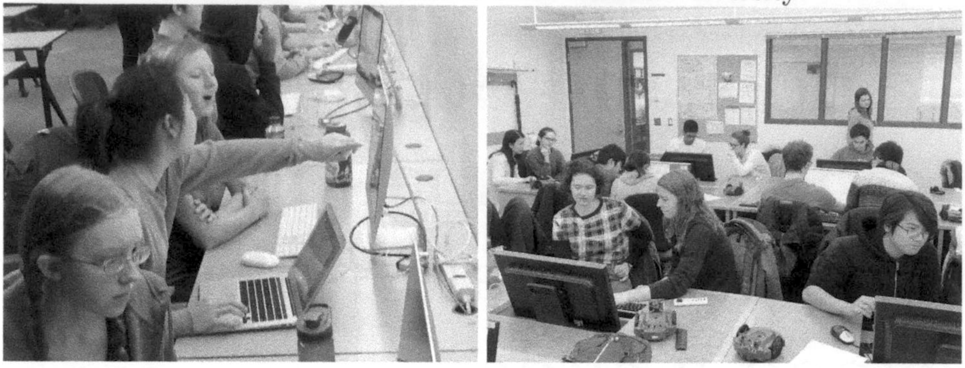

Combating Implicit Bias and Discrimination (e.g., Figure 26.1)

1. Do be careful in praising (or criticizing) only some Figure-26.1 Two Lab-based sections of CS2—Celebrate Diversity class participants.

   Commentary: Selective praise in class may be seen as a bias toward one student or group and against another. For example, if one person provides a nice answer to a question and receives a "well done" comment, then others providing helpful answers should receive similar praise.

2. Do and publicly advertise blind grading, where the instructor does not know whose solution is being scored at any time.

   Commentary:

   - A well-observed phenomena, sometimes called the "halo effect", is that graders have a tendency to give the benefit of the doubt to students who generally perform well, but are less forgiving and sympathetic to students who generally perform less well—independently of the answer actually written.
   - When students take a written test, I ask them to write their names on the first page only. Then I grade all question 1 responses, then all question 2 responses, etc. After grading each question, I record the score and turn the paper to the next problem before going on.
     - ◇ This allows me to grade each answer fresh—without knowing whose paper I am reading. (I might be able to guess somewhat according to handwriting, but even then a similar handwriting might be common to several students.)
     - ◇ This approach also ensures that I do not know students' scores on previous problems when I am grading the next one—again reducing the possibility of bias when grading.

3. Don't show bias in interacting with all students.

   Commentary:

   - As an extreme, do not address some as "Mr. Walker" and others as "Donna".

- Avoid use of terms, such as "guys" that some may consider a general reference, but others will consider sexist and exclusionary. For example, avoid "these guys", "those guys", and "you guys".
- With increasing cultural sensitivity to matters related to gender, one must be careful with the use of pronouns—what is appropriate when referring to a specific student: "he", "she", "ze", etc.? Some faculty recommend asking each student to complete a note card on the first day to clarify such matters.
- Personally, I call all students by their first names and avoid pronouns. For me, this is a simple approach that avoids may potential pitfalls.

4. Do be careful to avoid bias in examples, pictures, etc.

   Commentary:

   - Not all professional people are white males, so one must be careful that examples involving professional people include a diverse range of names and references. (Of course, similar comments apply to any career group.)
   - Images send messages regarding who are included in a group, profession, club, etc., and images also suggest which groups might be excluded or not welcome.

Standards

5. Do maintain standards, but be creative in how students demonstrate mastery—reflecting different personality types and learning styles.

   Commentary: Years ago, I had the same student for calculus I and calculus II. Throughout the student did well on homework and in class discussions. However, in the first course, the student consistently performed poorly on tests. By the end of Calculus I, the student was frustrated in that test scores did not seem to reflect the student's knowledge. Immediately after the first test in Calculus II, the student came directly to my office and indicated, "It happened again. I did not show what I knew". The student then indicated a wish to show me that the student knew each answer and proceeded to write a completely correct solution to every test problem on my blackboard. Thereafter, I allowed the student to take the test in my office—on the blackboard, and the student aced every test. The point is that, at least for me, grades should reflect actual mastery of the relevant material, but different students may flourish in one environment rather than another. I am open to students demonstrating their understanding in various settings, but the course grade must be based on actual evidence, not on assertions of "I panicked".

Presentation Basics

6. Do speak loud enough, without overwhelming some in the front of a room.

   Commentary: Speaking clearly and at an appropriate volume does not guarantee a conversation will be effective. However, mumbling or speaking too quietly or too loudly or utilizing a monotone voice will likely guarantee that other qualities of a presentation will not be effective.

7. Don't fall prey to distracting mannerisms:

   A. Avoid verbal ticks, such as "like", "uh", "basically", "you know", and the like.

   B. Avoid unnecessary physical gestures and movements that sidetrack a listener's attention from the matter at hand.

   Commentary: Students need to focus on content, logical connections among ideas, techniques, etc. Idiosyncrasies in presentation can capture a listener's attention, undermining the examination of the intended subject matter.

8. Do ask an observer to periodically attend a class session to note what is working well, what details might be distracting, and what improvements might be possible.

   Commentary: As an instructor engaged in classroom interactions, it is very difficult to objectively determine how behaviors and practices are being received by students. Are notes clear, can students read slides/board notes, is the instructor too loud or too quiet, etc.? An observer can provide worthwhile feedback without having to lead the class at the same time.

9. Don't pretend you know everything.

   Commentary: The discipline of computer science changes at a rapid rate, and few (if any) can keep up with all current developments in the field. Thus, it is fine to say you do not know an answer, but then either think it through with the class or tell them the following class (after doing your homework).

10. Do articulate your thought process, when your have to think about the answer to a student question.

    Commentary: Thinking quietly, by yourself, may seem efficient in getting an answer, but talking about the process provides insight about how you approach a problem. Further, students can observe both false starts and successful approaches, and you are modeling the problem-solving process.

Class Structure and Organization

11. Do start each class by presenting an outline of what will be covered, so listeners can follow the various topics being covered and can put the pieces together as the class proceeds.

    Commentary: Within an oral presentation, listeners cannot go back to previous material to check connections with what happened earlier. Rather the speaker must make those connections. Beginning class by outlining topics to be covered helps identify an initial structure for what follows.

12. Do encourage students to bring their books to class, and then refer to graphs, tables, etc. rather than reproducing extensive content on the board.

    Commentary: Since it takes time to read and digest large amounts of data and long passages of text, classroom activities can easily become bogged down when extensive details are presented. When handouts or the textbook are available for reference, class time can be devoted to highlighting important elements and making connections among topics. When students see the data or descriptive material in the book, they will know where to go for in-depth study after the class is over.

13. Do define terms when they are first used, so all in the room will know the meaning of what is being said. Similarly, avoid excessive use of acronyms, as being largely unnecessary.

    Commentary: Lack of definitions can represent an inherent bias against under-represented groups who may not have encountered the terms previously, but could excel if they were brought into the discussion.

14. Do show the payoff of each topic presented.

    Commentary: For example, covering the mechanics of program verification without giving experience on developing proofs and code together give the mechanics without any insight about why anyone would care.

Slides and Notes on a Whiteboard/Blackboard

FIGURE 26.2 Well-designed Slide, with index at left and concise, clear notes in body

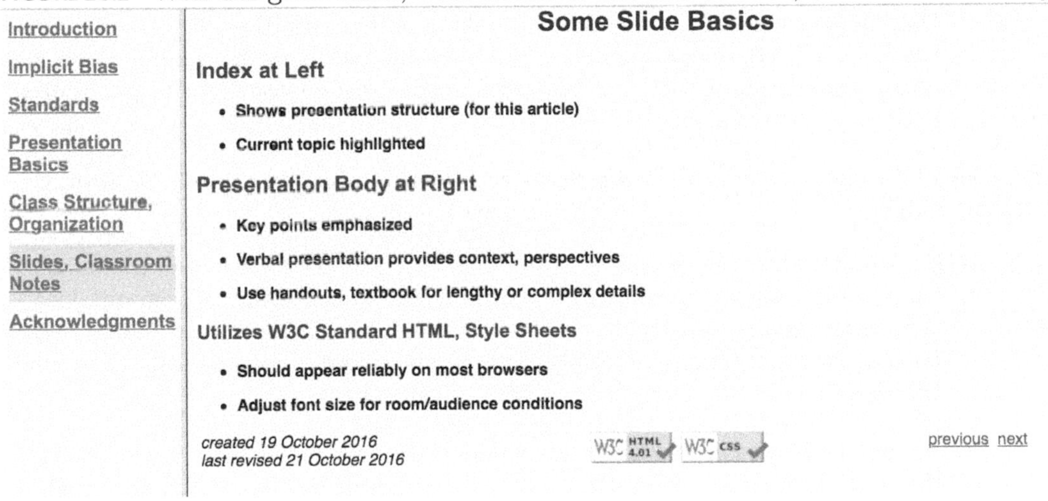

Introduction
Implicit Bias
Standards
Presentation Basics
Class Structure, Organization
Slides, Classroom Notes
Acknowledgments
Some Slide Basics
Index at Left
• Shows presentation structure (for this article)
• Current topic highlighted
Presentation Body at Right
• Key points emphasized
• Verbal presentation provides context, perspectives
• Use handouts, textbook for lengthy or complex details
Utilizes W3C Standard HTML, Style Sheets
• Should appear reliably on most browsers
• Adjust font size for room/audience conditions
created 19 October 2016
last revised 21 October 2016
W3C HTML 4.01
W3C CSS
previous next

15. Don't copy the book to slides or your notes, and then copy the slides or notes to the black board/white board.

    Commentary:

    - If the book already provides information, there is little gain in repeating it in the same way—that is already done.
    - A common goal of a degree program is to provide students with the skills they will need for a career requiring lifelong learning. Copying the book and reading it to students seems counter to this need for lifelong-learning skills. In contrast, helping students to learn on their own (with support and guidance) can be an important element in preparing students for life after graduation.

16. Don't read slides word-for-word.

    Commentary: Attendees will read the slides—teacher commentary can provide additional perspectives.

17. Do visit the classroom ahead of time to determine what parts of a slide are easily visible from all locations in the room.

    Commentary: In some rooms, the bottom third of a slide may be obscured by furniture or otherwise not visible. In other rooms, lighting fixtures near the ceiling may obscure the top part of a slide. Knowing about such obstacles is important in creating slides that are clearly visible by everyone in a classroom.

18. When writing on the board, read aloud what you are writing, so students can take notes as you are writing; then when done writing, repeat what you said (likely in different words) as you are looking at the class.

    Commentary:

    - Make eye contact with your students throughout the classroom as much as possible.
    - Not only does reading aloud fill dead air while writing, it causes listeners to hear the material once while they are taking notes. Restating the material after the writing is completed reinforces the material and helps drive home the main ideas: one instructor writing is bolstered by one student copying and two spoken reviews of the material.

19. Don't fill a slide or the board with extensive content in small print (e.g., Figures 26.2 and 26.3).

    Commentary: Students cannot read and digest vast amounts of material in a short time—even if they can read all the details from the back of the classroom. Don't talk to the board or a monitor. Do write large enough to be seen easily at the back of the room (and periodically check the board after class by walking around the classroom).

    Commentary: Sometimes in the excitement of the moment, it is easy to rush through the writing process. At least for me, the result can be sloppy and too small—unless I consciously work to maintain legibility for all (even those in the back of the room).

20. Do plan what you will put on the board where, so the presentation will be logical and easy to put into students' notes.

    Commentary: Since slides or board notes will likely be transcribed into student notes, the board needs to make sense as the session progresses. Inserts, erasures, arrows, etc. can make notes very hard to follow after the fact, and students may have difficulty reconstructing the ideas presented.

21. Don't write haphazardly on the board.

Commentary: Expect that your notes (from slides or the board) will represent exactly what students write in their notes. When writing on the board is sketchy or randomly placed, student notes likely will be similarly unorganized, and students likely will have difficulty determining how the pieces fit together.

22. Do assume that what is written on the board or on slides will be transcribed into students' notes (but assume nothing else will go into students's notes).

    Commentary: As I have looked at student notebooks during office hours over the years, I almost always can observe copies of what I have written on the board (not always copied perfectly, but usually representing a valiant effort at copying). Although some students add their own comments during note taking, my experience suggests that added comments are relatively rare, and when present, these comments are only sometimes on target.

FIGURE 26.3 Cluttered Slide with Excessive Color and Graphics

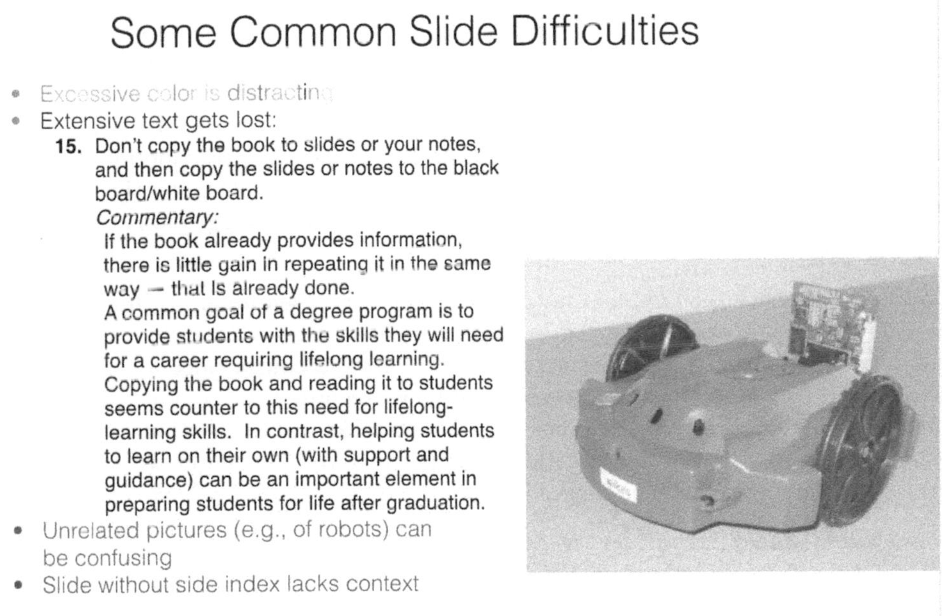

Some Common Slide Difficulties
Excessive color is distractin
Extensive text gets lost:
15. Don't copy the book to slides or your notes, and then copy the slides or notes to the black board/white board.
Commentary:
If the book already provides information, there is little gain in repeating it in the same way — that is already done.
A common goal of a degree program is to provide students with the skills they will need for a career requiring lifelong learning. Copying the book and reading it to students seems counter to this need for lifelong-learning skills. In contrast, helping students to learn on their own (with support and guidance) can be an important element in preparing students for life after graduation.
Unrelated pictures (e.g., of robots) can be confusing
Slide without side index lacks context

Acknowledgments

Many thanks to Charlie Curtsinger, Grinnell College, and Titus Klinge, Iowa State University and Grinnell College, for providing feedback on a draft of this column and suggesting additional discussion points for this column. Some ideas regarding presentations grew out of materials prepared by Kent McClelland, Department of Sociology, Grinnell College.

26.2 SUBSEQUENT REFLECTIONS

I consider teaching as an evolving enterprise. When one begins teaching, all aspects of the work may seem new, different, and perhaps intimidating. An individual may try to cover some basics, but the person likely cannot pay attention to various details and refinements.

As one gains experience, some parts of teaching may become familiar; the instructor has prepared several daily class sessions and provided some leadership within the classroom. Depending upon the course format selected, the individual may have given several lectures, organized discussion groups, engaged students within a lab setting, or otherwise facilitated student learning. With some elements of teaching under control, an instructor can consider adjustments and refinements—what parts of teaching seem to be going well, what seems to work less well, what approaches work for one group of students or another, etc.?

As semesters progress, a faculty member may try experiments, adapt ideas from colleagues, apply approaches mentioned at conferences, etc. In my experience, a desire to improve yields a life-long challenge. Regardless of what teaching successes I have had in the past, I continue to explore new ways to do even better.

As an example, years ago, I watched a new instructor at the beginning the person's teaching career. On the first day of the semester, the person obviously was extremely nervous, stood next to the board (as far away from the students as possible), spoke with hesitation, and had difficulty articulating topics to be presented. By the end of the semester, however, a remarkable transformation had occurred. The person showed self confidence, moved easily around the room (a lab), engaged the students, spoke naturally, and clearly had developed a rapport with all in the room. Many basics of teaching had become natural, although the faculty member was still working on various elements of teaching. Over the following years, the individual continued to develop and became an exceptional teacher.

With this evolutionary process in mind, Chapter 25 presents 24 teaching suggestions, and Chapter 26 presents 22. For new teachers, such lists may provide initial guidance, but even this modest collection of basic practices has considerable potential for intimidation. Together, Chapters 25 and 26 cover some basics, but more suggestions or lists could seem overwhelming.

Suggestion: Rather than expand upon the suggestions given here, beginning instructors are encouraged to ask an experienced instructor to sit in on a course from time to time—just last year, a new faculty member asked me to observe a course once a week for a month or two. (Some faculty prefer the classes to be video recorded, although I am neutral on this approach.) After each visit, the observer might make 1-3 suggestions—not a long list of refinements, but rather a few practices that the new teacher might work to incorporate in the next week's classes. Over time, incremental adjustments may have a dramatic impact!

As a final analogy, consider the process of learning to write. Children or adults learning a new language begin by writing simple sentences with a basic vocabulary and syntax. Over time, their vocabulary expands, they master additional grammatical forms, and they learn how to convey complicated ideas. Eventually, they learn how to structure blocks of writing (e.g., well-organized paragraphs, papers structured into sections), and they polish their work to yield effective and efficient materials. Although those learning to write start with basics, their writing develops and matures. Through this process, feedback and experience can have a substantial impact, if authors pay attention and work to improve.

Similarly, effective teaching may begin with basics (such as those identified in this and the previous chapter). With practice, effort, and on-going feedback, teaching can evolve from basic to quite satisfactory to quite good and perhaps to outstanding. But one must start simply and try not to be intimidated.

CHAPTER 27

Lab layouts for individual and collaborative class sessions

This column first appeared in ACM *Inroads*, Volume 8, Number 3, September 2017, pages 17–19[271]

27.1 ORIGINAL COLUMN

IN SEPTEMBER, 2010, I wrote a *Bits and Bytes* article for ACM Inroads on "Configurations for Teaching Labs" [231]. Since then, I have had numerous conversations about this theme of lab layouts with colleagues nationally. Recently these conversations seem to have increased in frequency, raising four common questions:

1. During a lecture or class discussion, some students may type at workstations. How can I minimize distracting workstation activities that might interfere with others?

2. How can lab equipment be arranged to allow an instructor to circulate easily during lab activities?

3. How can seating be arranged to promote collaboration (e.g., pair programming)?

4. With high enrollments, how can a teaching lab be arranged to accommodate a relatively large number of students (some of whom may have their own laptops), while addressing questions 1, 2, and 3.

This column reviews three laboratory layouts that highlight alternative configurations that support instructors in addressing these commonly-occurring questions:

- Workstations arranged in rows, with students facing forward
- An open concept, in which aisles are perpendicular to the front of the room
- A mixed layout, with a traditional classroom arrangement in the center and workstations around the outside

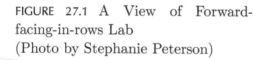

FIGURE 27.1 A View of Forward-facing-in-rows Lab
(Photo by Stephanie Peterson)

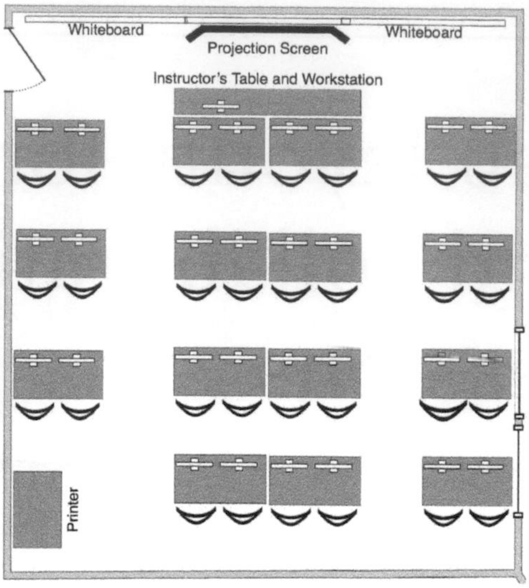

FIGURE 27.2 Floor Plan for Forward-facing-in-rows Lab

The following discussion describes each of these lab layouts in some detail. The September 2010 article provides broader commentary and additional detail.

For purposes of comparison, each floor plan considered describes a 28' by 31' teaching lab, with two 5'-wide windows at the back of the classroom on the right and with a 3'-wide door near the front on the left. Also in what follows, all tables for students are 5' wide and 2.5' deep, which seems adequate to support two students.

A Forward-Facing-in-Rows Lab Layout

The photo and floor plan in Figures 27.1 and 27.2 show a typical classroom at one college: fifteen 5' tables have two workstations for individual work, arranged in four rows, with two aisles providing access. All seats face a 24' whiteboard, with two additional sliding boards that move side to side. A projection screen may be lowered to cover one 8' section of the whiteboard; other parts of the whiteboard are visible at all times. The instructor utilizes a table and workstation in the middle of the room, in front of the first row for students. If all workstations function properly, this room's capacity is 30, but a capacity of 26-28 is more realistic to allow for 2-3 workstations being down.

Unfortunately, this room does not support any of the above questions very well.

- The instructor cannot easily see student monitors from the front of the room.
- Tables have two workstations, providing relatively little space for student notes, discouraging collaboration, and leaving little space for student laptops.
- Rows are 3' apart, restricting instructor movements (particularly with backpacks on the floor).
- Side aisles are about 4' wide, making movement from front to back in the room reasonably easy.
- The instructor's workstation may interfere with some sight lines for students.

Common Variations

- The projection screen may be placed on the right of the whiteboard (Figure 27.1), in the middle (Figure 27.2), or on the left.
- Tables may contain one workstation rather than two, to encourage collaboration and support pair programming.
- Workstations may be omitted at a few tables, to make room for students with laptops.
- If a printer is not needed to support the class, an additional table with workstation(s) and chairs might be included.
- Several approaches are in use at different schools to limit student use of workstations during class discussions.
  - ◇ The instructor may control a monitor on/off switch to disable monitors on student workstations.
  - ◇ The instructor may leave monitors in place but turns off Internet routers, so only local work is allowed.
  - ◇ The instructor may lower monitors during discussions to improve sight lines to the front of the room.

Disabling monitors during class discussion can keep students focused, but also prevents online note taking.

Commentary

Practices vary widely from school to school and from one part of the world to another. In this author's experience visiting dozens of schools, this configuration is by far the most common. I think of this layout as being "traditional" due to its widespread use at schools I have encountered, but this approach clearly is not universal. (For example, one reviewer notes that this approach is emphatically not used at the reviewer's school.)

In the environments which I have seen, this approach provides easy instructor access from front to back. However, the instructor likely must go to the front of the room to move from one aisle to the other. Overall instructor access to students is somewhat restricted. Also, depending on instructor controls, students may be able to work (or play or explore the Web) during class discussions, with the instructor not being able to observe what the students are doing.

An Open Lab Layout

Figures 27.3 and 27.4 show a different layout for the same room, based upon two key observations for Figures 27.1 and 27.2:

- Much space in Figures 27.1 and 27.2 is devoted to relatively narrow aisles between tables for students, and
- When tables with workstations face forward, an instructor cannot see student monitors, and students must look over the monitors in front of them to see the board.

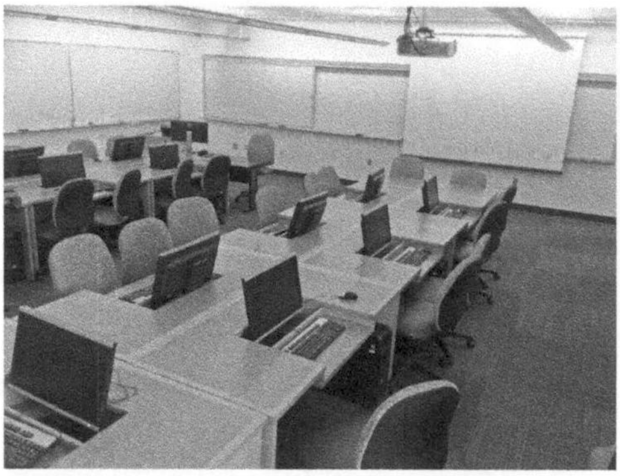

FIGURE 27.3 A View of an Open Lab (Photo by Stephanie Peterson)

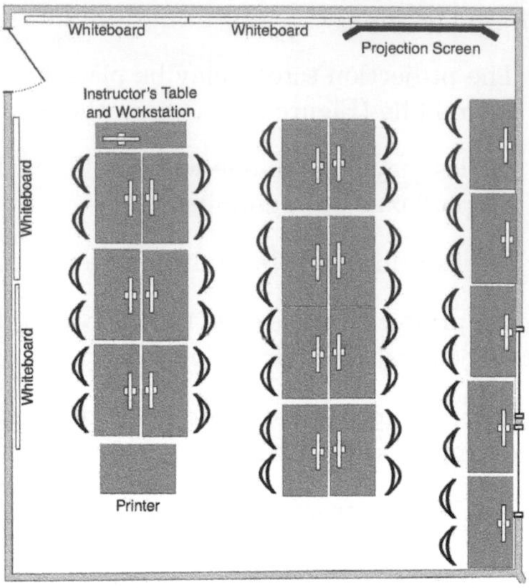

FIGURE 27.4 Floor Plan for an Open Lab

Figures 27.3 and 27.4 resolve these matters by turning tables 90 degrees and placing tables back-to-back. The front of the room contains the same whiteboards with projection screen as found in Figures 27.1 and 27.2. This configuration also allows another row of whiteboards along the left side of the teaching lab. With this arrangement, students work in pairs at 19 tables, so the room could seat 38 students; allowing for a few workstations to be down from time to time, 34-36 students is a reasonable capacity.

Common Variations

- Options for the projection screen are the same as in the floor plan for the facing-forward-in-rows plan.
- Tables may contain two workstations rather than one, to provide a workstation for each student rather than each pair. In my experience, one workstation per table is more common for teaching labs with this configuration.
- Workstations may be omitted at a few tables, to make room for students with laptops.
- If a printer is not needed to support the class, one or two additional tables with workstation(s) and chairs might be included.

Commentary

In my visits to a range of campuses, this approach seems reasonably rare, although those schools using this approach seem enthusiastic. With this arrangement, most issues presented at the beginning of this column are resolved: Instructors at the front of the room can see all student computer monitors. Also, students turn 90 degrees for lecture or in-class discussion, and then turn back for lab work?helping students focus on the class activity at hand.

FIGURE 27.5 One View of a Mixed-mode Lab Layout

FIGURE 27.6 Another View of a Mixed-mode Lab Layout

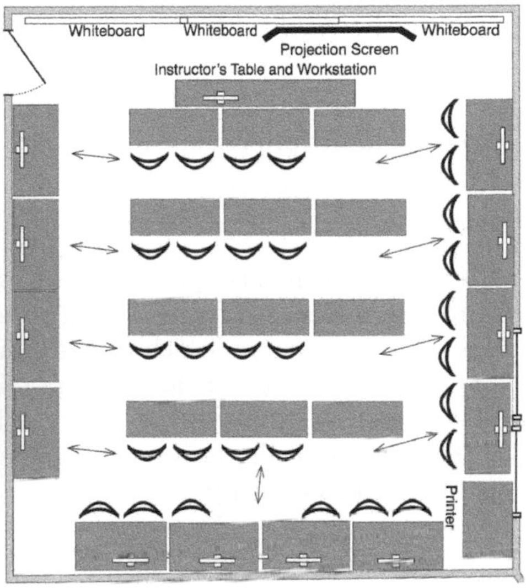

FIGURE 27.7 Floor Plan for a Mixed-mode Lab

A Mixed Mode Layout

A third layout for a teaching lab includes separate spaces for group discussion and for collaboration at workstations. In my experience, this approach seems particularly popular with social studies faculty, but also is used in some STEM courses. Figures 27.5, 27.6, and 27.7 illustrate some of the variety found with this approach for various programs. All implementations devote the central region of the room to class discussion. Students sit in the middle area, away from workstations, for lectures or group discussions. The front of the room contains the same whiteboard and projector screen options discussed previously. The sides and rear of the room contain tables and workstations for lab activities. Since students are sometimes seated in the middle and sometimes at workstations, some chairs may move between central seats and workstation tables to avoid some clutter. In Figure 27.7, the central section contains four rows of three tables, accommodating 24 students, and the workstation areas seat a similar number. Allowing for 1-2 workstations to be down from time to time, 20-22 seems a reasonably reliable capacity for this approach.

Common Variations

As Figures 27.5, 27.6, and 27.7 illustrate, many variations are in widespread use, choosing different options for the central section and varying configurations for workstation tables.

- The central section may contain tables in rows (Figures 27.6 and 27.7), round tables for student cluster (Figure 27.5), individual chairs (without tables) for small and large-group activities, etc.
- Workstation tables on the sides and in the rear may seat one (Figure 27.5) or two (Figures 27.6 and 27.7) students. Tables along the sides of the room may face forward (Figures 27.5 and 27.7) or be back-to-back facing each other (Figure 27.6).
- In Figure 27.7, the printer resides in a corner that is not easily utilized by workstations, so removing the printer might not allow additional seating.

Commentary

Personally, I have seen this approach on several campuses?-not as often as the forward-facing-in-rows approach, but more often than the open lab concept. Faculty report this layout works well in courses which utilize contrasting class formats, as students can move easily and quickly from discussion (in the middle) to lab (at the sides). Sight lines work well, and students are unlikely to be distracted by workstation activities during discussions. However, this approach allocates separate spaces for discussion and for lab for each student and thus limits room capacity.

Summary

Table 27.1 summarizes several features of these approaches for a 28' by 31' teaching lab, in which a 5' by 2.5' table supports two students with one (or two) workstations.

Table 27.1: Summary of Lab Layouts
(28' by 31' teaching lab; 5' by 2.5' tables with 2 students each)

	Forward-facing-in-rows Layout	Open Lab Layout	Mixed Mode Layout
Capacity (with printer) (max/with 1-2 machines down)	30 (26-28)	38 (34-36)	24 (22-24)
Capacity (without printer) (max/with 1-2 machines down)	32 (28-39)	40 or 42 (36-38 or 38-40)	24 (22-24)
Student sight lines	Reasonable	Good	Good
Instructor circulation to help student	Sometimes limited	Easy access	Easy access
Instructor can see student monitors	No	Yes	Yes

Also, all configurations can encourage collaboration and pair programming, if one workstation is placed on each table, whereas two workstations per table encourages tasks by individuals rather than pairs.

Acknowledgments

Many thanks to Stephanie Peterson, Academic Technology Support Specialist, Grinnell College, for her photographs that appear in Figures 27.1 and 27.3. Thanks also to the reviewers who raised several points.

CHAPTER 28

Lab-based courses with the 3 c's: content, collaboration, and communication

This column first appeared in ACM *Inroads*, Volume 8, Number 4, December 2017, pages 24–29 [270]

28.1 ORIGINAL COLUMN

AN OVERALL CURRICULUM INCLUDES much technical material (sometimes called "hard" skills) and people-oriented material (sometimes called "soft" skills). Thus, within an undergraduate computing program, one or more courses must address at least three curricular elements, as shown in Figure 28.1:

- content: courses facilitate student learning of specific technical content
- collaboration: students need practice collaborating (e.g., working in pairs) in preparation for graduate work and/or careers in industry
- communication: students preparing for technical careers must learn to
  - ◇ read technical materials, as they hone their skills for life-long learning
  - ◇ develop oral skills, in interacting with team members, clients, and management
  - ◇ sharpen their ability to write for clients, peers, and publication

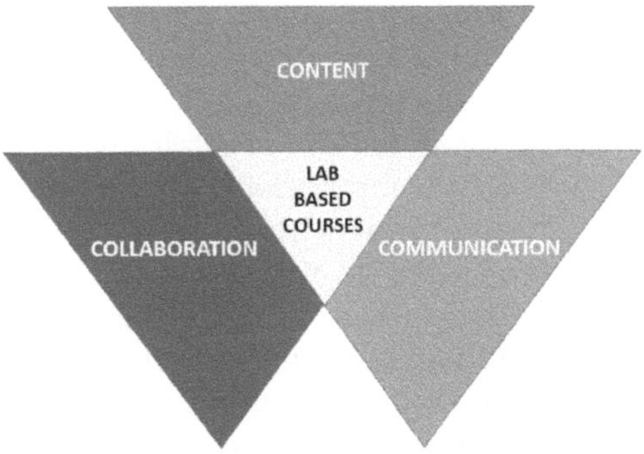

FIGURE 28.1 Lab-based Courses Combine Content, Collaboration, and Communication (Image by Theresa P. Walker)

This column describes a style of course that combines all these elements within a lab-based environment; students are actively engaged throughout almost every class session.

To stay within space constraints, this narrative focuses on seven main course components:

- basic approach
- instructor's role
- practice with many communication skills
- student buy-in
- assessment
- staffing
- start-up issues

In the past, this type of pedagogy has been named a "workshop" style; more recently, this course format may be considered one type of a "flipped classroom". The following notes describe this pedagogy. Much material in this column has been reported previously, in considerable detail, in [203, 237, 50, 255]

Basic Approach

For over 25 years, computing educators have advocated the inclusion of laboratory work within courses (see, for example, [8, 146]). Although practices vary widely, faculty at conferences often report introductory courses meet in a classroom three times a week for 50-60 minutes and once a week in a lab for 1, 2, or 3 hours. Typically, classroom activities involve lecture, discussions, or group activities, whereas lab work involves students working at a workstation individually or in pairs. Altogether, students interact with an instructor (or assistant) 4, 5, or 6 hours per week, with the time split between the classroom and lab.

In the lab-based format, classroom and laboratory activities are combined and integrated. Personally, I have been using this approach since Fall, 1992, basing my courses on earlier pedagogy developed by mathematician, Eugene Herman, for linear algebra. In this format, all course meetings take place in the lab, and students work collaboratively in pairs most of the time. Thus, if a course normally would involve three 1-hour classroom sessions and one 1-hour lab, this revised class format would meet 4 days a week for 1-hour in the lab. Ideally, if a traditional course met for three 1-hour sessions in the classroom plus 3 hours in the lab, the total of 6 hours might be divided into four 1.5 hour class sessions in the lab. (Since four 1.5-hour classes violates scheduling guidelines at some colleges, an alternative just moves all class sessions to a lab. With this schedule, the three 1-hour sessions allow students to begin various topics; the 3-hour session introduces some new material, but also provides time for students to catch up.).

Figure 28.1 illustrates the start of a [largely] hypothetical introductory course on Imperative Problem Solving with C, utilizing this lab-based format. An actual course following this format is available at [267].

This format emphasizes active-learning, and lectures are rare. Before each class, students prepare by studying a reading, watching a video, or engaging some other material. Each class session begins with announcements and answering student questions from either previous labs or the preparatory material. Thereafter, students immediately start working in pairs on lab exercises. When each lab session ends, students may have finished, but if not they are well underway and should finish the lab for homework.

Several additional comments follow:

- I assign lab partners and change the assignments each week. Each week, students learn to collaborate, but they cannot develop over-dependence upon a partner—the partner will be different next week.

- Although all in-class labs are done collaboratively, students also are assigned several programming assignments (for homework) and tests (in class) to be done individually.
- I distribute the full schedule of readings, labs, assignments, etc. on the first day of class, so students can plan their work.[1]
- In one variation of this course format utilized by my colleague, Jerod Weinman, a few clicker questions might be included in the student-question segment at the start of a class to highlight elements of reading and to help students begin thinking about what they have read.
- Interacting with students during a lab is particularly helpful in identifying and correcting misconceptions or missed concepts early, so this approach can be particularly helpful in keeping students moving in productive directions.
- Logistics do not allow me to collect lab write-ups daily. However, labs build upon each other, so collecting and grading one of every 3-5 labs often is adequate to check student progress.

Instructor's Role

A lab-based environment with collaboration is student centered. With students engaged in lab activities, an instructor largely has relinquished control. The instructor assigns readings and shapes instructions for labs and assignments, but students take responsibility for their own work once a class begins. In this setting, a faculty member serves as a mentor, guide, coach, and provider of suggestions and feedback. An instructor responds to questions and helps students if/when they become stuck. Unlike a lecture-oriented format, however, the instructor is not leading the class from one topic to the next.

In my courses, I circulate through the lab regularly. To encourage interactions, I try to check in with each pair of students about every 30-minutes: "are you getting started ok," "any questions about ?," "what issues are arising on the current step in the lab," etc.? In principle, students may ask questions at any time, but practically my presence often prompts a question. Also, if the same question(s) arise several times, I may interrupt the

[1]See Chapter 33 for additional comments regarding a daily schedule.

FIGURE 28.2 Start of a Hypothetical Introductory Course on Imperative Problem Solving with C

Week 1	Lab Exercise 1 Course Overview Linux/Mac OS X	Lab Exercise 2 C Basics	Lab Exercise 3 More Practice in C	Lab Exercise 4 Basic Input
Week 2	Lab Exercise 5 Intro. to Program Organization	Lab Exercise 6 Project 1: (Program to sing a song)	Lab Exercise 7 Elements of C Types and Declarations	Lab Exercise 8 Conditionals
Week 3	Lab Exercise 9 More Conditionals	Lab Exercise 10 Loops	Lab Exercise 11 Simple Simulations	Lab Exercise 12 More Prog. Mgmt: Functions, Assertions

lab for a 5-10 minute mini-lecture to clarify an idea, correct a misconception, provide an example, or otherwise move the students forward. (Since these mini-lectures arise based on student comments, an instructor likely cannot control what questions might arise and thus cannot prepare careful lectures. Such dynamics highlight an instructor's lack of control in this style of lab-based course.)

Practice with many Communication Skills

Peer pressure often encourages students to prepare for class as students may not want to seem unprepared before their lab partners! Further, much class time is devoted to working with a partner and answering questions from the lab. Such activities provide constant practice sharpening communication skills.

- *Reading:* Personally, I designate pre-class readings, so students have daily experience reading technical material. (Some faculty use videos or other media.) At first, students may have trouble delving into technical resources, but students improve over a semester?developing important skills for life-long learning. And, students can ask about confusing passages during the next class!
- *Collaboration:* Each pair of students works at one workstation, and I direct students to change who is at the keyboard at least every class session. To help students develop collaborative skills, some class time may be devoted periodically to techniques, practices, and responsibilities related to pair programming. In my experience, students may feel awkward working with a partner during the first week, but the second week with a new partner goes better?they apply what they learned in week 1 to the week 2 partnership. Similarly improvements are common in weeks 3 and 4. After a month or two, students often have developed solid abilities for working effectively in pairs.
- *Writing:* Although numerous lab exercises involve writing a program or code segment, I also ask students to describe an algorithm, compare one construct or algorithm with another, explain why an approach might have been used in one context but another approach in a second context, etc. The point is that students need to become comfortable expressing their ideas—not only in code but also in prose to partners, clients, and others. The inclusion of writing exercises in labs also sends the message that computing is more than coding—students need to think and communicate![2]

Altogether, a lab-based course can include reading, teamwork, and writing in fundamental ways. Communication skills are not considered an after thought, but rather an integral part of the course.

Student Buy-in

The lab-based pedagogy of this course relies upon student buy-in to the class format. The course framework places primary responsibility for learning with the students, and the instructor works to facilitate students' development. The schedule sets a moderate, but steady pace for material, and students who work regularly on readings, in labs, and with homework have strong track records of success. In my experience at Grinnell College and the University of Texas at Austin, students showed excitement and enthusiasm about this type of learning. As might be expected, not all students came to class prepared for every class, but a substantial majority were ready to start each new lab on schedule.

[2]See Chapter 16 for additional thoughts on writing within the curriculum.

However, local cultures vary, and in some environments, many students may come to class unprepared. For example, I am aware of one course offering with this format in which a substantial group of students did little reading and expected to be spoon fed. When a lab writeup was required, they reported they had searched the Web for 2-3 hours and had difficulty finding supporting material. When they were shown the reading, the lab went smoothly. However, for the next required lab, they again expressed frustration in searching the Web before being directed to the reading. This pattern continued through the semester.

Some faculty have suggested having a short quiz or other graded exercise before each lab to encourage preparation for class. Although I have not found this necessary at Grinnell or UT-Austin, additional incentives might be considered if students to not accept the need for class preparation.

Assessment

Although I have taught lab-based courses regularly since Fall 1992, sections of this course have always been reasonably small (e.g., mostly below 35). Such numbers do not support extensive analysis of this pedagogy's effectiveness or comparisons with other approaches. Some results, however, are possible:

- When I moved to this lab-based pedagogy, the course covered about 20% more content than a traditional course with separate lecture and lab. Both formats had identical contact hours, but the lab-based course covered more topics than were possible using the traditional format. [203, p. 212]

- In comparing student performance on a final exam, students in the lab-based course scored as well as those in the traditional course on questions related to traditional material, but also scored well on material beyond the scope of what could be covered in the traditional course. [203, p. 212]

- At Grinnell, drop rates are always quite low, so the low drop rates in the course with a lab-based pedagogy are lovely, but not exceptional. At UT-Austin, drop rates in the CS1 course using the lab-based pedagogy seemed comparable to the best sections using traditional pedagogy, and the lab-based drop rates were substantially better than the averages for traditional sections. [203, pp. 212-213]

- At UT-Austin, the fraction of students starting CS2 after taking the lab-based CS1 course seems higher than from traditional sections.

- At Grinnell, beginning computing courses have adopted this lab-based pedagogy, and the percentage of women taking these courses and becoming computing majors has been about 3-times the national average. Likely many factors contribute to reasonably strong enrollments among women, but it seems likely that the lab-based pedagogy with collaboration may be one helpful factor.

Altogether outcomes from students taking lab-based courses seem promising, although relatively small enrollments in various sections prohibit extensive comparative analysis.

Staffing

Within this lab-based environment, students largely work independently in pairs. For the most part, students engage with the material and make progress. However, assistance is needed when questions arise, clarifications are required, or bottlenecks surface. At a beginning level, a difficulty may be as simple as what key to push or what a compilation error means. In later courses, difficulties may arise less often, but may require considerable discussion with an instructor.

In my experience, when students collaborate in pairs, an instructor can handle 20-24 students within a lab section. For beginners, student questions may have short answers and require little instructor time, but students may become stuck and frustrated without quick help. With classes of no more than 24 or so, one instructor can provide reasonably quick feedback, and a lab session can progress reasonably smoothly.

When enrollments exceed 25, my experience suggests that an undergraduate assistant can provide adequate help for each additional 8-10 students in the course. (A graduate teaching assistant might be able to support somewhat more student.) Further, although my experience is limited, this level of support seems to scale reasonably well with classes up to 70 or so. In my classes, I have been able to support up to about 40 students with the help of one undergraduate assistant at every lab session, and one additional assistant attending lab every other day. Additional scaling might work well for larger classes as well, but I can only speculate on the support needed for large lab sections.

With relatively large enrollments, note that an assistant need not be expected to answer all student questions. Rather, if an assistant can handle a range of relatively straight-forward queries, then the instructor can focus on topics that require considerable background, experience, or insight.

Start-up and Maintenance Issues

Once a lab-based course has been established, maintaining and refining the course requires only modest effort. Personally, after students work on a lab during one term, I often edit the current materials, so they will be updated for the following year. Such revision rarely requires extensive instructor time.

Getting started, on the other hand, requires extensive effort, as both readings and labs must be identified for each class session. If the course utilizes a textbook, then the daily reading schedule may reference specific sections or pages Development of new readings, materials, etc. requires much more work.

Turning to the development of in-class labs, when I first switched format to a lab-based approach, I expected to rewrite traditional labs to fit the daily-exercise format. However, I quickly realized that simple editing was inadequate; in a traditional format, labs only cover selected topics?other material is discussed in lecture or discussion. In a lab-based format, students need guidance through the entire course, so labs must be comprehensive.

Rather than undertaking course development individually, a collaborative effort can work well. If two instructors teach parallel sections, each can develop half the readings and labs, with feedback from the other. When my colleague, John David Stone, and I followed this approach, the work was far from trivial, but we were not overwhelmed either, and the first semester provided a reasonable foundation for subsequent semesters.[3]

Acknowledgment

Many thanks to Theresa P. Walker for her help designing and creating the image that appears at the start of this column.

[3]Chapter 37 outlines a different development approach using student collaborators.

Conclusions

A lab-based course using collaboration can be an extremely effective course format. The course places responsibilities upon students, as it deepens their technical knowledge, sharpens their communication skills, and provides experiences working in teams. In many contexts, students cover more material than in traditional forms (with separate lecture and lab) . When students buy into this pedagogy, they perform well on tests. Further, although extensive comparative studies are not available, drop-out rates generally seem low, and the recruitment and retention of women strong. Start-up for this course format requires substantial effort, but on-going refinement of courses with this pedagogy requires only modest updating.

CHAPTER 29

Active learning and/or flipped classrooms

This article appears in this book for the first time, demonstrating that new topics might be considered as subjects for future columns.

29.1 NEW ARTICLE

IN THE 1990s, PHYSICS AND MATHEMATICS FACULTY were particularly active in exploring new pedagogy to facilitate student learning. This chapter provides an historical perspective on active learning in the 1990s in physics, mathematics, and computing, followed by developments in computing education that have led to today's flipped classrooms and other active pedagogy.

Introductory Physics in the 1990s

Within physics in the 1990s, faculty observed relatively high dropout rates, poor grades, and high student frustration. In describing Sheila Tobias' landmark research reporting student reactions to traditional science courses [185], Priscilla Laws wrote that students "paint a devastating portrait of introductory courses as uninteresting, time consuming, narrowly fixated on the procedures of textbook problem solving, devoid of peer cooperation, lacking in student involvement during lectures, crammed with too much material, and biased away from conceptual understanding." In response in 1991, Laws developed workshop physics, in which class time utilized computer simulation tools and active student experiments as primary elements of a "Calculus-based Physics without Lectures" [107, p. 24].

More generally, many physics faculty focused on actively engaging students within class, called "interactive engagement (IE)" by R. R. Hake. Between 1992 and 1996, in an impressive study involving 62 introductory physics courses with 6542 students, Rake collected data and compared student performance in lecture-oriented classes and those using IE. He concluded, "48 courses ... which made substantial use of IE methods achieved an average gain ... almost two standard deviations ... above that of the traditional courses" [80, p. 64].

Altogether, throughout the 1990s, physics faculty were experimenting with alternative forms of pedagogy for introductory courses, and by 1999 McDermott and Redish compiled a listing of 224 articles about teaching and pedagogy in physics [123].

Introductory Mathematics in the 1990s

Within mathematics, conversations about first- and second-year courses included both content and pedagogy. For example, as early as 1990, Eugene Herman was working to integrate lecture and labs within the same classroom for a second-year linear algebra course. Eventually, this led to a textbook and software, *Visual Linear Algebra*, published in 2005 [87].

In considering the teaching of calculus, some mathematicians throughout the United States formed a calculus reform movement with at least two themes:

- Some faculty viewed introductory calculus as having too many formulae at the expense of connecting concepts. Some described the traditional calculus course as following a "formula of the day" format. In an extreme case, each section of one textbook presented a formula and 2–4 examples. Students then were to solve 50-80 similar problems—mostly changing the numbers to be plugged into the section's formula.

- In parallel with physics instruction of the time, calculus instructors experimented with the use of labs, active-learning exercises, and cooperative learning. Supporting this innovative work, the Mathematical Association of America (MAA) published several pieces within its MAA Notes Series, including *Learning by Discovery: A Lab Manual for Calculus*, (MAA Notes Series #27, 1993) [179], and at least three monographs discussing "Cooperative Learning" (MAA Notes Series # 37, 44, and 55 [79, 61, 180].

Introductory Computing through the Mid-2000s

In the 1980s and early 1990s, computing educators were in remarkable agreement regarding key elements of introductory computing. Most colleges and universities utilized the programming language Pascal in their introductory courses, and the content of the first year typically included such topics as imperative problem solving with a top-down methodology, structured code with functions and procedures, data structures (e.g., arrays, linked lists and trees), abstract data types (e.g., strings, stacks, queues), files, and elementary algorithmic analysis. For example, the first Advanced Placement Computer Science Course and Exam, in 1984, indicated a description for an introductory year-long course that matched introductory sequences at a substantial majority of schools in the United States.

With this common content, students were expected to write programs to solve a range of problems. To support this student work, courses offered one or two general types of labs.

- *Open labs:* Computer laboratories often contained workstations that allowed students to work on their own to complete their assignments. Sometimes, a lab assistant might be available to answer questions, but often, little or no assistance was available.

- *Closed labs:* Some introductory courses included scheduled lab time, during which students were given a series of structured steps to provide practice with concepts and techniques recently covered in class. During these lab sessions, an instructor or assistant was available to answer questions and provide guidance.

In the 1980s, computing faculty discussed the relative merits of open and closed labs. Closed labs had the advantage that students could gain practice and receive help, but closed labs also required a substantial investment of instructor time—both to prepare materials and to attend the lab sessions. For example, for those who considered computing to be evolving from a mathematics tradition, some argued that mathematics (e.g., calculus) traditionally did not have structured labs, so why should computing?

By 1990, several articles, such as "Laboratories in the computer science curriculum" by Parker, Cupper, Kelemen, Molnar, and Scragg [146], promoted the use of closed labs, and over time closed labs for introductory computing courses became common.

Further, with the acceptance of closed labs, faculty experimented with both the content and the pedagogy for these labs, eventually finding a balance between traditional classroom activities and labs. For example, in Fall 1992, I reorganized an introductory computing course to follow a lab-based pedagogy [203], based on Eugene Herman's integration of lecture and lab for linear algebra [87].

Another important factor in encouraging experimentation centered on Pascal's inability to scale well for large-scale program development. Originally, Pascal was designed as a language for beginners, and another language was expected later for large software packages. As software applications became larger and more complex through the 1990s, an increasing number of schools shifted their introductory course from Pascal to another language that supported either separate modules (e.g., Modula 2) or object-oriented problem solving.

For the Advanced Placement Computer Science (APCS) Course and Examination, the Development Committee determined that Pascal could not continue as the language for the course, but there was no national consensus regarding what language to choose. With this challenge, APCS switched from Pascal to C++ in 1999 amidst considerable controversy.

Experimentation throughout colleges and universities continued however, and shortly after the switch to C++ discussions for APCS began that would yield, in 2003, a change to the newly-emerging Java programming language.

With this exploration of problem-solving paradigms and introductory programming languages, experimentation with pedagogy continued, and in 2004, Clement issued a "Call for action (research): Applying science education research to computer science instruction" [37], including the "use of manipulatives" [p. 354], the use of spreadsheets, experiments and prediction, and exercises that span several levels of learning.

Between 2005 and 2006, Jeffrey J. McConnell compiled several approaches in a 4-part series of articles on "Active and Cooperative Learning: Tips and Tricks". Part 1 provided general background and considered the possibility of using "dramas or kinesthetic learning activities that get students to physically act out a concept or algorithm" [120, p. 27]. Part 2 "discusses different levels of risk in class exercises and how activities can be designed to minimize that risk" [119, p.34]. Part 3 "discusses the characteristics that make groups effective as well as techniques for the formation and evaluation of groups" [121, p. 24]. Finally, Part 4 covered "the design and evaluation of classroom exercises" [122, p. 25]. Altogether, this series presents a very nice review of many ideas for active and cooperative learning that had emerged in the 1990s and through the mid 2000s.

Altogether, computing educators had not reached a consensus regarding either the content or pedagogy for introductory courses, and experimentation proceeded in many directions.

Two Faculty Populations at Many Conferences

Interest and activity in computing education and computing educational research have continued and expanded from the 1990s through today. Several ongoing events highlight the current level of interest and faculty involvement.

- The annual Symposium of the Special Interest Group on Computer Science Education (SIGCSE) within the Association for Computing Machinery attracts progressively more attendees from year to year, and the attendance for 2017 surpassed 1,400.

- SIGCSE's conference on Innovation and Technology in Computer Science Educaiton (ITiCSE) began in 1996 and continues to have a strong following.
- Beginning in 2005, SIGCSE has sponsored an International Computing-Education Research Conference (ICER) which alternatives between the United States and outside North America.
- Within the United States, regional conferences sponsored by the Consortium for Computer Science in Colleges (CCSC) continue in over a dozen venues annually.

As I reflect upon the work presented and discussed at these events, computing faculty continue their experimentation with computing pedagogy, building upon earlier work from the 1990s. Pragmatically, conference sessions often serve two distinct groups of attendees.

- Computing-education researchers seek to engage in formal studies of effectiveness, working to conduct controlled explorations of various pedagogical techniques. This work strives to collect data with substantial student populations to document what approaches are effective under specific circumstances. The ICER conferences are clearly aimed for these researchers.
- Many computing teachers focus on the day-to-day elements they encounter in their classrooms, and this group often presents experience reports that document their pedagogical approaches, challenges, and successes. Often these faculty teach at schools in which enrollments are too small to allow multiple sections of courses and extensive statistical analysis. Many CCSC conferences seem generally oriented to this constituency.

The SIGCSE Symposia and ITiCSE Conferences seem to provide a balance between these two types of sessions, with both the formal researchers and the practitioners finding some sessions for them, but also wanting more options for their interests.

In reviewing many presentations and sessions over the years, I have noted that many experiments involving active learning have concluded that students indeed learn more with various active-learning techniques, but also that courses using these techniques do not cover as much material within a given quarter or semester. Upon further investigation, it seems that some faculty continue to utilize traditional lectures, just as had been included previously, but the courses then add active learning activities as well. That is, courses may be duplicating class time for lecture-based presentations and active-learning activities. In listening to some of these conference sessions, three points come to mind:

- If a course includes both lecture and active-learning activities to cover the same material, then students likely are seeing the same material twice, so they are likely to have an enhanced mastery of this content.
- If active-learning activities are added without reducing time devoted to lecture, then time spent on covered topics will expand, and fewer topics can be included.
- If active learning is identified as an effective mechanism to enable student learning, but a course utilizes lecture as before, then *instructors are not trusting active-learning pedagogy*. If active learning is believed to foster learning, then a course should trust that active learning will yield student learning, and traditional lectures can be reduced or eliminated.

Interestingly, in recent years, educators and educational psychologists are refining the notions of active and passive learning. For example, Chi and Wylie identify four levels of cognitive engagement with an "ICAP Framework" [34]:

- **I** or **Interactive** Behaviors: In engaging presented material, a student interacts with one or more others, with each individual contributing to an on-going discussion.
- **C** or **Constructive** Behaviors: Students respond to material by creating something new, beyond the initial material—perhaps a rationale or a further explanation.
- **A** or **Active** Behaviors: Students engage in an "active mode of engagement" "if some form of overt motoric action or physical manipulation is undertaken" [34, p. 221].
- **P** or **Passive** Behaviors: Students may listen to a lecture, but they need not take any action. For example, in passive learning, students might hear a presentation, but they would not be expected to take notes.

Altogether, Interactive, Constructive, Active, and Passive modes represent a sequence of decreasing student engagement with a topic. With these distinctions, Chi and Wylie describe "the ICAP hypothesis (that $I > C > A > P$), showing the relevance of the hypothesis to classroom learning" [34, 220]; students are expected to learn more with increasing levels of engagement.

Computing in the Late 2000s and Beyond

Even a cursory review of proceedings from the SIGCSE and CCSC conferences documents computing faculty's ongoing interest in experimentation with both content and pedagogy.

One area of development involves the utilization of emerging technology to aid teaching and learning. Two examples illustrate many of the experiments and efforts in this area.

- Several conference reports describe the use of online bulletin boards or discussion groups to support student-student and student-faculty conversations between class sessions. Formats vary, but the general idea is to allow students to post questions, suggest approaches, brainstorm with others, and collaborate. Sometimes a bulletin board only allows students to ask questions, with answers supplied by the instructor or teaching assistants. Other bulletin boards promote more general discussion. With the range of activities possible, reported experiences seem mixed. For example,
  - ◇ In one experiment, Mihail, Rubin, and Goldsmith report, "Our findings suggest that, overall, making more posts, posting more questions and engaging in Devil's Advocacy have positive effects on learning, while making more informational posts, explaining to others and making longer posts do not" [127, p. 409].
  - ◇ Whaley describes the use of blogs to encourage students to delve into suggested readings, report their reactions, and then read other students perspectives. [284]
  - ◇ Stone utilized weekly student blogs as a useful mechanism to obtain regular feedback about student perspectives on a course in progress [181].
- In all-sized classes, but particularly in large classes, involving students can be challenging, as individual one-on-one student-instructor interactions consume considerable time. One popular approach asks students to press buttons on a remote device, called a clicker, to respond to instruction questions. After initial polling, both the instructor

and students can view a graph that shows the number of students submitting each answer. At this point, approaches vary: sometimes an instructor will respond by giving the proper answer with explanation, and sometimes students will break into small groups to discuss the alternatives and then re-vote. Porter and Simon provide a nice explanation of several possible alternatives with this approach. [149]

When clickers are unavailable (perhaps due to cost or technical constraints), large, inexpensive square cards might display contrasting colors along each edge on one side (the other side remains plain). In response to a question, each student raises the card with the proper color on top, and the instructor views the results. Also, since the reverse side of a card is plain, students cannot see the responses of those in front.

Another direction in the development of computing education pedagogy relates to past experiences with workshop pedagogy or lab-based courses, in which student labs or other activities replace lectures (e.g., [87, 107, 237] and chapter 28 in this book). The more general perspective, now often called a "flipped classroom", moves the presentation of content from an in-class lecture to a before-class activity, and homework activities shift to in-class work.

In a traditional lab-based format, student preparation often involved reading, although now faculty may produce a video of their lecture for viewing before class. With the "lecture" now done before class and class time devoted to the previous "homework" assignments, the elements of a class session are thus "flipped" or "inverted". Rutherfoord and Rutherfoord provide additional details and perspectives in [158].

With strong interest in approaches that actively engage students, it is hardly surprising that several specific methodologies have emerged, each of which provide a framework for course formats and activities. A short description of three of these approaches follows.

- *Peer Instruction (PI):* "PI is characterized by asking challenging, in-class conceptual questions of students. For each question, students individually respond, discuss the question in small groups, and respond again based on their new understanding" [150, p. 358] Often, PI utilizes clickers, through which students can record their conclusions, although the general concept of peer instruction is clearly more general.

- *Process-Oriented Guided Inquiry Learning (POGIL):* "In a POGIL classroom, teams of 3-5 learners work on instructor-facilitated activities. Through scripted inquiry and investigation, learners discover concepts and construct their own knowledge" [104, p. 159]. Teamwork is essential, as students collaborate on activities that suggest broader principles and concepts. Although not strictly required, many instructors assign specific roles to students in each team. Instructor training and support is available from The POGIL Project [151].

- *Team-Based Learning (TBL)* Following the general approach of a flipped classroom, students following TBL first encounter material before class. After an initial question/answer session in class, students may take a test to assess initial understanding. Next students work with a team to apply the new material and gain understanding. Classwork may conclude with assessment of each teams' success. Based on test results, class time may be devoted to clarifying ideas or correcting misconceptions. [106]

At present, PI, POGIL, TBL, a workshop format, and a lab-based pedagogy all have a strong following and documented success. Also, experiments continue with additional formats or combinations of approaches—all with the common goal of actively engaging students within a learning context. However, reports also indicate that strong learning outcomes may require student buy-in within a local setting, suggesting the local environment may have a substantial impact on what seems effective at one school or another.

CHAPTER 30

Combining technical depth, social/ethical issues, and active student involvement

The article in this chapter is newly written for this book, providing coverage of material that has not yet appeared in my columns.

30.1 NEW ARTICLE

TO BEGIN WITH WHAT MAY SEEM OBVIOUS, a course or course module on technology/computing and social/ethical issues likely requires at least three elements:

- *The understanding of some basics of technology and computing and their consequences:* Contemporary life involves a wide range of technological successes and products, the popular news press typically presents a stream of capabilities, and the entertainment industry sometimes includes fictional accounts of hypothetical situations that may or may not be possible. Students, particularly at the beginning levels, may have little experience separating fact from fiction and the capabilities of technology from its limitations. For example, in reviewing student work about technology and society, I have encountered papers reporting "facts and technical capabilities" with citations to the movie "2001: A Space Odyssey" and a personal Web page on Facebook. A realistic study of technology and society must clarify current capabilities, reasonable expectations for future technology, and what is known to be not feasible or impossible.

- *A framework for the consideration of social and ethical matters:* When asked how to evaluate the advantages, disadvantages, consequences, and ethical implications of potential actions, students need to have a framework to guide their analysis. In practice, people involved with science, technology, engineering and mathematics (STEM) may appreciate techniques and processes to solve problems technically, but may have little background in determining consequences, social impact, and ethical principles.

- *An environment that requires students to actively engage vital concepts and issues:* I have talked with many fourth graders who can recite principles of "love your neighbor", but who also report on-going teasing (or bullying) of a student on the playground.

> By the time students have enrolled in college, they likely can provide easy answers to simple questions about proper behaviors. Challenges arise, however, in practice, when different principles lead to contrasting actions, and computing professionals need to make choices. Within a classroom, students should confront complex issues to deepen and test their understandings and perspectives.

Over the years, I have tried to include comments about social and ethical issues within a range of computing courses (e.g., CS1/CS2, analysis of algorithms, introduction to artificial intelligence, etc.). Since these courses have a technical focus, students are developing technical mastery throughout the course, addressing the first essential element listed above. However, these courses typically do not devote substantial class time to ethical principles and foundational approaches for the evaluation of social and ethical issues. Thus, these courses allow some discussion of computing, its impact, and its consequences, but without a foundation these courses cannot delve into matters of social and ethical implications and analysis. (Also, if this foundational background were provided in technical courses that were electives or for which students had choices, then the same background would have to be included in each course. One of these courses could not assume students had engaged this material elsewhere, so each course would have to repeat the same foundational material.)

The remainder of these notes consider the above three elements in the context of full courses rather than modules within other courses. Topics for discussion include:

- getting started with social/ethical foundations
- choices in designing a course
- the role of case studies
- inclusion of current events
- engaging students

Getting Started with Social/Ethical Foundations

As students begin consideration of the social and ethical dimensions of computing, they need an understanding of basic principles and philosophies.

One approach for classroom discussion builds upon a general textbook introducing principles of ethics. Two widely-used books are:

- Sara Baase and Timorhy M. Henry, *A Gift of Fire: Social, Legal, and Ethical Issues for Computing Technology, Fifth Edition*, 2018 [18]

- Michael J. Quinn, *Ethics for the Information Age, Seventh Edition*, 2016 [152].

Another approach immerses students in cases and legal discussions at an early stage. For example, M.I.T.'s course, 6.805/STS085/STS487, *Foundations of Information Policy*, asks students to read a sample legal brief on one subject and write a brief on a second topic, using the sample as an example—all before the first day of class.[1]

However students study basic social and ethical principles, they also should relate general principles to considerations and practices within the field of computing. Three sources seem particularly helpful in making initial connections.

- The *ACM Code of Ethics* [3]
  The basic statement of ethical principles for computing professionals within North America.

- Diane Whitehouse, Penny Duquenoy, Kai K. Kimppa, Oliver K. Burmeister, Don Gotterbarn, David Kreps, and Norberto Patrignani, "Codes of Ethics and Cloud Computing", *SIGCAS Computers and Society Newsletter*, Volume 45, Number 3, September 2015. [285]
  A fascinating study of the evolution of codes of ethics for the computing community internationally, including variations in cultural expectations. The application of ethical principles to cloud computing also highlights differences in perspectives throughout the world and the need to interpret standards within local environments.
- Bo Brinkman, Don Gotterbarn, Keith Miller, and Marty J. Wolf, "Making a Positive Impact: Updating the ACM Code of Ethics", *Communications of the ACM*, Volume 59, Number 12, December 2016, pages 7–13 [29].
  ACM's Committee on Professional Ethics work on updating the 1992 ACM Code of Ethics through 2017. This report of the development of computing and the process for reviewing codes of ethics provides a worthwhile perspective when considering how social and ethical issues might be developed and updated.

Choices in Designing a Course

Every course requires instructors to make choices. For courses connecting STEM fields to social and ethical issues, three choices may have special significance: the target student population(s), the balance between breadth and depth, and the selection of one or more instructors.

Target Student Population(s)

Planning a course involving technology/computing and social/ethical issues likely requires making assumptions regarding the students that will be enrolled. Courses with a wide range of audiences could be very successful, but course details likely will depend upon the backgrounds of the enrolled students. Some possibilities include the following:

- *Audience: Upper-level computing majors:* If students have taken 2-3 years of undergraduate computing courses, they likely have considerable technical background. A course would need to provide a framework for analysis of social and ethical issues, but students likely have experienced challenges of software development, systematic testing plans, etc. In this setting, after providing a philosophical foundation, a course could move quickly to case studies and implications of technical consequences.
- *Audience: Upper-level philosophy majors:* If students have taken several philosophy courses, including an introduction to ethics, they still may need to experience writing some programs to gain a basic understanding of how a development process works, what can go wrong, how testing might be done, etc. An introduction to programming might start a semester, or basics of code development could be interspersed with considerations of other themes.
- *Audience: A mixture of upper-level computing and philosophy majors:* If some students have a technical background and others an understanding of philosophy and ethics, a course might draw upon both strengths by combining these students into groups that include both backgrounds. In this approach, the course might introduce case studies relatively early, with groups having to draw upon various students experiences to address underlying issues that grow out of different stakeholder perspectives. With a mixture of students, each can help others learn and provide missing background.

- *Audience: General students with no assumed background:* When students can be assumed to have little experience in either computing or social/ethical principles, the course likely will need to begin with a general introduction to both areas. Discussions and/or labs may begin at a reasonably general level; the course may need to emphasize breadth as students need to develop background in whatever they study.

Breadth versus Depth

A review of syllabi from several schools suggests that courses assuming either technical background or an understanding of philosophy or ethics tend to explore some range of topics in reasonable depth. The start of a course can supply additional background, but students generally are prepared to delve into cases and challenging settings in which stakeholders' interests conflict.

However, when offered with little prerequisite, much class time likely must be devoted to basic technical topics and an ethical framework. Toward the middle or end of a semester, small groups might explore one or two areas in some depth through their investigations, but this type of course likely would have to highlight general themes at an introductory level.

Selection of One or More Instructors

Yet another decision regarding a course relating STEM fields with social and ethical issues is who should teach the course? In particular, many computer scientists earned advanced degrees in other subfields of computing; they may be interested in social and ethical issues, but their work on these issues may have developed later. Further, when computing enrollments are high, computing faculty may need to teach courses within their original competences (e.g., CS1/CS2, algorithms, architecture, graphics, artificial intelligences, etc.).

On some campuses, however, some non-computing faculty may have interests, even research, related to the impact and ethical challenges of technology and computing.

Altogether, discussions of social and ethical issues within STEM fields are interdisciplinary in nature, and thus may open various staffing options, depending upon local interests and enrollment constraints. The following list suggests some of the possibilities.

- The course may be taught by a faculty member within a computing or engineering department. As an example, for many years, Donald Gotterbarn, now retired from the Computer and Information Science Department, at East Tennessee State University, taught numerous courses, including Ethical Issues in Computing [74]. After his official retirement, he has continued to work within ACM and internationally in this field.

- The course may be taught by a faculty member housed in another department, but whose interests and scholarship activities address topics related to technology and computing and their impact and implications. Two prominent examples are:

  ◇ Jeffrey Koperski, Department of Philosophy, Saginaw Valley State University, teaches "Philosophy 210B Online—Engineering and Computer Ethics" [100].

  ◇ Chuck Huff, Department of Psychology, St. Olaf College, has a long history teaching courses related to computers and society, most recently "Ethical Issues in Software Design" [91].

- The course may be team taught, often with faculty from several departments. As an example, M.I.T.'s course, "MIT 6.805/STS085/STS487: Foundations of Information Policy" is taught by six faculty members from various departments and groups. [1].

The Role of Case Studies

Although some courses encourage discussion of general principles and approaches for considering social and ethical issues arising from technology and computing, many courses utilize case studies. In some cases, these examples illustrate general themes that have been discussed previously. In other cases, study may begin with the case studies, and principles arise as generalizations from these specifics.

In reviewing courses and repositories utilizing case studies, several formats seem in common use. However, many materials seem to follow a similar overall approach that reviews facts, identifies consequences, considers alternatives, and articulates a reasoned conclusion.

- MIT/s course, 6.805/STS085/STS487, *Foundations of Information Policy*, prescribes "six key sections" for a legal brief: an identification of the issue, a history of the parties involved in the case, a statement of relevant facts, the legal finding of the court, the rationale explaining the legal finding, and any dissenting statements from other judges hearing the case. [1, Assignment 0: Briefing writing and policy analysis]
- The Applied Ethics in Professional Practice Program of the National Institute for Engineering Ethics (NIEE), with archives at the Murdough Center of the Texas Tech University, structures cases with an initial set of facts, a discussion of "Alternate Approaches to resolving the issues presented", survey results from parties involved with the case or alternatives, and a forum of other interested parties. [133]

Although these and many repositories seem available on the Web, my experience suggests that URLs and links change with moderate frequency. For example, in searching for references for this article, over half of the links I found were no longer valid; in some cases, a page indicated dozens of links to interesting-sounding case studies, but only a handful were active. However, even with such setbacks, a simple Web search for

"case studies" computing social ethical issues

yielded over one million results, many of which seemed potentially interesting and useful for inclusion in courses.

Inclusion of Current Events

Many courses seem based on general readings (e.g., texts on social/ethical principles) and case studies (e.g., law briefs or case descriptions as described earlier). However, some faculty and students appreciate in-class discussions of current events—many of which may relate to codes of ethics or social implications of computing.

- Some faculty require each student to locate and present in class a designated number of current events (e.g., 3 events). For each report, a student must identify the case, provide basics of what happened and who were impacted, and the result of the current event. In-class discussion might also consider possible approaches for resolution. In several reports by faculty, it seems common for one day each week (e.g., Fridays) to be designated for the report of these current events or news stories.
- In some cases, an instructor may review daily or weekly newspapers and/or magazines to identify reports that highlight course themes. In this context, the teacher rather than students identifies and presents the current events stories. Some faculty using this approach report that initially they may have to work to find appropriate stories, but after some experience finding news reports becomes reasonably easy; and sometimes the difficulty is limiting the number of identified events rather than finding enough.

Faculty using current events in class often report that such stories and the resulting discussions keep students involved and reinforce the relevance of the material under study.

However, utilization of current stories can take substantial class time. (For example, if 20 students use only 4 minutes to report each of 3 events, then the total time devoted to this activity may be approximately 4 hours—likely over a week of class time.) Also, reliance upon current events can undermine overall course planning and instructor control. As a semester unfolds, the nature of current events may be difficult to predict: some topics may arise frequently and others rarely or not at all. Further, the timing of topics may seem random, possibly interfering with a logical progression of topics through a semester.

Altogether, use of current events has clear advantages, but inclusion of these stories may impact other parts of the course.

Engaging Students

As discussed throughout this book, teachers should select a pedagogy that actively engages students. Such a guideline is particularly important for courses related to social and ethical issues, because a cursory review of principles may not highlight the challenges and conflicting circumstances that arise in real-life situations. Overall, some mechanisms are needed that require students to delve deeply into cases or examples.

In reviewing several courses, a few approaches seem particularly common to encourage student engagement.

- Many courses require students to participate within the classroom. Students must come prepared to discuss a reading, and sometimes the instructor or the students themselves post questions to initiate in-class discussions.
- Most courses seem to require one or more written papers; many course are designated as "writing intensive." Students need to delve into a focused situation, explain the perspectives of several different stakeholders, identify possible conflicts, and suggest a plausible approach to resolve the problem(s).
- Some courses require students to present their findings (e.g., from current events or from their explorations). In my experience, students often have difficulty with their initial presentations—they think they are prepared, but may stumble when speaking or badly misjudge time, for example. Also, for each talk, I provide feedback, based on student comments and my own observations. In many cases, students want another chance, so I usually try to schedule at least 2 oral presentations (3 is better). Often, a second or third talk is dramatically better than the first!
- Some faculty organize debates on topics—perhaps dividing a class into multiple groups of 2 or 3. In a debate format, the groups may have to represent one stakeholder in one debate context, but then represent another stakeholder in a second. If a case includes several stakeholders, one debate might involve several different groups (not just 2).
- In some classes, students or student pairs must propose a resolution to a social/ethical issue, including both the recommended action and the rationale. Sometimes other students in the class may be asked to vote for the best one or two solutions.

In any of these formats, a central objective is to encourage students to take ownership of one stakeholder perspective and then to view an issue from multiple viewpoints.

Additional resources related to a course on STEM fields and social/ethical issues are identified in Chapter 31, "Selected/annotated references for course formats."

Selected/annotated references for course formats

DISCUSSIONS OF ACTIVE LEARNING SPAN A VAST LITERATURE, including theoretical frameworks, conferences and resources from supporting professional organizations, classroom techniques, the exploration of social and ethical issues, and related physics and mathematics resources. The following annotated list of references provides starting points for additional reading in each of these areas.

Theoretical Frameworks

- R. R. Hake, "A six-thousand-student survey of mechanics test data for introductory physics courses", *American Journal of Physics*, Volume 66, Number 1, January 1998, pp. 133–137 [80]
  This early report, comparing results from 6542 students from 62 introductory physics courses, demonstrated the profound impact of active learning techniques over traditional lecture formats.

- Michelene T. H. Chi and Ruth Wylie, "The ICAP Framework: Linking Cognitive Engagement to Active Learning Outcomes", *Educational Psychologist*, Volume 49, Number 4, 2014, pp. 219–243 [34].
  A careful description of four modes of student engagement: Passive, Active, Constructive, and Interactive, followed by considerable analysis.

Organizations with Ongoing Publications and Conferences

- The Special Interest Group on Computers and Society (SIGCAS) within the Association for Computing Machinery (ACM) maintains links to a range of programs and resources, accessible from their home page: http://www.sigcas.org [167].
  Resources include a link to materials on computer ethics and to the Pledge of the Computing Professional.

- The Special Interest Group on Computer Science Education (SIGCSE) within the Association for Computing Machinery (ACM) supports a range of conferences, projects, and funds related to computing education. Home page: http://www.sigcse.org [169]. The SIGCSE-sponsored conferences: the SIGCSE Symposium, ITiCSE Conference, and the ICER Conference are outlined briefly in Chapter 29; see SIGCSE's home page for an up-to-date schedule.

Classroom Techniques

Several different approaches and techniques have emerged over the years to encourage and support active learning. The following identifies some of the widely-reported efforts.

- *Lab-based Pedagogy and the Inverted Classroom:*
  - ◇ Henry M. Walker, "A lab-based approach for introductory computing that emphasizes collaboration", *Proceedings of CSERC '11, Computer Science Education Research Conference*, Open Universiteit, Heerlen, Netherlands, 2011, pp. 21–31 [237].
    A detailed description of lab-based pedagogy, including its application in several settings.
  - ◇ Diane Horton, Michelle Craig, Jennifer Campbell, Paul Gries, and Daniel Zingaro, "Comparing Outcomes in Inverted and Traditional CS1", *ITiCSE '14 Proceedings of the 2014 Conference on Innovation and Technology in Computer Science Education*, 2014, pp. 261–266 [88]
    An interesting analysis of the effectiveness of the inverted or flipped classroom, including the conclusion "while students in the inverted offering do not report increased enjoyment and are no more likely to pass, learning as measured by final exam performance increases significantly" [p. 261].
- *Peer Instruction (PI) and the Use of Clickers:*
  - ◇ Leo Porter, Dennis Bouvier, Quintin Cutts, Scott Grissom, Cynthia Lee, Robert McCartney, Daniel Zingaro, and Beth Simon, "A Multi-institutional Study of Peer Instruction in Introductory Computing", *SIGCSE '16 Proceedings of the 47th ACM Technical Symposium on Computing Science Education*, March 2016, pp. 358–363 [150]
    A description and analysis of peer instruction, based on experiences from seven instructors at different schools.
  - ◇ David Lindquist, Tamara Denning, Michael Kelly, Roshni Malani, William G. Griswold, and Beth Simon, "Exploring the potential of mobile phones for active learning in the classroom", *SIGCSE '07 Proceedings of the 38th SIGCSE Technical Symposium on Computer Science Education*, March 2007, pp. 384-388 [110].
    An account of clickers and their extensions to mobile phones for use in the classroom, particularly within the context of peer instruction.
- *Process-Oriented Guided Inquiry Learning (POGIL):*
  - ◇ The POGIL Project [151]
    This organization, with funding from the National Science Foundation, the Department of Education, the Hach Scientific Foundation, and the Toyota USA Foundation, provides resources and holds workshops for high school and college teachers supporting the POGIL pedagogy.

- ◇ R. S. Moog and J. N. Spencer, Editors, *Process-Oriented Guided Inquiry Learning (POGIL)*, Oxford University Press, 2008 [132]
  The original reference and manual for the POGIL approach.

- *Team-Based Learning (TBL):*
  - ◇ L. K. Michaelsen, A. B. Knight, and D. L. Fink, *Team-Based Learning: A Transformative Use of Small Groups in College Teaching*, Stylus Publishing. Sterling VA, 2004 [126].
    The foundational presentation of team-based learning, now utilized world-wide.
  - ◇ Team-Based Learning Collaborative (TBLC), Home page at https://www.teambasedlearning.org, 2017 [184]
    The home page begins "an organization of educators from around the world who encourage and support the use of Team-Based Learning in all levels of education." The site includes links to consultants, modules, materials, research, news, and conferences and workshops.

Exploration of Social and Ethical Issues

Two references provide a standard for ethical behavior for professionals within the computing field.

- The *ACM Code of Ethics* [3]
  The basic statement of ethical principles for computing professionals within North America.
- Bo Brinkman, Don Gotterbarn, Keith Miller, and Marty J. Wolf, "Making a Positive Impact: Updating the ACM Code of Ethics", *Communications of the ACM*, Volume 59, Number 12, December 2016, pages 7–13 [29].
  ACM's Committee on Professional Ethics work on updating the 1992 ACM Code of Ethics through 2017. This report of the development of computing and the process for reviewing codes of ethics provides a worthwhile perspective when considering how social and ethical issues might be developed and updated.

Several textbooks provide foundational background for a consideration of social and ethical issues related to computing. Two potential textbook candidates follow:

- Sara Baase and Timorhy M. Henry, *A Gift of Fire: Social, Legal, and Ethical Issues for Computing Technology, Fifth Edition*, 2018 [18]
- Michael J. Quinn, *Ethics for the Information Age, Seventh Edition*, 2016 [152].

A few additional references:

- Diane Whitehouse, Penny Duquenoy, Kai K. Kimppa, Oliver K. Burmeister, Don Gotterbarn, David Kreps, and Norberto Patrignani, "Codes of Ethics and Cloud Computing", *SIGCAS Computers and Society Newsletter*, Volume 45, Number 3, September 2015. [285]
  A fascinating study of the evolution of codes of ethics for the computing community internationally, including variations in cultural expectations. The application of ethical principles to cloud computing also highlights differences in perspectives throughout the world and the need to interpret standards within local environments.

- Donald Gotterbarn, *Software Engineering Ethics Research Institute*, http://csciwww.etsu.edu/gotterbarn/ [74].
  A wealth of materials from Don Gotterbarn, including articles he has written over his career, case studies, and links to materials.

- Cathy O'Neil, *Weapons of Math Destruction: How Big Data Increases Inequality and Threatens Democracy*, Penguin Random House, New York, 2016 [144]
  A popular account of the use and misuse of mathematical modeling and big data, and the impact of this work on contemporary society.

Related Physics and Mathematics Resources

Physics Resources: Physics faculty were particularly active in the 1990s in exploring pedagogical technique to encourage student engagement within introductory physics. Beyond Hake's work, cited earlie, two widely-cited sources follow.

- Priscilla W. Laws, "Calculus-based Physics without Lectures", *Physics Today*, December 1991, pp. 24–31 [107]
  A description of the application of discovery learning, with the use of computer tools and active student experiments to help students discover and clarify basic principles of introductory physics.

- L. C. McDermott and E. F. Redish, "Resource letter", *Computer Science Education*, Volume 67, Issue 9, September 1999, pp. 755–767 [123].
  An organized compendium of resources for teaching and learning within the field of physics.

Mathematics Resources: In the 1990s, the Mathematical Association of America published several monographs highlighting forms of active learning.

- Anita Solow, *Learning by Discovery: A Lab Manual for Calculus*, MAA Notes Series #27: Resources for Calculus, 1993 [179].
  Within the Calculus Reform Movement, faculty developed a range of small-group and laboratory-based activities. This monograph supports inquiry-based learning and contains 26 laboratory modules that highlight topics within a standard calculus course.

- Nancy L. Hagelgans, Barbara E. Reynolds, G. Joseph Wimbish, Mazin Shahin, Ed Dubinsky, *Practical Guide to Cooperative Learning in Collegiate Mathematics* MAA Notes Series #37, 1995 [79].

  Ed Dubinsky, Barbara Reynolds, David Mathews, Readings in Cooperative Learning for Undergraduate Mathematics, MAA Notes Series #44, 1997 [61]

  Staff of Project CLUME, MAA, Anthony Thomas, MAA Staff, Anthony Thomas, *Cooperative Learning in Undergraduate Mathematics: Issues that Matter and Strategies that Work*, MAA Notes Series #55, 2001 [180]
  Many mathematicians within the Calculus Reform Movement experimented with using small, collaborative student groups within calculus courses. These monographs present several perspectives on the use of pedagogy involving cooperative learning.

VII

Preparing a Course

IN CHAPTER III OF HER BOOK, "The Mind of the Maker", Dorothy Sayers describes the development of a creative work (e.g., a novel, a painting, a musical composition) as involving three separate, but interconnected, elements: [161]

- A "Creative Idea:" a vision or sense of the overall work. Although this Idea is unconscious and intangible, Ms. Sayers observes that it may motivate an artist to conclude that a specific phrase or visual detail or musical segment does or does not fit within the whole. Overall, the Creative Idea provides guidance, direction, and cohesion for the rest of the creative process.[161]
- A "Creative Energy" or "Activity": The working out of the Idea in time and space. In Ms. Sayers' words, "Everything that is conscious, everything that has to do with form and time, and everything that has to do with process, belongs to the working of the Energy or Activity ... ."[161, p. 38]
- A "Creative Power:" The manifestation of the Idea and Activity as the book that the author and reader read, the painting that the painter and viewer see, and the audible music that the composer/musician and listeners hear. [161]

With the interplay of the Creative Idea and Activity, the creative person works through many elements of the work. The Creative Idea provides a direction, and creative person's Activity provides substance as various pieces are developed and shaped. As the creative process, Ms. Sayers notes,

> A writer may be heard to say: "My book is finished—I have only to write it"; or even "My book is written—I only have to put it on paper." [161, p. 42]

At the end of the process, a reader, viewer, or listener may only experience the finished work (which Ms. Sayers calls, the "Creative Power"), but the end result is successful only when the Idea and Activity have come together.

Although the analogy certainly is not perfect, elements of this narrative seem to fit my experience with course planning and teaching.

- Initially, a faculty member may have a vision for a course, perhaps described at a very high level as a course title or as a course description. (analogous to the Creative Idea)
- Course preparation, before the course starts, requires an instructor to make many decisions—both at a conceptual level and at a detailed level. (analogous to the Creative Energy or Activity)

- Once a course starts, students experience prepared materials (e.g., a syllabus, day-by-day schedule, a statement of textbook and/or other resources, an outline of course work, and a plan for assessment. (analogous to the Creative Power)

As this introductory narrative suggests, I think about teaching a course as a long-term process that ends in the classroom. A great deal of my work for a course begins much earlier, when I first must ponder a vision or high-level sense for the nature of the course. From a hazy beginning, I can start to articulate learning objectives, possibilities for content and scheduling, an inventory of appropriate textbooks or software or other resources, a syllabus, and a day-by-day schedule. Pragmatically, some elements of planning may reasonably be decided while the course is underway, but many elements must be determined before the course can start. With the time and effort required for course preparation, one might consider paraphrasing Ms. Sayer's comments by noting "My course is finished—I just need to teach it [in the classroom]."

This part focuses upon an instructor's work from an initial idea or vision through the planning process. After an initial chapter (Chapter 32) on getting started, the Part focuses on specific elements of before-the-semester class preparation: a day-to-day schedule, reading(s) and class work, the syllabus, and textbooks and other resources, Often, such class preparation can be extremely time consuming, and traditionally all of the work is done by an individual instructor. The last chapter of this Part (Chapter 37), however, outlines another approach for class preparation through a collaborative, student-faculty partnership.

CHAPTER 32

Planning and organizing a course for the first time

This column first appeared in ACM *Inroads*, Volume 7, Number 4, November 2016, pages 12–17[264] Subsequent reflections on this subject appear later in this chapter.

32.1 ORIGINAL COLUMN

THIS COLUMN IS THE SECOND in a series designed to help new and inexperienced faculty get started in their teaching. The *Classroom Vignettes* columns for both September 2016 (Chapter 25 in this book) and March 2017 (Chapter 26 in this book) list do's and don'ts related to the classroom environment, including

- student-faculty interactions and general classroom parameters (September 2016), and
- combating implicit bias, details of slides, board notes, and similar day-to-day activities (March 2017).

This *Curricular Syncopations* column presents practical tips and techniques for planning and organizing a course for the first time. As with other columns in this series, the comments presented here are designed as a starting point rather than a definitive or complete statement regarding effective teaching. Although elements of effective course planning may vary widely among faculty, variations of the suggestions here are used by a range of faculty who teach in varying settings.

Getting Started: Course Goals and Learning Objectives

In considering a course, natural starting questions include, "What topics should this course cover?" and "What should students be able to do when the course finishes?" More generally, identifying course goals and student learning objectives represents a major first step in course planning. Typically, I draw ideas from multiple sources and possibly review draft statements from others. Then, to internalize my understanding of a course, I write my own statements—in my own words. If existing materials already are well formulated, my own work may involve rewriting or editing for my own benefit—but writing myself helps me understand what the course should cover and where it may fit within the broader curriculum.

When I am done, I need to be clear about both the list of topics to be covered and the level of mastery expected. To be reasonably complete, such a narrative should include statements of

- technical skills,
- expectations regarding skills for lifelong learning, such as locating relevant manuals, reading documentation, locating and analyzing articles, etc.,
- development of communication skills, and
- development of ability to work collaboratively.

For many courses, several resources may be available to help in the creation of a partial list of goals and objectives

- At a high level, a course description (e.g., in a college catalog or on the Web) may identify general topics and objectives. Although published course descriptions often are general (so courses can evolve over time with minimal paperwork and administrative overhead), at least they may provide a start.
- If the course is a prerequisite for other courses, then one can ask what students are supposed to know for those later courses. Starting competencies for later courses may provide an initial draft of the learning outcomes for the prerequisite course.
    - ◇ Sometimes, written descriptions of later courses may describe expected background.
    - ◇ Often, at least in my experience, actual expectations of instructors of later courses may differ from published statements, so chatting with instructors may be particularly helpful in determining desired goals and outcomes for the prerequisite course.
- If a course has developed from a grant proposal, the proposal itself likely contains relevant goals and objectives.
- If a course is designed to meet criteria set by employers or accrediting bodies (e.g., ABET), then goals and learning objectives may be drafted, in part, from those professional criteria.
- Although such sources may help, schools seem inconsistent in maintaining lists of goals and learning objectives for their courses.
    - ◇ If the course has been offered previously, and if previous documentation is available, an instructor already may have a good draft of topics and outcomes.
    - ◇ If the course was recently proposed to a curriculum committee or department chair, the proposal may provide a fine starting point.
    - ◇ Unfortunately, in my experience, some schools seem casual in their course-approval process, and documentation may be spotty (or worse)!

When a course has not been taught previously or when few narratives seem available regarding expectations, an instructor may need to be creative and resourceful in getting started.

- If a similar course is offered elsewhere, one might contact instructors at other schools.
- If a textbook covers much of the material envisioned for the course, then one might contact a publisher, author, or instructor that uses the book elsewhere.
- When attending a conference, one might ask attendees about their thoughts.
- *CS2013* from ACM/IEEE-CS [10] identifies about 80 "course exemplars"—innovative and successful courses that have been offered at various schools.
- *CS2013* also lists a large number of topics for consideration. (However, since this litany of topics is organized by knowledge areas rather than courses, this listing may or may not be particularly useful for initial course planning.)
- Assuming one is a member of ACM SIGCSE or other computing-education organization, one might send email to a listserv for ideas.

In what follows, the developed list of goals and learning objectives provides a target for the course. In future steps, we may find that the list is overly ambitious; the course may contain more material and/or articulate higher expectations than can fit into a single course. Thus, some flexibility may be needed as course planning proceeds.

Initial Deliverables

Overall, a well developed list of goals and outcomes represents an essential first step for course planning, requiring substantial time and effort. For many instructors—particularly new faculty, the next steps can seem equally intimidating and time consuming. Before going on, therefore, it may be helpful to determine some initial goals for the planning process; what must be done before a course starts, and what can/should be left until later.

Throughout my teaching career, I have found the development of three primary documents to be essential before the start of a semester:

- *Reading list:* Book stores and students need to know what, if any, textbooks and/or articles will be required. Typically, such information must be available a month or two before the semester begins, so stores have time for ordering materials and students can be prepared for the first class. If even more time is available, an author might talk to a publisher about putting together a customized textbook, with chapters selected from one or more existing books. (In my experience, publishers are quite willing to consider such tailored volumes for high-enrollment courses, but perhaps less interested in courses with relatively few students.)
- *Syllabus:* Students need to know the context for a course, any assumed background, course goals, intended student learning outcomes, course format, and expectations for course work.
- *Day-by-day schedule:* During a semester, I want to proceed at a measured and consistent pace. All material should be covered without being rushed. I have heard of courses in which an instructor assigns multiple books or several hundred pages of reading in the last week in an attempt to get through the expected material, and I want to avoid such problems.

In preparing these materials, an instructor likely will need to consider what resources students will consult, mechanisms for student feedback and assessment, and class format and pedagogy, as discussed later in this column. Such considerations may require substantial initial thought and effort. However, I try to separate this initial planning from the day-to-day activities that are required when the course is underway. In particular, in planning a course, I spend little time on such on-going details as

- Development of class notes (e.g., slides, lecture/discussion notes),
- Identification of specific day-to-day videos and animations that may support the presentation and learning of specific topics,
- Writing final versions of exercises or assignments, and
- Division of students into small or large groups.

In each of these areas, details likely will evolve as the semester proceeds. In class, for example, I may need to emphasize some topics, clarify misconceptions, address common mistakes, etc. Further, as a practical matter, when I prepare my class notes more than a few days ahead of time, I forget what I wrote, and I end up reading my notes to the class. The result can be remarkably dull and unmotivated class sessions!

The remainder of this column reviews these three main deliverables in turn and comments upon the planning required for each, and also provides some initial thoughts regarding eventual assessment of the course. Although discussion treats the reading list, syllabus, day-by-day schedule, and course assessment separately, decisions made for one may impact the others; development of one deliverable may require adjustments in the others—likely in an iterative process.

Reading list

A resource list informs students what materials (e.g., books, articles, notes, software, computing equipment) they will need for success in the course.

In many courses, whether utilizing a traditional format or a workshop (or flipped) style, textbooks provide foundational background. The course works through all or part of a book as the semester progresses. In practice, use of a textbook also may simplify planning: for example, developing a day-by-day schedule largely involves determining how much time to spend on each chapter and section. A reviewer of a draft of this column noted that representatives from textbook publishers can be helpful in identifying existing books or brainstorming the combination of selected elements from one or more books to yield a customized textbook for a course. (Some popular textbooks come with numerous ancillary materials, including quizzes, workbooks, etc. Although these resources may work very well in some settings, my personal experience has not found these particularly useful.) Some additional thoughts on the role of textbooks may be found in [239].

Even when textbooks are used, however, additional material may be needed regarding the use of a local computing environment, operating system, or software package. In such cases, a reading list should identify relevant resources, such as reference manuals, on-line documentation, or local notes.

Also, modern pedagogy may ask students to read material, view videos, work through animations, listen to lectures, or otherwise engage material before a class session begins. Some sources to consider include catalogs of existing videos, searches for YouTube videos, and reviews of recent SIGCSE proceedings and *Inroads* articles for relevant animation software.

Yet another factor influencing a reading list may involve course goals regarding student skills for lifelong learning. Beginning students may need considerable support and guidance; full textbooks, starting documents, and complete reading lists may be particularly helpful. However, eventually, students will need to learn how to work with manuals, find materials on their own, and analyze the correctness of on-line sources. When course objectives include the development of lifelong learning skills, a reading list might state explicitly what resources are required, what types of sources are encouraged, what materials are to be avoided, and what responsibilities students assume when searching for materials.

In considering required materials, *a new instructor is warned that the development of course materials is usually time consuming.* For example, in a lab-based introductory course I started in 2011, I worked with four assistants for 10 weeks (about 500 hours total) to organize the course into modules, supplement a textbook with system-based readings, write labs for most class sessions, and develop appropriate exercises and assignments. Although this may seem extreme, these statistics highlight that a new instructor, teaching a new course for the first time, likely will not have time to develop materials for an entire course before its first offering.

Similarly, development of videos, audio tracks, and lab manuals can consume vast amounts of time and effort.

Personally, when I first teach a course, I tend to select existing materials (e.g., a textbook), and I utilize materials of colleagues (with attribution, of course). Although I [usually] know the material, I may or may not anticipate student difficulties, and a local lab environment may exhibit unexpected idiosyncracies. During an initial course offering, I may develop some limited resources, but I largely rely upon existing materials—the reading list mostly identifies books, articles, manuals, colleague's notes, etc. that have been developed and tested elsewhere.

As I teach a course multiple times, I tend to write some of my own materials, identify videos or animation software, and rely progressively less on the materials of others. Initially, however, my reading list usually includes established resources that I largely can utilize "as is".

Syllabus

A syllabus serves as a contract between students and the instructor. One part discusses the context or background for the course, likely including prerequisites or other needed experience. Another part typically includes goals and expected student learning outcomes—what skills and insights a student can expect to acquire by the end of the course. Yet another part of the syllabus typically covers rules and procedures of the course, the department, and the school, such as policies regarding academic honesty, support for students with disabilities, class attendance, etc. Such elements might be considered largely separate from class format and pedagogy—perhaps covering a half of a typical syllabus.

The other half of a syllabus typically describes assessment and class format.

Student Assessment

Reasonably early in the planning process, an instructor must decide how student assessment will be handled: how will students receive feedback (formative assessment) and how will students be graded (summative assessment)? Of course, any type of assessment should be connected with course goals: To what extent are students mastering the material identified in course objectives.

Some types of activities might include

- writing programs,
- writing papers,
- quizzes or tests: in-class, take-home, or oral,
- labs,
- class presentations by students (individually or in groups),
- submission of questions for class discussion,
- journal,
- class participation, and
- projects.

Each of these vehicles has its own advantages and each can provide useful insights into student learning and mastery. However, each also requires both time and effort. For example, in non-majors' courses of 20 students, I may assign 2-4 page papers to support the development of writing skills as well as in-depth student engagement of a topic. Over a semester, feedback on papers can facilitate learning and make a difference in helping students write effectively, but I spend about 1 hour per student per paper assigned—clearly not an assessment vehicle that scales!

More generally, consideration of options may include personnel available for grading (e. g., teaching assistants), automated tools, enrollment size, on-line resources for grading and recording results. New faculty likely should talk to experienced teachers in their departments regarding what approaches seem effective, within time constraints, in their local environments.

Once determined, assessment options should be included on a syllabus: what artifacts will be submitted, when each is due, what format(s) are required, etc. Further, how will results from various activities be combined and weighted in a final grade? From a syllabus, students should be able to start planning their work schedule for the semester—how much time can be devoted to each course, employment, etc.?

Course Format

Within the classroom, a basic question is "what will students be doing and what will an instructor do?" Modern pedagogy promotes the active involvement of students during a class, and preparation for a class should include who will do what when.

- Lecture, based on a textbook, often is relatively easy for a beginning teacher, as it requires relatively little preparation of new materials, such as labs, clicker questions, small-group assignments, exercises, etc. Although modest student involvement may arise by encouraging student questions and/or calling on students individually, lecture generally does not promote extensive student involvement. (For myself, I tend to rely upon elements of lecture the first time I teach a course, since I have few materials available for individuals, pairs, or small groups. These materials, however, develop each time I teach a course, so I usually give fewer and fewer lectures over successive semesters.)

- Peer instruction, typically using clicker questions, encourages students to discuss material and to respond to focused questions throughout a lecture-based course, allowing an instructor to guide students based upon their responses and their interactions with those around them. Porter and Simon [149] provide additional background regarding some possible uses of peer instruction and clickers.

- Recitation sections provide a mechanism to divide a large class into groups, allowing students to ask questions and obtain direct feedback. When used, an instructor will need to consider what will be discussed in each section and if/how material will be presented.

- Small group activities can allow students to attack problems or discuss questions within a context where everyone is strongly encouraged to contribute. Note taking and group reporting sometimes are useful to gauge progress—particularly if responsibilities are rotated through a group. When used, an instructor will need to decide the group sizes, how groups are constituted (e.g., students choose or the instructor assigns), and how often groups are changed.

- Labs provide opportunities for direct student practice, either individually, in pairs, or in small groups. When used, decisions are needed regarding what equipment should be used (e.g., individual machines owned by the school or college-owned computers in labs). Whatever approach is used, materials are required to ask relevant questions and guide students through stages of learning.

As with approaches for assessment, new faculty may wish to consult experienced teachers about common approaches and possible resources in a local environment.

Once decided, the syllabus should document the main features of the class format, so students will have appropriate expectations, understand what preparation they will need each day, and plan to participate appropriately in class.

My December 2005 *Inroads* column [219, Chapter 35 in this book] provides a more detailed discussion of what might be included in a syllabus. One example of a reasonably complete syllabus may be found at

http://www.cs.grinnell.edu/~walker/courses/161.fa16/syllabus.php

Day-by-Day Schedule

From my very first teaching experience (as a graduate student), I have prepared a draft day-by-day schedule for each course, identifying what material will be covered each day. I strive to proceed through material at a reasonably constant pace through a semester, but I also must understand options as the semester evolves. For example, when students seem to require more time, I need to know what low-priority or optional topics I might omit. Overall, a day-by-day schedule allows me to make decisions as a semester progresses without losing sight of the big picture and overall course goals.

Even with years of experience, I find development of a day-by-schedule is a [long] iterative process. For each topic (e.g., technical components, sections from a book, exercises in a lab manual, etc.), I estimate how long each topic might require, and collect the pieces in an easily-modified format (e.g., a spreadsheet, collection of post-in notes on a whiteboard, lines in a text file, etc.) Within the schedule, I include tests, in-class discussions, etc., following the pedagogy determined and the corresponding class format. Following the lead of my colleague, Jerod Weinman, I may include some days for a "pause for breath" to allow some flexibility. Almost always, my first draft schedule extends weeks (occasionally months) beyond the designated end of semester. Since a school rarely seems willing to change its academic calendar, I then consider what topics might be combined, streamlined, or cut. Also, I explore whether a rearrangement of topics might allow greater efficiency. Eventually (often over multiple days), a day-by-day schedule emerges that fits within the time allotted and also covers the mandated goals and learning objectives. (Of course, if juggling a schedule continues to require too much time, then discussions with the department might be appropriate regarding whether the course is overloaded and how goals might be rethought.)

Once prepared, I distribute the draft-day-by-day schedule to students on the first day of classes, *with a strong statement that the schedule is tentative and likely to be adjusted from time to time.* Although I may change topic and reading details through the semester distribution of the schedule allows students to start planning their semester. However, in making schedule adjustments, I consider test dates to be fixed—again so that students can plan.

As an illustration, a current example of a day-by-day schedule may be found at http://www.cs.grinnell.edu/raisebox -4pt~walker/courses/161.fa16/schedule.php. Additional thoughts regarding a day-to-day syllabus may be found in [242].

Course Assessment

Yet another component of planning a course should involve the extent to which the course, when offered, can be considered successful. More importantly, course assessment can identify ways to improve weaknesses and to make good courses even better.

Opportunities for course assessment vary greatly by institution. Some schools have offices, staffed by professionals, to plan and implement assessment instruments and mechanisms; other schools provide little support, and individual faculty are left to handle their own assessment activities. Whatever the local environment, initial planning can begin the process of course assessment and improvement.

Course assessment is complicated, in that some outcomes may be short-term in nature, but others may require a long-term analysis. At a basic level, course assessment includes a determination of whether students are learning what they are supposed to learn and whether they are developing appropriately. Some observations follow.

Technical Content:

When course learning objectives include specific technical skills and knowledge, assessment may include tests, projects, or other formal instruments.

- results on test(s) or other vehicles can help evaluate student achievement regarding each learning objective.
- if this course represents a new approach for a previously offered course, one can compare student responses to identical problems on [final] exams or projects.
- If one course serves as preparation for a later course, one can analyze student performance in the later course— are students appropriately prepared?

Some outcomes may require a long-term view:

- Student progress for some high-level curricular goals, such as ability to read technical reports or student preparation for lifelong learning, might be addressed through exit interviews at graduation or through alumni surveys.
- Student preparation for careers might be determined through feedback from employers or from alumni interviews.

Other Considerations:

The value of some common indicators may depend upon the local environment and context.

- *Failure rates*: One session at ITiCSE 2016 focused upon failure rates in introductory courses, reporting rates as high as 50% or even 90%. At other schools, failure rates of 5% are viewed as shockingly high. Whatever the distribution of low grades, analysis is required to discover student preparation, comparison of failure in other courses, and other contextual matters. (A failure rate of 50% might be wonderful in an introductory course, if other sections of the course have a 75% failure rate—but the grading for the course with the lower rates also might result from simple inflation of all grades?)

- *Enrollment trends*: If a course is required (e.g., for a major), then students must take it, and enrollment trends may not indicate student sentiment regarding the course's success. If the course is an elective or one of several options, however, then strong enrollments might suggest some types of success.

Students themselves can provide useful feedback on their experiences with a course, but students will not know whether content is appropriate or whether other approaches are better or worse. Students can provide insight regarding the pace and rigor of a course, their understanding of the work load, the level of supporting resources they experienced, and the amount they believe they learned.

- When I first teach a course, I often distribute a mid-term evaluation after a month or two, to determine student perceptions of strengths and weaknesses and to help me refine the next part of the course to address identified shortcomings.

- Many schools require end-of-course evaluations of courses. At many schools, these may provide only high-level feedback (e.g., was the reading helpful), and I often distribute a much more detailed survey for details of labs, readings, assignments, etc.

Still another form of feedback can come from other CS faculty attending a course, outside observers (either faculty or members of a school's assessment office), faculty from other departments who are taking the course (I have had faculty from anthropology, physics, French, mathematics, etc. take my courses), or other outside evaluators. At some schools, faculty sit in on each other's courses to provide feedback and make suggestions—but unfortunately, such activities can be time consuming.

Whatever options might be available within a local environment, initial planning should include consideration regarding how the success of a course can be assessed and what mechanisms might be available to help improve it through several course offerings.

Acknowledgments

Thanks to Titus Klinge, Iowa State University and Grinnell College, and to Charlie Curtsinger, Grinnell College, for their feedback on an early draft of this column. Additional thanks to Marge Coahran for her suggestions and insights. Reviewers of a draft of this column also contributed several comments.

Summary

For every college-level course I have taught since I was a graduate student, I have distributed three documents on or before the first day of class: a reading list (often transmitted to a local bookstore or posted on the Web), a syllabus, and a day-by-day schedule. Together, these materials help students obtain needed resources and set expectations for the semester.

Behind the scenes each of these documents requires thinking through the course at a high level, as outlined in Figure 32.1. The reading list requires identification of resources, with consideration of what materials will serve as the foundation supporting the course. The syllabus requires consideration of class format and pedagogy—what will the instructor and students do to encourage learning. The day-by-day schedule provides a vehicle to put the pieces together within a logical structure that fits within a semester.

FIGURE 32.1 Some components of course planning

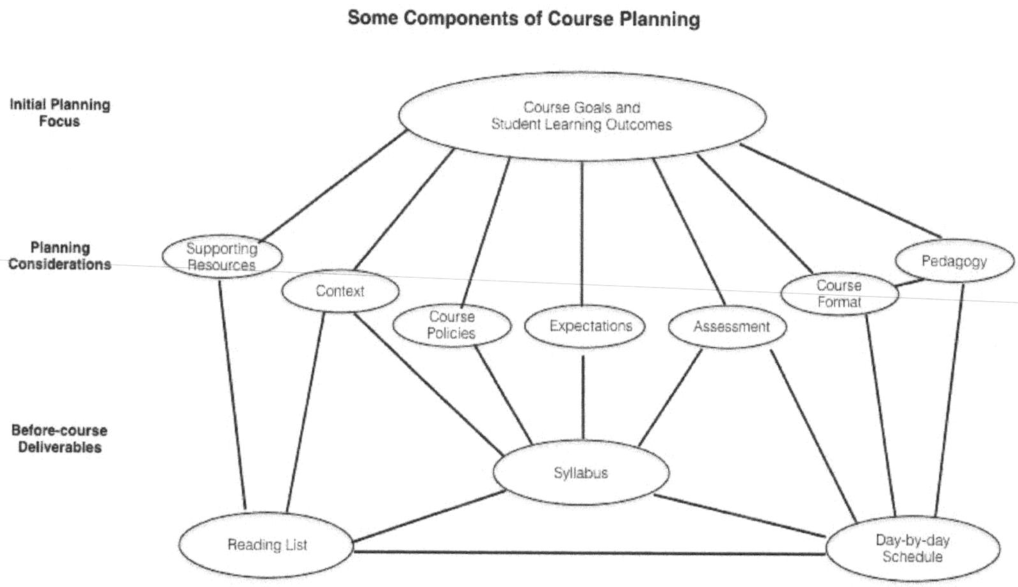

With these pieces in place, much of my planning is completed, and I can turn to the daily routine of teaching, assured that the pieces will fit reasonably within a coherent framework that supports the course's goals and objectives.

As a side effect, distribution of these items provides an aura of preparedness—students think I know what I am doing, adding to my credibility as the instructor for the course. I may have questions, but students usually believe the course is coherent and well constructed!

32.2 SUBSEQUENT REFLECTIONS

The opening remarks for this part (Part VII) discussed three interconnected elements of the creative process. For writers, artists, and musicians, Dorothy Sayers identified these elements as the "Creative Idea", "Creative Energy" or "Activity", and "Creative Power". [161, pp. 38-42]. When considered as part of the teaching process, Part VII reworked these elements to yield roughly "Course Vision", "Course Preparation", and the "In-class/student experience."

From this perspective, the original column, *Planning and organizing a course for the first time*, highlights the "Course Preparation" component of the course/teaching process. Three vital products of this preparation include a reading list, a syllabus, and a day-to-day schedule—all described in the original article and expanded in later chapters of this

Part VII. Before a course officially begins, "Course Vision" represents a complementary component.

In some respects, a Course Vision may include a general notion of a subject or course title (e.g., operating systems, analysis of algorithms, introduction to artificial intelligence, networking), together with some sense of the level of course (including such elements as the student background, course rigor, work load). Applying Dorothy Sayers analysis of the creative process, the success of a course depends, at least partially, on the consistent interplay of the vision with aspects of the course preparation (and subsequent activities within the classroom).

Often, faculty work to develop this synergy of vision and preparation—either consciously or intuitively. For example, an operating systems course at the senior (fourth-year student) level likely assumes more technical background, more rigor, a greater work load, and higher expectations for student capabilities than an operative systems course for second-year students. In my experience during external reviews, when assumptions regarding student background, capabilities, and motivations are consistent with both the course vision and course preparation, then many students can meet or exceed the learning objectives, and students typically react well.

In contrast, in a few of the my external reviews, I have encountered some faculty who seem to have an unrealistic vision. For example, an operating systems course with only a pre-requisite of introductory computer science might require substantially more background and sophistication than can be reasonably expected of second-year students. In this context, course preparation may seem to cover topics at one level, but the course vision may promote a substantially higher-level. Of course, a few top students might thrive in this environment, but the course may be quite overwhelming for most students—even very solid students with considerable long-term potential.

In my experience, a few of these courses within a curriculum can have a substantial negative impact on morale, student interest, enrollments, and motivations. When students believe a course is unreasonable, they may work to get through (if the course is a requirement), but they may have little incentive to dig into the material and explore new ideas.

Unfortunately, avoiding disconnects between vision and detailed preparation can be a challenge, because much of an intuitive vision may be vague and uncharted. However, a few questions may help instructors clarify their vision:

- What expectations of abilities are reasonable for the target student population?
- To what extent are the course goals and learning objectives consistent with the actual abilities of students expected in the class?
- To what extent is the pace set by the day-by-day schedule likely to be manageable for students with the expected background and ability?
- Is the background required for readings consistent with the background and experience of the targeted students?
- How much insight, time, and problem solving ability are required for the assignments and projects?

Of course, each course should help students progress in background, insight, problem-solving ability, etc., and students may be pushed as they development. By considering questions such as these, instructors might ponder whether the course they are preparing is consistent with their vision.

CHAPTER 33

Course planning: the day-to-day schedule

This column first appeared in ACM *Inroads*, Volume 3, Number 3, September 2012, pages 22–24[242] Subsequent reflections on this subject appear later in this chapter.

33.1 ORIGINAL COLUMN

WHEN I WAS IN COLLEGE, friends at various schools sometimes complained that scheduling in a course had gone badly. The instructor had planned to cover eight books, but had spent the first ten weeks on just two books. Thus, the instructor was trying to squeeze the six remaining books into the last four weeks.

Partially in reaction, I have prepared tentative, day-by-day class schedules for every college-level course I have taught. Although my initial motivation was to avoid problems of bad time management, I have learned the benefits of extensive and detailed course planning extend well beyond simple matters of pacing. This column identifies benefits and motivations for developing a detailed course schedule, potential pitfalls, and some practical approaches for constructing a schedule.

Motivations and Benefits

Initially, I developed day-by-day schedules as a mechanism to monitor the pace and sequencing of material in a course. Courses must cover specified material to prepare students for subsequent courses. Often topics have varying priorities: some high-priority topics must be covered, but other topics may be optional. Also, it may be important to provide examples, but the selection of examples is open. (If an instructor is not careful, extensive coverage of numerous interesting examples on one topic can interfere with the discussion of other topics. For example, five examples of recursion in CS1 may be helpful but fifty examples may prevent coverage of other essential topics.) Of course, a detailed schedule also helps clarify sequencing of topics. When topics build upon each other, a tentative day-by-day schedule allows an instructor to consider what background students will know when introducing each new topic.

A tentative, day-by-day schedule also can help in the selection of textbooks. For example, when I first came to Grinnell, my colleagues (particularly Charles Jepsen) would identify two to five possible textbooks for a multi-section course such as Calculus I and II; Charles would prepare possible day-by-day schedules for each book. Often, these schedules would show that some textbooks could work well within a course's time constraints, while other textbooks would not. Typically, potential daily schedules helped reduce an initial list of five books to just one or two candidates.

Over the years, I have added much detail beyond pace and sequencing to my day-by-day schedules.

- A schedule can show readings for each day. (I often ask students to submit "discussion questions" on each reading electronically before class, so they will be prepared and I know what to highlight in class.)
- The daily schedule can show due dates for assignments, projects, or other work. In a course that meets four days a week, I may plan a mix of labs and individual programming assignments to be due every third class day. The day-by-day schedule helps space these materials evenly throughout the semester.
- If I expect to be at a conference or on a consulting trip during the semester, I may schedule a test for a day I will be away. (My colleagues have been wonderful at covering for me on such occasions). A day-by-day schedule helps to work the test into the overall flow of a course.
- When considering due dates, I try to be sensitive to culturally-based events such as religious holidays. Planning ahead helps me avoid such dates for tests or other activities.
- Variety in a course may be encouraged with a range of planned activities. For example, several of my colleagues (e.g., Janet Davis and Jerod Weinman) specify Mondays as "news" days in which students present recent news articles they have encountered. Similarly, I might schedule group activities for Tuesdays, or "discussion questions" for Wednesdays. Each activity can add a valuable dimension to a course, but a schedule must include time for each activity.

After I have made a day-by-day schedule to organize a course, I distribute the schedule to students on the first day of class. Of course, the initial schedule is tentative, and I tell students that adjustments are likely through a semester. (Typically, an occasional topic may shift by a half day—a full day shift is possible, but rare. More often, I may drop an optional topic to keep the overall course on schedule. Dates of assignments and tests do not change, although the content for these activities may vary.) Students regularly remark that a day-by-day class schedule helps them in planning. Students know when assignments, projects, etc. will be due; and they are able to balance course work with other commitments.

Of course, distributing a tentative, day-by-day schedule at the beginning of a course also reassures students that an instructor is on top of the material and has the course well prepared.

Potential Pitfalls

Although preliminary course planning can be very helpful, each offering of a course is different—different students have different backgrounds, abilities, and interests. Some topics may require more time one semester than another; some students may ask more or deeper questions; technical elements of a lab environment may make work particularly efficiently or may add unforeseen difficulties. Altogether, a tentative, day-by-day schedule aids planning, but an instructor also must be flexible.

Several techniques can help an instructor adjust to variations from one course offering to the next.

- Some topics can be identified as "essential" or "core"—these must be covered.
- Some topics can be specified "as time permits" or "optional" or "for extra credit". A published schedule might or might not label these topics explicitly, but the instructor will know which topics can be abbreviated or dropped as needed.
- Some days may be labeled as "catch up", "questions", "review" or equivalent; my colleague, John Stone, instituted scheduled days to "Pause for Breath", and my colleagues, Janet Davis, Samuel Rebelsky, and Jerod Weinman, have adopted versions of this approach.

In practice, I consider a tentative, day-to-day schedule as a baseline reference, not a mandated rule. If interactions with students indicate more time is needed on a topic, I review the schedule to determine how to adjust upcoming plans or which topic to drop. If material moves faster than expected, the schedule suggests what might come next.

Some Approaches for Constructing a Schedule

In practice, development of a tentative, day-by-day schedule involves an iterative process. When I start planning a new course, I typically go through a textbook or list of topics, assigning time for each book section or topic. I consider what might be involved with each part of the material,and I estimate an amount of time. Typically, my initial schedule provides a lovely day-by-day plan that may extend to 20+ weeks—but Grinnell's semester only has 14 weeks of classes. The iterative process then examines each part of each topic and class: what can be combined, what might be omitted, could reordering make coverage more efficient, do I really need all five examples, etc.?

Throughout this process, I make an adjustment and examine the schedule again. While this iterative process may be accomplished in many ways, here are several well-tested approaches.

- Write topics in a text file, with a blank line between each day's material. Use a program to read the lines to yield a weekly table of days and topics. (Aside: this can be a great CS2 exercise, if you do not already have access to such a program.) After revising the material for a day, rerun the table generator to view the revised program. An example of such a table for a course on *Algorithms and Objected Oriented Design* may be found at
    - ◇ http://www.cs.grinnell.edu/~walker/courses/207.sp12/sched.pdf (TeX format) and
    - ◇ http://www.cs.grinnell.edu/~walker/courses/207.sp12 /lab-index.shtml (html format).
- Use a spreadsheet to identify topics, with one class per row. The spreadsheet tracks successive days; counting days gives a weekly or monthly schedule; and insertion and deletion of rows allows easy adjustment of the schedule.
- Use post-it notes to record each course topic, and make a large grid on a whiteboard or poster for each week. Put the notes on the grid to obtain a tentative course outline; move the notes from one position to another for course revision. Figure 33.1 shows this approach, as developed by colleague, Janet Davis, for two courses, CSC 105, *The Digital Age* (left) and CSC 364, *Networks* (right). The colors of the post-it notes identify types of activities or themes (e.g., the right column for Networks uses green to indicate lab topics).

FIGURE 33.1 Post-it Notes for Course Planning (with permission from Janet Davis)

- Divide the course into modules (e.g., one to two week blocks), giving a high-level organization of topics and materials. Then work within each module to obtain a daily schedule. An example of the module approach may be found at http://www.cs.grinnell.edu/~walker/courses/161.sp12/ semester-outline.shtml .

Over the years, I have found course planning forms the foundation for my teaching. Development of a tentative, day-by-day schedule takes significant time—particularly for courses I have not taught previously, but this initial work makes a dramatic difference during a semester. I know what to focus on during the stresses of a semester, and I have reasonable confidence each course will cover appropriate material in a timely way.

Acknowledgment

Many thanks to Marge Coahran for her suggestions on an early draft of this column!

33.2 SUBSEQUENT REFLECTIONS

I continue to utilize the day-by-day schedule as a central component of course preparation. For example, when I prepared recently to teach a new course, the day-by-day schedule helped me decide what topics might fit in the semester, how much time to spend on each topic and in what order, when assignments and tests might be scheduled, etc. Throughout, three elements seemed key:

- The schedule must highlight the overall organization of the course, so overall time allocation for major components of the course is clear.
- The schedule must contain reasonable detail for each day, so a review can identify the amount of material being covered and whether that level of content seems reasonable for the student backgrounds.
- The schedule must be flexible, as I work to arrange class sessions, add or remove topics, move material from one class session to another, determine due dates for assignments, etc.

In planning, the day-by-day schedule is central in putting many course elements together. Further, after the course starts, the day-by-day schedule remains valuable, when it becomes apparent that students may need more time than expected for one topic or less time for another topic.

The Low-Tech Post-it Note Solution

In considering mechanics to develop the day-to-day schedule, the post-it note approach (Figure 33.1) seems a particularly effective low-tech approach. With topics, examples, and other details on separate post-its, a schedule can show much detail. Further, adjusting or moving topics is easy. In addition, if the post-it notes are kept on a board or wall throughout a semester, then small changes in the schedule can be planned and recorded while a course is in progress. Of course, some logistics of the post-it approach require some work:

- Care and movement of many notes may be required, if work for an entire day is added or removed.
- Once a post-it-based schedule is determined, the schedule must be reformatted for distribution to students in the class.

Altogether, using post-it notes seems a lovely low-tech mechanism for developing and maintaining a day-by-day schedule.

A High-Tech Database/Web Approach

In the past few years, I have been experimenting with a database approach, coupled with Web interfaces, for developing and maintaining a day-by-schedule. This approach requires software to record, modify, and display the schedule; and course elements must be properly recorded in a database. With preliminaries done, several features become easily available:

- Multiple programs provide several views of the schedule. For example, one view might show the entire day-by-schedule, another details for a specific day, and another a listing of all assignment deadlines.
- Web-based interfaces allow students to view all elements of the class from any Internet-connected browser world wide.
- With a common database, the update of one element of the schedule will be visible in all views. All details of the schedule will be consistent.
- Since display software draws from the database, the construction of an up-to-date schedule requires no additional work. (Unlike the post-it approach, there is no need to transcribe the schedule from the daily post-in notes to a separate handout for students.)

On the downside, the database approach has at least two negative aspects:

- This database approach requires data entry of course elements (just as the post-it approach requires writing many separate post-it note pages).
- This approach requires the development of programs to manipulate and display database information. (Since this software development is not particularly difficult but requires care, this application might be a lovely exercise for a software project class!)

Currently, I continue to experiment with details of this database/Web interface approach that include these features:

- Separate programs display an overall day-by-day schedule (Figure 33.2), the details of resources and in-class activities for any specific day (Figure 33.3), a list of all assignment deadlines, a description of the overall structure of a course, and other perspectives (still being identified).
- A left-navigation section on both the overall day-by-schedule (Figure 33.2) and each detail for a specific day (Figure 33.3) gives links to class elements for the previous day, the current day, and the next class meeting, as well as the next deadlines for assignments or tests.
- The day-by-day schedule (Figure 33.2) presents a calendar-like view showing the days of the week when the course meets. For each day, this schedule provides a brief title of the work to be covered, together with a link to the details for that day.
- The view for any specific day (Figure 33.3) provides links for any examples or readings, together with a link (titled "details") for a lab or other in-class activity planned for that day.
- Behind the scenes, administrative programs allow an instructor to identify the start date of a semester, dates for breaks or vacations, specific testing dates, etc. Additional capabilities allow an instructor to allocate the number or days for each logical session and to designate if one session may be combined with another on a given day. With these parameters set, a new day-by-day schedule is created with a simple click of a button (combined with a relevant username and password).

FIGURE 33.2 A Database-driven Day-by-day Schedule and Interface

CSC 161

Grinnell College

CSC 161: Imperative Problem Solving and Data Structures

Fall 2016

Course Home Resources Course Details: Syllabus, Schedule, Deadlines, Topic organization MyroC Documentation Project Scope/ Acknowledgments

Tue, Sep 13, 2016

Previous Activity on Class Schedule: Mon, Sep 12, 2016

- module 001: Conditionals with the Scribbler 2

Today's Work:

- module 001: Loops and Scribbler 2 Motion

Next Activity on Class Schedule: Wed, Sep 14, 2016

- module 001: Simulations and More Prog. Mgmt.: Functions, Assertions

Close Deadlines:

- Today: none
- Next deadline(s):
  - Fri, Sep 16, 2016: Supplemental Problem 1

Course Schedule

Symbols at the start of segment and session names reflect the status of these materials.

Monday	Tuesday	Wednesday	Friday
Aug 22 *New Student Days*	Aug 23 *New Student Days*	Aug 24 *College Registration*	Aug 26 module 000: Course Overview module 000: Linux/Mac OS X Basics
Aug 29 module 000: Linux/Mac OS X Basics module 000: C Basics	Aug 30 module 000: C Basics	Aug 31 module 000: More C and eSpeak and Makefiles	Sep 02 module 000: MyroC, the Scribbler 2, and Basic Input
Sep 05 module 000: Intro. to Prog. Org. module 000: Song Program Project **(required)**	Sep 06 module 000: Song Program Project **(required)**	Sep 07 module 001: Elements of C module 001: Types and Variables	Sep 09 module 001: Types and Variables module 001: Conditionals with the Scribbler 2
Sep 12 module 001: Conditionals with the Scribbler 2 **Due: module 000: Song Program Project**	Sep 13 module 001: Loops and Scribbler 2 Motion	Sep 14 module 001: Simulations and More Prog. Mgmt.: Functions, Assertions	Sep 16 **Due: Supplemental Problem 1** module 001: Robot Actions Project **(required)**

FIGURE 33.3 A Day View from the Database-drive Schedule

CSC 161

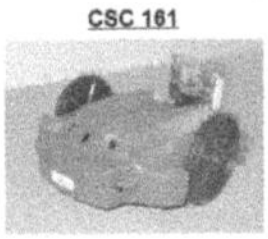

Grinnell College

CSC 161: Imperative Problem Solving and Data Structures

Fall 2016

Course Home Resources Course Details: Syllabus, Schedule, Deadlines, Topic organization MyroC Documentation Project Scope/ Acknowledgments

Tue, Sep 13, 2016

Previous Activity on Class Schedule: Mon, Sep 12, 2016

- module 001: Conditionals with the Scribbler 2

Today's Work:

- module 001: Loops and Scribbler 2 Motion

Next Activity on Class Schedule: Wed, Sep 14, 2016

- module 001: Simulations and More Prog. Mgmt.: Functions, Assertions

Close Deadlines:

- Today: none
- Next deadline(s):
  - Fri, Sep 16, 2016: Supplemental Problem 1

module 001: Conditionals, Loops, and Scribbler 2 Motion:: Loops and Scribbler 2 Motion

Available supporting resources

Resource category: Example

- robot-dance.c
- while-obstacle.c
- 3-loops.c

Resource category: Reading

- Reading on Loops

In-class Work

- Details

Scheduled dates

- Tue, Sep 13, 2016

CHAPTER 34

Utilizing student class preparation to promote active learning in class

This column first appeared in the *SIGCSE Bulletin*, Volume 39, Number 2, June 2007, pages 13–14[222] with the title, "Reading and class work". The revised chapter title highlights a theme throughout the column: various approaches are possible to motivate students to prepare for class by completing assigned reading or other activities ahead of time. Further, advanced student preparation presents several possibilities for active learning within the classroom.

Subsequent reflections on this subject appear later in this chapter.

34.1 ORIGINAL COLUMN: "READING AND CLASS WORK"

MANY YEARS AGO, I heard jokes about faculty transcribing material from a textbook onto the blackboard, students transcribing the board material into their notes, and the content going through the brains of neither.

Reflections

Over a decade ago, I developed a lab-based CS1 approach, in which students read material before class, worked in pairs on labs during class (I served as consultant and facilitator), and finished the lab as homework. When developing this approach, I initially devoted the first part of each class to reviewing the reading, but I soon realized this review was counter productive; students had little incentive to prepare for class or study a reading.

More recently, an instructor in another discipline commented that students lacked the skills to read technical material, so the faculty member repeated all relevant textbook material in class. The instructor agreed students should develop reading skills before graduation, but NOT in any of this instructor's introductory, intermediate, or advanced courses.

Altogether, these examples illustrate some tension between reading assignments and class activities.

- We assign reading to help students prepare for class, but many students have trouble reading technical materials.

- We want students to graduate with good technical reading skills, but classes may give students little incentive to practice.
- We want to help students learn specific topics, but lecturing about reading material can undermine students incentive to do reading assignments.

These observations raise questions of what a teacher should assign for reading and how class sessions should relate to material covered in reading. The following comments detail one approach to these issues and outline some other approaches.

Discussion Questions

One approach asks students to submit two "discussion questions" on each reading. As several of my syllabi indicate, these questions should seek to clarify points of confusion or to expand ideas just introduced briefly. For the most part, we should avoid simple factual questions (unless the facts are in dispute). The instructor will assume that students have a clear understanding of topics not covered in questions, and students should be prepared to contribute to class discussions on all areas not asked. To clarify further, the syllabi suggest the following types of questions might be appropriate:

- Questions concerning terms not defined in the reading,
- Questions to clarify ideas
- Questions about the relationships between ideas or approaches
- Questions regarding assumptions behind approaches
- Questions concerning the motivation or necessity of ideas or protocols

The following logistical details support this general approach:

- Students may submit questions via email or through on-line discussion software (e.g., Blackboard). I set a deadline for the night before a class, and employ a system that fairly enforces deadlines. I then collect and organize them, and hand them out at the start of class.
- After questions are organized, I distribute them to the class either in paper or through a web page. (If using web pages, I make the page accessible to the campus only, both to preserve privacy and to keep the materials away from prying eyes, such as parents.)
- I first distributed questions without attribution, but I have found that attaching names often results in questions that are more thoughtful. When using the web for distribution, I use only first names or first names and last name initials, so names will be clear to class members, but relatively anonymous to others on campus.
- Assigning points provides an incentive for students to complete the questions. For speed and convenience, I use a binary scale: 1 point for any reasonable question; 0 points otherwise.

Usually, I allow students a little flexibility. For example, students might submit 30 of a potential 34 questions, with any additional counting as extra credit. Counting these questions as 5%-10% of the course grade provides adequate incentive, so ignoring these questions might lower a final grade from A– to B+ or B. However, this percentage will not help a student pass without substantial work in other parts of the course (e.g., tests, projects).

I then base my lesson plans on the questions received. For example, if no students ask about a topic, I strike that material from class presentations; why go over what students already have understood? In addition, if student questions focus on a few sections, then class time can emphasize those areas. When questions arise, I might go over details in the text from a different perspective. In some cases, questions may suggest a basic misunderstanding or a lack of background knowledge; in such situations, a class might address misconceptions or provide background rather than explicitly answer the questions.

Sometimes, written material seems clear to the instructor, but not the students. In these cases, I may copy a few pages onto transparencies, so the students and I can read them together in class. For example, a logical argument may proceed with statements, such as "according to Theorem 5.7" or "by our previous discussions of efficiency"or "by an approach similar to Dijkstra's Algorithm".

Introductory students often just skip over such statements, just as they might skim material when reading fiction. When we read this material together, I can stop to ask "What did Theorem 5.7 say?"or "How does the discussion of efficiency here relate to previous discussions?" or "Describe the approach to Dijkstra's algorithm that applies here". Thus, student questions may indicate when it might be fruitful to model how to read technical material.

Some Additional Alternatives

The primary motivation for discussion questions arises from the desire to have students submit something tangible related to the reading. Other approaches can accomplish similar results. Some variations include:

- Have students submit a paragraph summary of each reading before the start of class.
- Ask students to answer a few questions or solve problems based on the reading before class, and base class discussion on questions that arise.
- Give periodic, short quizzes on reading at the start of a class. Questions may be simple – provided students have done the reading.
- Pick one or more students' names out of a hat at the start of each class, where each student summarizes one section or subsection from the reading.

Of course, many approaches are possible. The point is that students are encouraged to read technical material, and the class can serve as a safety net for those parts that seem difficult or confusing.

34.2 SUBSEQUENT REFLECTIONS

When a pedagogical objective involves students being prepared for class, an important challenge is how to encourage to do the necessary preparatory work. Since the original column appeared, common formats for preparatory study have expanded considerably, from traditional textbooks and paper-based notes to on-line articles, videos, and interactive resources. Even with new media available, however, the challenge remains: how to structure a course, so that students are likely to do the recommended work before a class session.

Unfortunately, today's students often seem remarkably busy, so they naturally make choices regarding how to allocate their time and energy. In this context, students may need incentives to complete readings and other preparatory work before class; intentions, even among the best students, may be strong and constructive, but demands of the moment can have an impact.

In response, an instructor may need to require explicit actions by students, either before a class starts or at the start of a class session. When some type of consequence or grade is associated with these actions, students may have particular incentive to complete before-class assignments and arrive prepared for class.

Several approaches, listed in the original column, provide clear incentives. As I have talked to colleagues at various schools, it seems that the effectiveness of various incentives may vary considerably from one campus to another.

- If students will be working on lab activities or other tasks with a partner during class, peer pressure can have a strong effect; students may not want to seem ill-prepared or ignorant in front of a partner.
- Discussion questions or paragraphs submitted before class require students to review material and compose a written response.
- Similarly, graded quizzes or other start-of-class activities may draw upon basic elements from preparatory material.

The wide availability of on-line tools may expand options for some class environments.

- Students might maintain a journal with their notes on readings, class work, etc. If an instructor has access to this student writing, then periodic checks might provide feedback on student work and also serve as input for a "preparation" grade.
- Students might complete an on-line, open-ended, feedback/reaction form for each reading.
- An instructor might post a few questions in anticipation of the next class, and students might be expected to fill out an on-line form for each question before class.
- Assuming the technology is available within a classroom setting, students might complete an on-line form to answer instructor questions at the start of each class.

The above thoughts identify some mechanisms that could provide incentives to students as they prepare for each class. Additional approaches abound and are only limited by an instructor's creativity and the capabilities of available technology.

indexstudent! class preparation—)

CHAPTER 35

What should be in a syllabus?

This column first appeared in the *SIGCSE Bulletin*, Volume 37, Number 4, December 2005, pages 19–21[219] Subsequent reflections on this subject appear later in this chapter.

35.1 ORIGINAL COLUMN

STUDENTS TYPICALLY RECEIVE A SYLLABUS at an early meeting of a course, Sometimes, schools provide faculty with guidelines or a template for the content of these handouts, but more often syllabi may display considerable variation in both the topics covered and the level of detail given. This article presents my perspectives on the purposes of a syllabus and its content. Other documents, of course, might cover some of this content. Traditionally, such syllabi were distributed in paper form, although distribution now may include e-mail or course web pages.

Goals of a Syllabus

A syllabus provides details pertinent to a specific course offering within three main categories. A syllabus can

1. Provide context or background to complement the general course description found in a college catalog,
2. Provide details that pertain to a specific offering of a course, and
3. Clarify cultural expectations that an instructor, institution, and many students may want to assume.

The first two of these categories expand upon the brief, high-level descriptions found in college catalogs. Further, since catalog statements typically require an extensive approval process, they often are sufficiently general that they need not be changed for small refinements and updates of a course. In contrast, a syllabus has space to explain how a course fits into an overall curriculum; what topics it will cover; what concepts and skills a student might expect to learn; what expectations and assumptions a course might make; and various mechanics.

In addition, instructors commonly make assumptions about procedures and expectations. For example, in some disciplines at various institutions, there is an understanding that students may work with each other on in-lab and small-group activities, but students are to write up their answers individually. Within this framework, difficulties can arise if students do not understand the boundaries for collaboration. Without clear guidance, students may believe they are following the rules, while instructors may believe that academic dishonesty has occurred. The syllabus provides a mechanism to make such expectations clear. Of course, this does not prevent students from behaving badly – particularly in stressful situations – but it does provide a clear framework for later discussions, perhaps with college committees on plagiarism or academic standing.

Suggested Content

Syllabi must fit within a local environment, and differences among departments and schools require adjustments of syllabi. The following annotated list identifies some common elements for course syllabi.

Main Course Goals and Objectives

Established faculty members often understand the goals of a course and how those goals fit within an overall curriculum. However, students and new faculty may not know this background. In addition, students often want assurance that a course contains valuable information and will proceed in a effective way. A syllabus can help put a course into perspective, explaining why it covers various topics, why assignments or discussions are structured as they are, and the like.

Basic Instructor/assistant Information

Students need to know where and when they can find an instructor or teaching assistant, how to contact this person, where to get help, and what process to follow when questions arise concerning assignments, tests, and grades. For example, it may be useful to indicate policies regarding contacting you through e-mail and telephone calls.

Textbooks (both required and recommended)

Courses often assign readings from textbooks or journal articles. A syllabus can identify which resources will be needed and how students are expected to access them (e.g., through purchase or the library).

Course Schedule (including likely times for assignments and tests)

Some schools require that notice be given for tests and other assigned activities. Even when not required, however, the posting in advance of anticipated due dates and in-class tests can help students plan their work (and give ammunition to faculty when talking to students about late work). In addition, after announcing the date of the final exam, some faculty members include the statement, "Do not buy airline tickets that overlap with your exams!"

Identification of Required Course Work (including tests and exams)

From time to time, students may complain that they did not know about an assignment or test. These circumstances seem to arise most often when assignments are given informally or when deadlines/test dates are adjusted without formal written notice. Placing such activities on the course schedule can have a significant impact in reducing problems related to the schedule. Once times are stated, an instructor is encouraged to abide by those dates; it may be better strategically to adjust the content of a test (to match what material has been covered recently) than to change the date of a test.

Policies Regarding Class Attendance

Faculty members usually expect students to attend class; students are responsible for material covered during class meetings. However, questions can arise regarding the appropriate procedures if a student is ill, must be away as part of a college-sponsored event (e.g., a sporting event or theatrical performance), or is working with a college office to resolve a crisis (e.g., a roommate experiences a medical or psychological emergency).

Policies Regarding Deadlines (including any late penalties)

A simple policy states that all work is due by stated deadlines. Practice, however, can complicate this ideal policy. For example, a syllabus can address such question as: Will late work be accepted? Will there be a late penalty? How will late work affect class participation? Similarly, consider a course that requires work in a computer lab. What happens if the equipment in that lab goes down the night before the deadline, or the day before, or the week before? A strict late policy might encourage numerous telephone calls to an instructor at home (perhaps late at night) if equipment fails. To resolve this, a due-date policy might include an automatic extension should lab- based equipment or the network fail during a period before a due date.

Expectations Regarding Collaboration

During experiences on committees that oversee cases of suspected academic dishonesty, I have noted numerous instances where faculty members made assumptions that they did not convey to their students. For example, a faculty member might expect that students complete homework assignments individually, but this is never stated in any course materials (and other courses might allow collaboration). Since discussions of possible cheating are rarely pleasant, it can be particularly helpful to clarify expectations explicitly in a written syllabus.

Expected Time Commitment

Although conventional wisdom suggests that students should devote 3-4 hours outside class for each hour in class, workloads may vary dramatically from one course to another or from one part of a semester to another — particularly if the course involves a project. A statement of realistic time expectations clarifies what should be considered the norm for a course and it helps students plan their time. For example, a student might reexamine her/his course schedule if four courses indicated a major paper or project was due the last week of a semester.

Grading Policies (with expected processes for determining the final grade)

Many students want to know the elements that go into a course grade and any formulae for computing that grade. From one perspective, such detail may emphasize grades excessively and encourage nit picking. However, grading policies also send messages about what is important in a course and guide students about how to organize their time. Ideally, students would devote full attention to every aspect of every course. Realistically, however, students take multiple courses, have job or family commitments, and address such mundane matters as lodging, cleaning, and food. The articulation of grading policies and schedules can help students make reasonable choices when real events conflict.

Setting the Tone for a Course

Statements and policies on a syllabus give students an initial perspective on a course. For example, at one campus, I recently read a syllabus that stated that an average between 80.00% and 89.99% corresponded to a B and that no rounding would be done; a score of 89.99% would yield a B, not an A. This raises a philosophical question about grading accuracy, because science and statistical methodology require that care be taken to ensure that data and results are presented within the level of accuracy of an experiment. Thus, the above course policy likely sends at least two messages:

1. Every point is vital; students should argue about every partial score on every exercise and test.

2. Experimental error and instructor variability never occur in grading.

Accommodations for Students with Disabilities

Schools often have established procedures, so that accommodations in assignments, lab work, or tests can be made for students with disabilities. However, introductory students may not know these procedures, and established students may be embarrassed to ask. A section on the syllabus reminding students of school practices helps resolve such problems.

Conclusion

In summary, a course syllabus can place a course within a curricular or programmatic context. Further, since a course exists within a culture and collection of expectations, a syllabus can clarify assumptions for that culture rather than hoping that "everyone understands that ...". Finally, a syllabus can clarify details, so that students understand how a course will proceed and can plan their work accordingly.

Acknowledgments

Thanks to Jonathan Chenette, Chris French, Shonda Kuiper, and Karen Shuman, who provided helpful feedback on drafts of this column.

35.2 SUBSEQUENT REFLECTIONS

The 2005 article provided a reasonably comprehensive list of topics that commonly appear in course syllabi, including

- Course goals an learning objectives
- Basic instructor/assistant information
- Textbooks (both required and recommended)
- Course schedule (including likely times for assignments and tests)
- Identification of required course work (including tests and exams)
- Policies regarding class attendance
- Policies regarding deadlines (including any late penalties)
- Expectations regarding collaboration
- Expected time commitment
- Grading policies (with expected processes for computing the final grade)
- Setting the tone for a course
- Accommodations for students with disabilities

All of these areas continue to be widely described in many [most?] contemporary syllabi.

However, in the intervening years, some changes can be observed—at least on some syllabi at a range of schools. In particular, it seems that syllabi are becoming more formal, sometimes serving as a type of contract between the instructor/course and students.

- *Objectives and Outcomes:* Modern syllabi often describe in some detail anticipated objectives and learning outcomes. In selecting courses, students may review these content statements to determine if a course will meet their educational and career goals. To date, I am not aware of a student suing a well-established, not-for-profit school and instructor, because the actual course did not cover the advertised content. However, in today's litigious society, it would not be surprising for a law suit to arise if an actual course did not focus on its stated objectives and outcomes.
- *Textbooks:* The 2008 Higher Education Opportunity Act (HEOA) asks schools to make textbook lists for courses available during registration. In this way, students will know likely textbook costs for courses they might take, and they may adjust their course work accordingly. Also, by knowing textbooks well before a class, students can comparison shop to locate needed books at the best prices.
- *Required work:* In discussions with colleagues around the country, a common excuse from students seems to be that they did not know about an assignment, test, or other required activity. In reviewing such complaints, a primary question is "how should a student have known about this work?" To anticipate such arguments, the syllabus should be as explicit as possible about course requirements.
- *Class attendance:* As with required work, a syllabus should indicate course expectations regarding attendance. Today's administrators commonly emphasize that if a course grade will be impacted by lack of student attendance, then the syllabus should indicate how attendance will be determined and what penalties will be applied if a student misses one or two or three or ... class sessions. Similarly, if a student is consistently late for class or consistently leaves early, the syllabus should indicate what, if any, consequences might arise.

- *Collaboration and academic honesty:* Course policies must be clear regarding collaboration, consultation with outside sources, use of answer books, discussions with other students, etc. If a student claimed, "I did not realize I was not allowed to talk to my roommate about this assignment," the question would arise whether the rules were clearly stated.
  - ◇ If course policies are the same for all work (e.g., tests, programs, written assignments, etc.), then the syllabus is an excellent place to make these rules clear.
  - ◇ If rules for tests differ from the rules for labs or other course activity, then the syllabus should clarify the overall rules, but it also is likely that each activity should explicitly state the rules for that activity. For example, if labs allow collaboration but written homework does not, a student must be certain whether a particular programming assignment should be considered a lab or written homework.
  - ◇ If students must work together for part of an activity (e.g., to collect data), but then they are to reach their own conclusions and write their own individual reports, then instructions must clarify what is and is not allowed regarding collaboration. For example, when they are collecting data together, are they allowed to talk about the data's implications in anticipation of the individual write ups?
- *Required institutional language:* At some schools, specific language is required regarding statements concerning accommodations for students with disabilities, students who become ill or who must address family or personal difficulties.
- *Publicizing support systems:* Beyond specific legal announcements (e.g., regarding disabilities or illness), some schools encourage inclusion of statements on support services, such as tutoring programs, writing centers, library resources, student affairs counselors, etc. Although paragraphs on various services can provide helpful information and encouragement, students may skim over syllabi that extend for several pages.

Since each institution may have its own policies and expectations regarding specific topics contained in a syllabus and specific language for certain sections, faculty should consult a dean or department chair regarding an individual school's requirements. In the past, syllabi may have been considered as informal documents that had little formal stature. However, current practice suggests that syllabi must be reasonably complete and formal. Students need to know policies, expectations, and rules clearly; and syllabi provide a natural place to explain course details.

On the positive side, syllabi provide the opportunity to clarify rules and expectations—helping a class progress smoothly. On the negative side, an instructor may have difficulty lowering a grade for specific behaviors, if a syllabus does not indicate specific penalties when designated rules are violated.

CHAPTER 36

The role of textbooks and multimedia

This column first appeared in ACM *Inroads*, Volume 2, Number 1, March 2011, pages 14–16[239] with the title, "The Role of Textbooks". This revised title highlights contemporary practice that may include a wide range of media in supporting student learning. Comments on some of these elements may be found in the "Subsequent Reflections" subsection that appears at the end of this chapter.

36.1 ORIGINAL COLUMN

MANY COLLEGE AND UNIVERSITY FACULTY expect students to acquire one or more textbooks to support course work. Over the years, however, many textbooks have become quite expensive: one widely-used calculus textbook is priced about $230.00 (U.S.), although an alternative paperback edition sells at about $150.00 (U.S.); several books for upper-level undergraduate computing courses are listed at about $150.00 (U.S.); and numerous computing textbooks are marked over $80.00 (U.S.). Further, at least one publisher's Web site advertises the possibility of renting textbooks or materials for limited time intervals (e.g., 60, 90 or 120 days). High costs, of course, can create hardships for students, so the question naturally arises, "Why might an instructor want to use a textbook at all?" This column considers possible roles of textbooks; in many cases a compelling case can be made for this resource, but faculty likely will want to have clear goals before requiring these potentially-expensive materials, and faculty will want their course practices to reinforce those goals.

Some Possible Uses of Textbooks

Textbooks vary substantially in their scope, design, intent, and ancillary materials. Uses of these materials may focus on students, faculty, or both.

- Uses primarily supporting students
  - ◇ Books may provide students experience reading technical materials (e.g., articles, manuals) that are essential for professional work in the field of computing.
  - ◇ Books may serve as reference works, allowing students to look up ideas and details that are covered in class or elsewhere.

- ◇ Books may provide a wealth of exercises, challenging problems, and research problems that allow students to practice and apply ideas.
- ◇ Some books come with student guides that provide outlines, review guides, worksheets, etc.

- Uses primarily supporting instructors
  - ◇ Some books provide slides to aid in lecture preparation.
  - ◇ Some ancillary materials may include test banks and other possibilities for assessment.
- Uses for both students and instructors
  - ◇ Books may explain ideas, techniques, and approaches in detail. Reading assignments guide students through new material, connect new ideas with previous concepts, and provide in-depth discussions of a subject. And, with explanations already available, instructors do not need to create new materials will all of those details.
  - ◇ Books may include lab exercises that support formal lab sessions.
  - ◇ Some books are particularly large and weighty and thus make wonderful door stops.

Some Constructive Practices

When textbooks are adopted for a course, a variety of practices can help utilize these resources effectively. Here are several examples.

Over the years, I have taught both calculus and the theory of computation several times. In each case, I identified a text that worked well; it was clear, readable, and complete. Thus, I did not have to cover all details of the material in class; I could provide examples to illustrate main points, give alternative approaches, provide contrasting perspectives, etc. My role was to amplify, clarify, and explain; but it seemed pointless for me to repeat the book in class. However, conversations with students indicated they often had difficulty reading technical material; their experience working through a text was limited, since past course work often allowed them to rely upon lectures. Thus, in both calculus and the theory of computation, I typically devote one class meeting to the collective reading of a book section. For example, during the class one on "reading", I initially focus with the class on the first few paragraphs and a statement of a theorem; what does a specific phrase mean, why might a condition be needed, how does the theorem relate to past work, what might motivate the introduction of the material at this point in the book, etc.? The class and I then review at the first proof: what is the idea of the argument; is this approach similar to anything we have discussed earlier; how is this material structured; when the proof references an earlier result, what does that result actually say, and why does it apply; why might one think this approach would yield a satisfactory proof; would another argument be simpler; etc.? Then, after a few rounds of discussion with the entire class, I present several questions on the next page or two of the book, and these are discussed in small groups. After hearing some small-group responses, I pose questions for the next page or two of the section. Overall this type of exercise models active reading; one normally reads a technical reference differently than a short story or novel, and some students may need initial guidance in tackling technical material.

In other courses, several colleagues and I ask students to read a passage for homework and then submit questions for discussion before the class meets. I then collect the questions, organize them into a logical framework, and make the questions available to the class (either on paper or, more commonly, on a Web page which can be read only by the class or on campus). Pragmatically, I have found questions tend to have reasonable depth when the questions are distributed with attribution. Questions are graded simply: acceptable or unacceptable, according to whether they demonstrate some connection to the reading; and the submission of questions constitutes a small part of the final grade. This type of activity serves at least two purposes. First, students must read the material in advance of class in order to complete the questions, so the students come to class prepared. Second, submitted questions can guide class discussion; an instructor need not discuss areas that seem clear, but rather can focus on areas of confusion, misconceptions, related topics of interest, etc.

As a variation, an instructor might pose questions on readings and expect student responses to be submitted before a class meeting. A review of these responses clarifies what parts of a reading have been understood, what requires clarification, and how students are thinking about the ideas.

As another variation, when a textbook is organized into short subsections, a student can be assigned to lead class discussion for the material for each subsection. This might involve a brief summary, but then should branch into a discussion in which the designated student offers initial probing questions. In my experience, all students are motivated to read material in this context, because they do not want to be embarrassed when it is their turn. Thus, student summaries can be short, and class time can focus on the development of ideas.

As yet another variation, in a small class of 4-6 students, I may organize coverage of the material into day-by-day topics normally based on sections of the book; and in rotation, each student presents material assigned for the day. Thus, the student must master the material, determine what was important, organize a lecture, prepare any needed handouts, etc. A student presenter has complete control of a class session, but the next class will move to the next topic. Since students understand they must provide any needed background for what follows, and since they do not want to be embarrassed before their peers, they consistently show outstanding motivation and preparation. Throughout the semester, I am available as a resource, but I lead the class only on the first and last day. This approach not only requires students to learn material well, but it also helps students develop their communication skills. In one guided-reading course in which I used this approach, students also developed a sense of wanting to surpass the others; by the end of the semester their appeared in formal attire, had written multi-page handouts, and had clearly practiced their presentations several times. Note that when students are presenting new material at least every week, assessment beyond the presentations may not be needed; the material builds significantly from one section to the next, and all listeners know whenever a topic is muddled.

A different use of a textbook arises in support of lab-based courses. For example, several years ago, I changed the pedagogy of a course from having separate lectures and labs to having a fully lab-based format. Initially, I wrote about 45 labs to support the course. To reduce development time, I coordinated the labs with a textbook; I had to write instructions for the lab, but the discussion of content was left to the book. In this setting the book had a clear purpose: it supplied material that was not available elsewhere.

Continuing this lab example, over the next several years later in the same course, I supplemented the materials for each lab, so that the labs had become largely self contained. In this context, students observed, and I agreed, that a textbook was no longer needed for the course, and the course no longer requires a separate book.

Of course, textbooks also often provide a wealth of exercises to provide students with practice on material. Traditionally, it is common for a course to assign problems to be turned in, and the resulting scores constitute some portion of a grade. While this approach can work well, logistics and time constraints limit the number of exercises to be turned in, since grading can be time consuming. Thus, in addition to regular problem assignments to be turned in, I commonly identify "supplemental problems" on textbook sections. These problems are not to be turned in, although I invite questions about them during class. As an incentive to students, I also announce early in a semester that about a third of the questions on each test will come from these supplemental problems – perhaps with numbers changed or text slightly edited. My reasoning is that if students are encouraged to work through several dozen exercises, the students will engage with the material and learn. Of course, a test itself may include only 2 or 3 of the numerous supplemental problems identified, but knowing some possibilities in advance can help relieve some stress among test takers.

Note that many of these approaches also work when an instructor organizes a class with a collection of articles rather than a formal textbook. Developing of a reading list can be time consuming for an instructor, and collecting these articles may require student access to the ACM Digital Library or an effort to obtain copyright permissions for a packet that can be duplicated. However, once compiled, a collection of articles can be utilized in much the same way as a textbook—particularly for upper-level courses that explore research in a given field.

Some Questions in Utilizing Textbooks

Although many of these uses of textbooks may be valuable in principle, instructors should consider how to utilize these resources and not undermine the actual value and effectiveness of the books. The following questions may further help brainstorming how textbooks might be used and how class sessions might reinforce goals and student learning.

- If an instructor digests the content of the text and repeats that material in question, why should the students bother to read the book, and what value is the instructor actually providing?
- If students have difficulty reading technical material (e.g., in the book), how does class work help them improve their skills?
- If the instructor covers material at the same level of depth as the book, why should students purchase the book at all?

In summary, this column makes two points: Faculty should consider the role that a textbook actually plays within a course, and faculty should examine how other course elements support the textbook's role.

Acknowledgments

Many thanks to Marge Coahran for her wonderful suggestions regarding both the structure and content of this column; her comments led to a substantially improved article.

36.2 SUBSEQUENT REFLECTIONS

Many of the themes discussed in the original column seem relevant today; as part of course planning, faculty need to consider what resources might be available and how those resources might be used. The original column focused upon printed textbooks, and courses now may draw on a wide range of support materials, including textbooks but also spanning other media.

The following four additional factors illustrate further issues that faculty should consider as they review materials that might support courses.

- Contemporary resources often involve a wide range of media, including textbooks, videos, CDs, Web sites, and other electronic media.
- Some colleges and universities routinely purchase copies of all required textbooks (sometimes supplementary books as well), and these books may be placed in a resource center (e.g., a reserve desk in the library).
- The 2008 Higher Education Opportunity Act (HEOA) asks schools to make textbook lists for courses available during registration.
- When materials are available primarily in electronic form, access may require students to view these resources via the Internet, so the cost, speed, and possible format for viewing may depend upon Internet/data plans and details of the Internet infrastructure in a local area.

The following notes provide some perspective on each of these factors.

Multimedia

Conference discussions and numerous articles report the use of video to introduce students to new material. Although this medium may reflect a change in the form for student assignments, assigned study before class does not . For example, many discussions of flipped classrooms describe videos (often video lectures) that students view before class, so that class time can be devoted to various active-learning tasks. In my own teaching, since fall 1992, I assign readings for study before class, so they would be prepared for lab activities throughout regular class time. In my classes, students read book sections or articles or specially-written handouts, but the pedagogy would work equally well with videos or other media. Altogether, contemporary resources may have expanded options for helping students prepare for class, but many of the purposes of these resources would seem to remain the same.

Lending/Reference Libraries

Some schools purchase a copy of the textbooks for every class and place them on reserve in the library or other resource center.

In principle, this policy provides access to required books for each student, and all students can be expected to have completed any assigned reading before each class. As one instructor at such a school commented, "No longer do I hear excuses that the bookstore was out-of-stock for a textbook or that a student had ordered a book online, but it had not arrived yet." Of course, students may develop other creative excuses, but book availability no longer should be an issue.

In practice, however, implementations at various schools may differ. Typically, a library or resource center might place a book on a 2- or 3-hour reserve, so that students must return the book within a prescribed time interval. The idea is to set a time interval to be long enough for a reading to be done, but short enough so the book can be used by several people. Although this may work fine in some settings, difficulties can arise in at least five ways:

- If a course is new or adopts a text that has not been used previously, then it may take the school some time to acquire and process the book before it is available to students.
- If a moderate number of students want to read the book at the same time, all copies might be checked out at some points.
- If students with a book are slow to return it, other students may be impacted.
- If students have prescribed work schedules or other commitments, they may need to use the book at specific times, and waiting for a book to be returned may not be feasible.
- If students live some distance from campus, travel to and from the resource center may not be practical.

Individual colleges and universities many address such matters in a variety of ways, but it seems appropriate for instructors to check on local practices before making assumptions regarding textbook availability.

Of course, if all required books are guaranteed to be available to every student, instructors may feel they have wide latitude in selecting the best book(s) for their courses.

The 2008 Higher Education Opportunity Act (HEOA)

The original column noted that instructors may wish to include the cost of textbooks as part of their decision making in selecting books. The 2008 Higher Education Opportunity Act (HEOA) takes this issue one step further: schools must make textbook lists for courses available during registration. In this way, students will know likely textbook costs for courses they might take, and they may adjust their course work accordingly. Also, by knowing textbooks well before a class, students can comparison shop to locate needed books at the best prices.

Dependence upon the Internet

On-line resources may provide wonderful insights, background content, and supplemental materials for a course—assuming all students have adequate Internet access. Many schools (at least in North American and Europe) may high speed Internet access on campus for students, faculty, and staff, so utilizing on-line resources may not be a problem when students are physically near their classes.

However, what conditions will students encounter when they go home—outside of class?

- If a student does not have a computer, or if a family has one computer to be shared among several people, then extensive Internet access for a course might require purchase of a computer and further charges for an [upgraded] Internet connection, if available.
- In some communities, telephone wires in parts of town are old, and Internet speeds are remarkably slow. In one college town, speeds on campus operated at remarkably high speed, but five blocks away, it took 20 minutes to upload a 1 megabyte picture. Such speeds make use of video streaming problematic. (Students might be willing to pay for high-speed Internet at home, but the service provider may not be able to supply it.)

- If students are using wireless connections (or perhaps other types of Internet access), costs of service may depend upon the amount of data transmitted and received. In this context, reliance on Internet materials (e.g., videos) may yield costs to students—even if the Internet source does not charge for viewing or access.

Altogether, the Internet can be a wonderful resource for materials, but those materials may or may not be available equally to all students in a class.

To summarize these reflections, options for media have expanded considerably over the years beyond traditional formats involving textbooks and other paper forms. From this perspective, forms have evolved, and details may be different. However, in many contexts, the underlying uses and constructive practices for textbooks might be reviewed more generally for today's classes.

CHAPTER 37

Course development utilizing student-faculty collaboration

For over five years, I worked with teams of students to create, develop, and refine a lab-based CS2 course that introduced imperative problem solving and programming in C while utilizing Scribbler 2 robots as an application theme. Overall, this work involved

- developing C-based infrastructure to support various robot-oriented commands over Bluetooth communication,
- organizing course topics into a logical semester schedule, and
- writing readings, examples, labs, and projects to support daily class sessions,

Two papers, written by student teams and myself, have described the curricular development and the infrastructure for this project [23, 50], and two papers, written by two other student teams and myself, have focused upon the development process involving student-faculty collaboration [92, 194].

This chapter draws upon these past papers and generalizes this experience.

37.1 NEW ARTICLE

TEACHING A NEW COURSE or reworking an existing course typically involves considerable time and effort. Typically course preparation requires many of the following activities:

- identifying course goals and student learning outcomes
- determining topics/content to be covered
- identifying writing resources
    - ◇ textbooks
    - ◇ articles
    - ◇ videos

    - ◇ supporting infrastructure
    - ◇ handouts for local computer environments

- organizing topics into a logical sequence and developing a course schedule
- writing course materials
    - ◇ syllabus
    - ◇ examples
    - ◇ labs
    - ◇ projects
    - ◇ assignments

Traditionally, instructors tackle all of these elements themselves.[1] However, this chapter suggests that much of this work can be achieved through a student-faculty collaboration.

Since the instructor usually is the subject expert, describes intended levels of student mastery, and understands pedagogical approaches, the instructor naturally must take a leading role in setting course goals, student learning objectives, and identifying appropriate topics and course content. However, these elements comprise only a small fraction of the long lists of tasks given above. This observation leads to the possibility that students might be able to collaborate with the instructor on many of course-development activities. Such student-faculty collaboration may be particularly helpful when a course is being reworked, in which case upper-level students already may have learned the material within the past few semesters.

Advantages of Student Involvement

Although an instructor knows the subject and understands issues of teaching and learning, students have both experience and perspectives that can greatly enhance the course-development process.

- Students who recently have taken a similar class may remember what they did and did not know when they first learned the material.
- Since upper-level students typically have learned material rather recently, they likely remember what topics seemed difficult, what insights helped them over hurdles, and what approaches led to dead ends.
- Students in the new class will largely be peers of the student developers, so the upper-level students likely have a good sense of what examples and language will resonate with the target audience.
- Students who have recently learned material may have insights regarding how much time might be required to master a new topic and how long assignments might take.
- When materials are developed, upper-level students may anticipate how the new students might respond to wording and examples.

[1]If several faculty will be teaching multiple sections of the same course, some collaboration may be possible, but often all of this work is done by one person.

An Instructor's Role

In my experience following this approach of student-faculty collaboration, an instructor's role is [at least] five fold.

1. Providing initial background and perspective
2. Organizing students into productive teams
3. Mentoring and coaching
4. Establishing a review process
5. Reviewing final products

Discussion follows for each of these points.

First, student developers have little experience in organizing and developing materials and in pedagogy. For example, in considering how to organize content into a logical sequence, the instructor likely will need to provide background; How might one approach the organization of content, so material at one stage can build safely upon previously covered material? Beyond specific content prerequisites, do some topics require more experience or practice than others? What pitfalls might be anticipated in sequencing topics, so work can progress in a natural progression?

Second, with a development team of four or more students, all students likely need an overall understanding of course development, but all students working together on all details likely will be inefficient. Rather, content might be organized into modules, and these pieces might be placed into a general sequence. For each module, a subgroup of two or three students might focus upon details. Frequent group meetings might allow presentation of several modules to review connections, continuity, omissions, and overlap. With coordination, groups can proceed in parallel, as long as they check in frequently with the instructor and report regularly to the overall team.

Third, when working on individual modules, students can and should brainstorm the flow of material. However, students likely have little sense of pedagogy, material development, use of examples, and other issues of teaching and learning. Thus, as development progresses, the instructor will need to explain teaching principles and respond constructively to student suggestions. As work shifts toward the writing of examples or labs or other materials, the instructor will need to help the student groups get started, perhaps outlining a reading or lab exercise or project. Students can then flesh out the outline, working with the subgroup as materials emerge. As students gain experience, the level of instructor interaction may decrease, but some ongoing discussions should be expected.

Fourth, once student groups have prepared draft materials, the instructor might organize a review process in which the work of each student group is reviewed in detail by another group. Feedback should include detailed comments of what works well, what is hard to understand, what reinforces a central point or detracts from that point, what working might be improved, etc. With this feedback, the group which prepared the draft can work on a revision. Naturally, the develop/feedback/revise cycle can be iterated as needed until a solid result has been achieved.

Fifth, once student groups have developed materials, with considerable review by other students, the instructor can review and approve materials before materials are deemed ready for classroom use.

Altogether, multiple student groups, working in parallel, require an instructor to guide, mentor, and review. However, since development proceeds in parallel, my experience indicates that substantial materials of strong quality can result—at a much faster rate than an individual instructor (even an experienced instructor) can produce without help.

Some Suggested Procedures

In my own experiences of student-faculty collaboration for course development and refinement, I have been most successful working with groups of four students. (Some instructors may be able to coordinate more than four students, but four is my practical limit.) The work then proceeds in phases:

- Since I know the material and the scope of the target course, I typically supply an initial list of topics and learning objectives for the course.
- Early in the development process, I review this list with the student team in some detail, so that all students understand the scope of the work. As with all activity in this collaboration, students are encouraged to ask questions, suggest additions and modifications, and brainstorm connections among the topics.
- To help the students become comfortable with the content, they initially work in pairs to develop examples for some or all of the topics. Typically, the development of examples continues for a couple weeks, in which case I assign one pairing of students for the first week and a second pairing for the second week.
- As the students gain experience, I suggest they work as a group of the whole to organize the full list of topics into coherent modules that might require 1-2 weeks of class time. The idea is that modules should be focused and coherent (hence not too long), but of sufficient content to require some integration of topics (hence not too short). Sometimes I join the full group as this organization starts, but getting out of the way for the first several days often yields innovative and interesting approaches that can work particularly well as the course develops.
- Once a draft schedule has emerged for the entire course, with topics organized into 1-2 week modules, I divide students into pairs to flesh out full details of each module.
    - ◇ Two pairs may work on the first two modules in parallel. For each module, one group develops, and the other provides feedback.
    - ◇ After two modules have been developed and reviewed, a new pairing of students can develop the next two modules.
    - ◇ Work progresses two modules at a time, until course development is completed.
- To ensure coordination among efforts, I typically meet with the entire group every week. In the language of agile programming, each week becomes something of a scrum.
    - ◇ Each group reports on its work.
    - ◇ After determining progress of the past week, we set detailed targets for the next week, and assign which pair will do what.
        - ∗ My students were enthusiastic about utilizing a Web page to chart progress, with the detailed tasks for each week identified.
        - ∗ Initially, each targeted task appeared in red on the page.

    * As substantive progress was made (e.g., a draft became available), I recolored the task as orange.
    * When the task passed its final review, the task color changed to green.

- Between the weekly meetings, I typically meet individually with each group every day or two to provide mentoring, coaching, and technical assistance.

As this outline suggests, my personal style is very much hands on. I set short-term goals, I coach, I give feedback, and I provide technical and pedagogical assistance. I expect instructors with different temperaments might allow greater flexibility and more latitude for each student group, as it works on its materials.

Some Results of One Case Study

As might be expected from an extensive student-faculty collaboration, such course development can be time consuming for the instructor, but it also is a deep learning experience for the students. Further, students working in pairs in parallel can produce a substantial body of material in a form sufficient for use in the class under development.

To be more specific, in following this approach for 10 weeks during one summer, working with four students (with funding from Grinnell's Dean's Office), the team produced an impressive quantity of solid materials:

- about 70 examples (C programs in this case)
- about 35 labs (each for use in a 1-hour lab session)
- about 8 projects (each integrating 1-2 weeks of material)
- about 15 selective readings on topics not covered in the textbook.

Although I devoted substantial time to this project over the 10-week period, I am confident I could not have come close to this productivity.

Beyond quantity with solid quality,

- the student team demonstrated considerable creativity in examples and constructing labs—with more variety and innovative connections that I likely could have done working alone.
- although several examples seemed corny to me, these examples were particularly well received by the students in the targeted class (the student team anticipated reactions of the targeted students particularly well).
- the student team developed and matured considerably, in technical mastery, in communication skills, and in their ability to work in teams.
- the students in the targeted class were particularly excited to see credits on the materials (e.g., examples and labs) that included the student team members. The credits helped connect the development with the targeted students and encouraged a sense of a learning community for the course.

Further, students in the revised course clearly had higher motivation than in the course before revision, and the students in the revised course did as well or better on exam questions taken from tests before the revision.

Overall, student-faculty course development allowed me to completely rework a course from scratch, adding an application theme, reorganizing content, rethinking labs and projects, etc. The effort required my attention, but the process

- yielded more course materials than I could have by myself,
- anticipated student needs, and
- provided a substantial learning experience for the student developers.

In the end, I spent less time in course development for a completely reworked and revised course, the students in the target course gained knowledge and insight, and the students in the development team developed skill and insight.

Acknowledgments

Many of the ideas presented in this article have evolved from my work with students on the MyroC project that brings imperative problem solving and C programming to a CS2 course. Student developers in this effort follow.

Student Development Teams		
Summer 2011 David Cowden April O'Neil Erik Opavsky Dilan Ustek	*Fall 2011* Erik Opavsky Dilan Ustek	*Fall 2013* Spencer Liberto Dilan Ustek Jordan Yuan
Fall 2014 Vasilisa Bashlovkina Anita DeWitt Anqing Liu Nicolas Knoebber	*Fall 2015* Marija Ivica Sara Marku Thu Nguyen Ruth Wu	

VIII

Instructors' Roles, Inside and Outside the Classroom

WHAT SHOULD AN INSTRUCTOR DO in the classroom? Traditionally, the instructor has been considered as "the sage on the stage", but that perspective now is well-known to be quite ineffective.

But if an instructor is not the central actor and active party, what role(s) might be appropriate? The chapters in this part consider several dimensions to this question.

FIGURE 37.1 Instructors Word Cloud

- Within the classroom itself, an instructor seeks to facilitate learning. From time to time, I quip that it is not important what material I cover in a class; rather what is important is what material students learn. From this perspective, a teacher's mission involves promoting and enabling student learning. Chapter 38 describes three related roles that support this mission: coach, mentor, and listener.
- Although teachers work to promote learning, ultimately students are the people who must achieve understanding and mastery. Learning is not achieved by students swallowing a pill, eating a magic elixir, or getting a shot. In the educational process, teachers can strive to create an environment that motivates, teachers can make resources available, and teachers can try to guide students through content. However, students also have a responsibility within the learning process. Chapter 39 explores boundaries for instructor activities and responsibilities. Some activities may seem common to many or all educational environments, but others may depend upon goals, philosophies, and local constraints.
- Numerous difficulties are well documented for class formats involving an instructor as a "sage on the stage", and yet lecture-based classes continue at many schools. Particularly when enrollments outstrip available faculty, highly-enrolled classes utilizing a lecture-based format may be common. Chapter 40 provides some perspectives on lecturing, reviewing both its advantages and disadvantages, and considering approaches that may encourage active learning within a lecture format. i

- Few people are likely to consider class sessions to be the same as theater. Students typically attend classes to learn something, while they go to the theater for entertainment. However, students also are unlikely to attend class to be bored or put to sleep. Such considerations lead to discussions on how class activities might engage students. What techniques might an instructor employ to make an activity memorable? In this regard, the realms of teaching and theater may overlap, and an instructor may use ideas from a theater setting within a class setting. Chapters 41 and 42 explore this connection between teaching and theatrical elements.

- As a semester progresses, a course moves forward. Classes meet, the instructor facilitates and/or presents, students [hopefully] engage with the material, etc. But how might an instructor determine whether the course is going well and what, if any, refinements might make the course even better? One way to proceed is to ask students, and Chapter 43 reviews approaches for gaining and utilizing student feedback.

- Finally in this part, each chapter explores elements of classroom teaching, but a few chapters cannot explore all research studies, experience reports, and best practices. Chapter 44 therefore identifies some additional resources for readers who wish to explore teaching in the classroom in greater depth.

CHAPTER 38

Teacher as coach, mentor, listener (part 1?)

This column first appeared in ACM *Inroads*, Volume 7, Number 1, March 2016, pages 18–21[265]. Subsequent reflections on this subject appear later in this chapter.

38.1 ORIGINAL ARTICLE

A STUDENT WALKS into an instructor's office and says (choose one):

- "I have a mental block about — — ; I don't get what I'm supposed to do or how to do it."
- "This course is too hard, because problems are difficult, and I have to think too long."
- "I had trouble with this homework problem; can you show me how to do it?"
- "My grade does not reflect my real knowledge!"
- "I work and work and work, and I still am doing badly in the course!"

No, this is not the start of a joke. Rather, this column considers how an instructor might respond in a constructive and honest way.

Last summer, circumstances caused me to change offices, and now I am fortunate to work next to the office of Minna Mahlab, Director of Grinnell's Science Learning Center (SLC). Over about two decades, Minna and I have talked about her work organizing student mentors, peer instructors, and one-on-one tutoring for students in biology, biochemistry, chemistry, neuroscience, and physics—I extrapolate to computer science. With my recent move, I can observe first hand the many facets of her coaching, mentoring, listening, and guiding students. This column combines selected observations of the SLC with my conversations with Minna and my own reflections on the role of a teacher in working with students. Of course, a single column can only begin a discussion; let me know if you have interest and/or suggestions for more discussion as Part 2 or even Part 3 in future columns.

As I consider how to respond to the above student questions, some initial thoughts include:

- My professional training does NOT include counseling, psychiatric analysis, or medical diagnoses.

- Although I want to be helpful, my insights and capabilities are limited. I can work to clarify expectations and provide context.
- Although I may be able to help students by providing background and guidance, unilaterally I cannot make them think or learn.
- I can, at times, coach students with their own self reflection and realistic self assessment.

In addition, Minna suggests a starting baseline when she tells students: "I can only work as hard for you as you are willing to work for yourself."

The remainder of this column examines each of these thoughts in some detail.

Boundaries on my Role as an Instructor

- "I have a mental block about — — ; I don't get what I'm supposed to do or how to do it."

Periodically, I hear broad and vague laments, such as "I can't do math", "the reading made no sense", "I get anxious when I try to work with my lab partner", etc. Pragmatically, I can explore with a student what such statements might mean and whether troubles might suggest possible patterns. Are difficulties with math related to computation, abstraction, translation of words into mathematical statements, etc.? How does a student approach reading, does the student have difficulties comprehending text in all subjects or only in some, and does the student follow the same approach in reading computer science as short stories or novels? Is the student anxious working with a particular person or type of person, or is all group work intimidating, etc.?

Sometimes, when issues relate to my specific course, it seems helpful to examine a specific example. In a recent reading or assignment, what part of the mathematics caused trouble, where did the student get stuck in an assigned reading, or what made working with a specific lab partner challenging? Rather than lamenting in general, it can be helpful to work with a student to identify specific strategies for the next assignment, program, or test.

However, when an issue seems to extend beyond a local environment—perhaps to several courses or subjects, then I need help. I cannot diagnose handicaps, learning disabilities, psychological or medical challenges, reading disorders, etc. In such cases, I can be supportive and sympathetic, but I have no training to uncover root causes.

Fortunately, I can be aware of resources available for referrals, and I can be pro-active in encouraging a student to consult professionals. In math, a student could talk to a mathematics laboratory or tutoring center to discover the student's proficiencies and gaps, and I can work with the student and the laboratory or center if corrective actions are needed. In reading, I can encourage a student to work with a reading lab to assess both reading speed and comprehension. For social interactions, I can encourage a student to talk to an academic counselor—perhaps even walking with the student to a counseling center to consider possible resources and approaches.

In summary, I find it useful to explore broad and vague statements to begin to understand the issues at hand. If the difficulties relate to my professional training in mathematics and computer science, I may be able to suggest approaches, provide background, and guide study. However, my background is obviously limited, so for serious investigation in other areas, the best I can do is to get the student to the professionals who can help.

Clarify Expectations and Provide Context

- "This course is too hard, because problems are difficult, and I have to think too long."

For the first 2-3 semesters in my college career, I was convinced I was going to flunk out. The issue was not if, but rather when. Although I had done well in high school, college was different. I did not know how college was different, but I was scared. In actuality, I worked much harder my first year in college than my second, and I worked much harder my second year than my third. During those years, however, I also learned how to study and work more effectively, and my grades went up substantially each successive year. On the other side, in my teaching, I also have met numerous students who expected college to be just like high school.

For many students, academic expectations change from secondary school to introductory college classes to upper-level college courses. Further, students at one level may have little experience regarding work at the next level, and many discover that what was successful yields mediocre or poor results in later courses. For example, after students proclaim that problems in my course are impossible, I ask how long they worked on their solution. When they report spending a full 25 minutes on mathematics homework or a programming assignment, I often respond that college typically expects students to work 3-4 hours on homework for each hour in class. In this context, 25 minutes represents only a respectable start. As course levels increase, problems and assignments become more complex, and students must work through multiple steps—each requiring insight and care. This takes time, patience, and creativity.

But why should expectations change and work get harder through a curriculum? Here are two parts of an answer.

- Problems in real life are often complex and difficult, and academic programs naturally seek to prepare students to address these challenges. Thus, students should expect to encounter progressively harder problems through their academic work—in many cases, assignments should be hard and challenging, as students prepare to address complex societal and business needs.

- Life would be dull and boring if we had to do the same work year after year. Rather, insight and excitement expand as we engage new ideas, new problems, and new approaches; and new knowledge can change how we think. For example, in his book, *Science of Programming* [76] , David Gries shows how the development and use of loop invariants can lead to simple, correct, and elegant code. To illustrate, consider the problem of reversing a linked list ([76, p. 215]) or sorting an array of 3 colors (the Dutch National Flag Problem, [76, pp. 214-215]). Initially, students often attack such problems with nested loops that require moderately complex logic and run in $O(n^2)$, and off-by-one errors are common. However, when code is developed in conjunction with carefully-identified loop invariants, solutions turn out to be simple and clear, run in O(n), and often run correctly the first time.

In summary, students have limited backgrounds. The content, approaches, and depth expected at each level will likely require students to develop and mature as they progress through a degree program. In response, students may express surprise and/or frustration. Instructors may be able to help—not by changing expectations, but by articulating new expectations, suggesting approaches that may be helpful, and encouraging students to stretch themselves.

Provide Background and Guidance, but Students Must Think and Learn

- "I had trouble with this homework problem; can you show me how to do it?"

From time to time I observe to students that if the real purpose of an assigned problem was for me to determine the answer, then I would be most effective if I did the problem myself. No need for all students to prepare and submit solutions and for me to grade their work. I could write a solution quickly, and save everyone time and effort. Instead, assigned problems provide opportunities for learning—integrating ideas, applying concepts and techniques in new contexts, and constructing polished solutions with care.

When students have trouble with a problem, opportunities arise for learning! Rather than give immediate and full answers to a student question, I often find it helpful to review what the student has done. Does the student understand the question asked? What techniques or ideas from readings or the class have been covered, and do any seem to have potential to help in at least part of this problem? Did the student write down anything to get started on the problem? Where did the student get stuck, and why did further progress seem blocked?

Often when starting a new problem, students look for a previously-solved problem from the textbook or class to serve as a guide. As a general strategy, this approach can motivate reflection of what known techniques might be helpful, and how one might bring past experience to bear on new situations. This approach, however, depends upon the nature of questions being asked. If students are asked to do the same activities repeatedly, then past examples may provide a fine guide—one can edit one solution to get the next. However, such examples may motivate little thinking and even less development of problem solving. If a new problem is reasonably different from past examples, then considerable thought may be needed. Past problems may illustrate a range of techniques, but thought is needed regarding how those techniques might be combined creatively to address new challenges.

Further, everyone—faculty and published authors as well as students—make mistakes from time to time and start attempts to solve a problem that do not work out. Of course, published articles and textbooks rarely describe their many failed attempts—only polished successes appear. However, in the background, students should understand that making mistakes can provide rich opportunities for learning! In trying to solve a problem, students need not fear trying an approach that turns out not to work; rather try something, and then consider what worked, what did not work, and what can be learned from the experience.

Perhaps the key point in talking to students is that getting stuck is a common experience—we all get stuck when working on problems, and we all try approaches that turn out to fail. The challenge is what to do when the current approach does not work as anticipated. This is where creativity, insight, and perseverance comes in. As a teacher, one can ask what a student understands, what the student has tried, and why the student might think one approach or another could be fruitful. However, since assignments are [usually] intended to encourage thinking and learning, just giving the answer can be counter-productive. Asking questions, providing guidance, and presenting strategies for possibly getting unstuck can require time and effort, but also pay long-term dividends.

Encourage Self Reflection and Realistic Self Assessment

- "My grade does not reflect my real knowledge!"

Over the years, I have observed that student perceptions of their abilities often do not seem well correlated with the insights and solutions they express in their work. Some top students comment that they know they have less background and experience than others, and they think they have just been lucky in their grades. Other students are vocal about their strengths, but consistently perform poorly. And, of course, still other students express self assessments that fit well with their work. (A reviewer noted that the difficulty of students with self assessment is an example of a more general pattern, called the Dunning-Kruger effect. The interested reader can find numerous reports and perspectives on this effect through a Web search.)

When talking with students about their performance on assignments and tests, listening and constructive feedback can be particularly helpful. When students do well, celebrate, congratulate them, AND reflect with them about what they did that might have contributed to their success. Perhaps they reread the textbook, rewrote their class notes, worked through examples on their own, started an assignment early, organized their time to allow both adequate attention to the work and a healthful amount of sleep, made up similar problems for practice, joined a study group, etc. In short, if they were successful, they should consider doing more of whatever helped. Further, an instructor might boost a student's self-confidence by writing, "nice job", "great insights", etc., on papers or saying something encouraging directly to the student.

In contrast, if a student's performance does not meet their expectations or their rhetoric, an instructor might explore how the student approached the exercise, and how that approach might be adjusted. For example, when students use their study habits from high school within a college setting, Minna Mahlab reminds students that college has different expectations. If the old habits no longer are working, there is no reason to believe that doing more of the same will yield better results next time. Rather, one might try new techniques for taking notes while reading, more careful planning and time management, summarizing the main points of a class session or reading, making up one's own examples, getting adequate sleep to be fresh for a quiz or test, etc. Of course, an instructor also should be prepared if discussions with a student suggest patterns of difficulty that might be addressed by referrals to counselors, medical personnel, or other resources.

In short, instructors can encourage self reflection and conclusions based on actual evidence, while paying attention to matters of self esteem, study habits, time management, and other aspects of life. Evidence can help build self confidence in those who may be overly nervous, but evidence also can help provide a reality check for the over-confident who may be in denial of the need for change.

Of course, this self assessment should not be considered a one-time activity. Rather, students should practice self reflection and evidence-based self assessment regularly—refining this skill of self analysis, just as they hone their other academic skills.

Learning Requires Constructive Student Effort, Perhaps with Mentoring and Guidance

- "I work and work and work, and I still am doing badly in the course"

I hear such complaints from time to time, and some of my first reactions focus on whether the work followed a logical plan, what work did a student actually do, how long did they spend on that work, and to what extent was the work sidetracked by distractions.

Within office hours, I especially appreciate a student who has done [or tried to do] the reading, who has tackled part of the assignment [perhaps getting stuck], and who arrives prepared with a list of questions or observations. Minna Mahlab notes that "a well-prepared student is never annoying to an instructor." The student is engaged with the material and has made progress (although perhaps not very much). In this context, the student's work seems productive, and we can discuss the issues at hand. Perhaps we should correct some misconceptions, fill in gaps, or brainstorm on what next steps might be productive.

On the other hand, I can do little when a student arrives without having tried the reading, attended class, reviewed class notes, or thought about forthcoming homework. When such cases arise, one approach can be to outline specific steps a student needs to take. "Before we can talk, you need to be prepared by doing X, Y and Z. When working on those tasks, either outline what you have learned or prepare specific questions for us to discuss."

In short, I can be encouraging, and I may be able to provide suggestions on how a student can be more effective in studying and attacking assignments and tests. However, since I am not a neurosurgeon—I cannot perform a brain transplant, I cannot learn for the student. Instead I need to review specific activities and brainstorm next steps.

Acknowledgment

Many, many thanks to Minna Mahlab, Director of the Science Learning Center at Grinnell College, for her many insights, creative approaches, helpful feedback, and constructive sharing over the past two decades. Particularly thanks for her thoughts and suggestions regarding the writing of this column! Thanks also to the reviewers for their feedback.

Conclusions

When working with students, active listening is vital. As students talk, an instructor can gauge the level of understanding, underlying misconceptions, mastery of both concepts and details, approaches to problem solving, etc. Further discussion can uncover how a student studies and approaches class activities. With such interactions,

- An instructor CAN mentor and coach, building upon professional background and subject-matter expertise.
- An instructor CAN encourage a student to move from broad and abstract statements (e.g., "I can't do this") to more specific statements that can be addressed (e.g., "I need to think about how to proceed when I get stuck").
- An instructor MIGHT help a student discover patterns that yield referrals to resource centers, such as counseling, mental health, and physical well-being.
- An instructor MIGHT be able to help a student become more effective and efficient in learning, by encouraging reflection and realistic self assessment.
- An instructor CANNOT learn for a student—an instructor can try to facilitate, but the student has to work on the learning itself.

38.2 SUBSEQUENT REFLECTIONS

The original article briefly considers the value of self reflection and realistic self assessment by students. In some cases, students may do well in reviewing their own strengths and weaknesses, and they can properly evaluate their status in a course. For others, however, self assessment can be hindered by psychological considerations or the lack of a realistic frame of reference. Two factors that can interfere with self reflection and assessment include the imposter phenomenon and the Dunning-Kruger effect. When either of these factors arise, an instructor may need to be particularly creative in supporting strong students and in identifying mechanisms to help weaker students.

The Imposter Phenomenon

Some outstanding students may believe that their success is an accident. From their perspective, they are not really very intelligent or insightful or effective, but their excellent performance on papers or assignments or tests depends upon luck or accident. Yes, their grades may be wonderful, but deep down they feel this achievement is not well founded. In time, they believe, their lack of ability will be discovered, and they will be shown to be the imposters they think they are. As an example, when I started college, I was convinced that I would flunk out; the question was not "if", but rather "when". Yes, I was getting reasonable grades, but such positive results just put off the inevitable. In short, such feelings illustrate the *imposter phenomenon*, sometimes called the *imposter syndrome.*

Originally Clance and Imes described this experience in their work with women: "Despite outstanding academic and professional accomplishments, women who experience the imposter phenomenon persist in believing that they are really not bright and have fooled anyone who thinks otherwise" [35, Abstract]. In recent years, discussion has expanded to other groups that are under-represented within the context of computing. Those subject to the imposter phenomenon may have outstanding records; grades and professional accomplishments may demonstrate considerable achievement and success, but the individuals still may believe they have been lucky and will be found out as incompetent at some time in the future. As an example, Harvey Mudd President, Maria Klawe, has written a particularly insightful article in *slate* about her experiences [99].

In addition to lowering motivation and self confidence, individuals may live in fear of being discovered as phonies. Further, as others are seen as being successful, those affected by the imposter phenomenon may feel that they do not fit into the successful group, and this may be a contributing cause for relatively high drop out rates from STEM areas for certain women and students from other under-represented groups.

As instructors seek to support and encourage strong students, the imposter phenomenon can undermine confidence and feelings of success. For instructors, connecting with these students can be particularly challenging; the students may feel they have just been lucky, raising a fear level that they may be more likely to be discovered as phony as they work more with faculty. Combating such perceptions can be challenging, and I make no claims that I have all of the answers. The following list indicates some possible approaches, but faculty may need to read and experiment to discover a range of approaches that can help.

1. Clance and Imes, in their pioneering article, identify elements of group therapy that may be helpful in overcoming this view of an individual being less capable than classmates. [35, pp. 246-247].

2. Maria Klawe promotes the value of talking about the imposter syndrome to prospective and incoming students, at conferences, and in conversations. She also suggests some

    > ways you cope with impostoritis (mine were "practice, practice, practice," "surround myself with support," and "look back as well as ahead"). [99]

3. Miyake et al report another approach that involves two short, in-class writing exercises, in which students reflect upon values that are important to them. [128].

    > Values affirmation, in which people reflect on self-defining values, can buffer people against such psychological threat. When they affirm their core values in a threatening environment, people reestablish a perception of personal integrity and worth ....

> The values-affirmation intervention used in this study involves writing about personally important values (such as friends and family). The writing exercise is brief (10 to 15 min) and is unrelated to the subject matter of the course. Nevertheless, it has been found effective in improving the grades of ethnic minority middle-school students and closing the racial achievement gap [40].

Miyake utilized this approach within an introductory physics course, with 15-minute writing exercises in weeks 2 and 4 of a 15-week course, just before a test in week 5. In describing the three-page exercise given to students, Miyake et al indicate

> The first page listed 12 values: *being good at art; creativity; relationships with family and friends; government or politics; independence; learning and gaining knowledge; athletic ability; belonging to a social group (such as your community, racial group, or school club); music; career; spiritual or religious values; and sense of humor.*[129]

The second page of the exercise provides prompts that ask students to describe "why the selected values were important to them (affirmation condition)", and the third page encourages further reflection.[129]

Although not completely closing a performance gap between men and women on test 1, this modest intervention did narrow results, the impact was observable later in the semester as well, and final letter grades for women improved from an average C to and average B for those involved in the intervention.

Also, since the exercises in this intervention were not related to physics, similar results seem reasonable for performance in computing courses as well. With this level of success with simple classroom activities, teachers may want to explore the imposter phenomenon and possible interventions in more detail as they plan their courses.

4. Personally, toward the end of an introductory course, I may strongly encourage a capable student to apply to be a class mentor, grader, or tutor for the next semester, and I may suggest that the student apply for a summer research experience with a faculty member over the following summer.

   Such encouragement may help bolster their feelings of success in a course, but it may not address the imposter phenomenon directly. On the other hand, if students can interact with others in the classroom or by grading or by tutoring, then the students may start to understand better their own strengths and insights.

   Similarly, when students join a research/development team, they may start with considerable nervousness, but they also can experience solid success over the long term.

The Dunning-Kruger Effect

Another factor influencing a student's self assessment within a course may be the Dunning-Kruger Effect. In a 1999, Justin Kruger and David Dunning observed that unskilled and low-ability individuals may overestimate their skill level, and "their incompetence robs them of the metacognitive ability to realize it" [101]. In a series of exercises, students took tests to determine their abilities at determining humor, performing logical reasoning, and analyzing American grammar. During this work, students also were asked to determine their success on these tests and their relative rank among their peers.

Throughout these studies, low-ability students consistently over-estimated their abilities, with students performing in the bottom quartile often indicating they were performing above average. Further, when shown a range of tests by others, the low-achieving students had difficulty evaluating the quality of those tests and determining their own performance relative to the others. As Ehrlinger, Johnson, Banner, Dunning, and Kruger summarized, "An additional meta-analysis showed that it was lack of insight into their own errors (and not mistaken assessments of their peers) that led to overly optimistic estimates among poor performers" [63, Abstract].

As one consequence of this effect, "College students who hand in exams that will earn them Ds and Fs tend to think their efforts will be worthy of far higher grades" [62].

At the other end of the ability spectrum, top-performing students often rated their work as being just slightly above average as well. Dunning and Kruger noted that some of the best students seemed to assume that everyone else was also doing well, so their strong scores were neither unusual nor note worthy.

Overall, the Dunning-Kruger Effect may suggest that some students may have stunningly incorrect views of their work and progress. From an instructor's perspective, a challenge is to convey to weak students that their work is not very good, and they may need to change their ways (e.g., study habits, approaches to homework, test preparation, etc.) Similarly, the best students may not realize their talents and successes.

As with possible responses to the Imposter Phenomenon, specific approaches for instructors for the Dunning-Kruger Effect may be difficult and require experimentation. Three initial ideas follow, and teachers are encouraged to consult the literature for additional ideas.

1. Dunning and Kruger, as part of their 1999 study, found that giving students some additional background and analytical skills may help self assessment; that is, by giving even a minimal level of additional background, student self-assessment may improve substantially.

2. Some anecdotal comments from conference attendees suggest that providing feedback on student test papers may be a start, indicating what responses are incorrect and why. When students learn what they have done is wrong and why, the students may better assess their current status in a course. However, the Dunning-Kruger Effect also indicates that some low-performing students may be clueless that they are not doing well. Instructors may need to consider how to identify such students, how to convince those students that they are not doing well, and how to encourage them to obtaining tutoring or other help.

   On the other side, instructors might consider how to communicate to top-performing students that they are indeed doing outstanding work, and encouraging them to continue and move forward.

3. In my own experience, I find that providing status reports for classes can help clarify to students their actual performance and how their work compares to that of others. In particular, I distribute a grade report every 2-3 weeks in a course, starting about mid-semester. This report indicates all grades that I have recorded, including test results, lab grades, programming scores, and any extra credit obtained. In addition, the report displays the current semester averages of all students in the class (with no names, of course). Together, this type of report serves at least three purposes:

- Students can check whether my records agree with their understanding of scores. Over the duration of a semester, I have to enter hundreds (or thousands) of grades, so typographical errors are a definite possibility. If students show me an exercise that they have gotten back, there is no trouble with me fixing a grade that has been miss-recorded.
- Students receive clear feedback on their current semester average. If that is low, for example, the report might encourage them to focus better on handing in assignments and completing exercises carefully.
- Students can determine how their semester average compares with others.
  - ◇ If their semester average is low and near the bottom of the class, they know clearly that what they have been doing is going badly. In my experience, such a report does not motivate everyone, but it does motivate a reasonable number and challenge their false sense of coasting along. For example, it is not uncommon for some students to talk to me about changing their approach: if their current activities are not going well, we can brainstorm what they might do differently.
  - ◇ If students are doing particularly well, they may develop a better sense of their abilities and talents.

As an example, the sample status report in Figure 44.1 illustrates a form that I typically distribute during a CS2 course. Note that the course allows several options for extra credit, so that some averages may exceed 100%.

Overall, continued education, careful feedback (showing strengths and errors), and forthright status reports may provide some help to instructors in helping students to understand the quality of their work—both positively and negatively. Faculty may need to consider additional study and experimentation to address the Dunning-Kruger Effect more completely.

Figure 44.1: A Typical Student Status Report

Status Report for CS2 - 02:57 PM on Tuesday, April 04, 2017

Name: Student's Name
Current Semester Average: 81.6% (Unverified)

Tests — Average 76.93%

	Original Score	Original Max.	Revision Score	Revision Max.	Adjusted Average
Test 1:	89	108	112	135	82.7%
Test 2:	89	108	-	-	82.4%
Test 3:	71	108	-	-	65.7%

Projects – Average: 92.0%

Required:	000	001	010	011	100	101	110	111
	25	21	24	24	25	-	18	24
Out of:	25	25	25	25	25	-	25	25

Supplemental Problems – Average (including extra credit): 72.8%

Required:	1	2	3	4	5
	15	13	21	23	19
Out of:	25	25	25	25	25

Optional Supplemental Problems:	6	7	8	9	10	11	12	13	14
	-	-	-	-	-	-	-	-	-

Labs – Average (including extra credit): 87.2%

Required:	Funcs	Floats	Lists	Movies	Queues
	23	15	19	25	17
Out of:	25	25	25	25	25

Optional:	Loops	Ints	Picts.	Movie	Music	Bash	Files
	-	-	-	-	-	-	10

Contact me immediately if you believe any of these scores are incorrect.

The following lists the approximate, current semester averages for this course.

102.9	102.2	101.9	99.9	98.4	97.9	97.5	97.1
96.5	96.3	95.2	92.9	91.3	90.6	89.5	89.5
88.9	88.3	87.6	87.1	87.1	84.1	83.2	81.6
80.9	80.8	77.1	76.3	75.3	55.5	38.0	25.3

CHAPTER 39

What teachers should, can, and cannot do

This column first appeared in the SIGCSE *Bulletin*, Volume 36, Number 2, June 2004, pages 20–21[217] Subsequent reflections on this subject appear later in this chapter.

39.1 ORIGINAL ARTICLE

AT A BASIC LEVEL, learning requires physiological changes within students' heads, and education seeks to accomplish appropriate neurological changes effectively and efficiently. While schools and faculty cannot change physiology by themselves (thank goodness), what roles should teachers play in the process? Here are three extreme models.

1. *In loco parentis:* schools act as substitute parents. While this includes teaching academic subject matter, schools also are responsible for social, ethical, and personal behaviors. As local parents, schools impose discipline, foster moral codes, develop leadership, and supervise personal choices. Schools also pay attention to personal needs and circumstances, looking out for the well-being of each student. For example, preschools largely act *in loco parentis.* Preschoolers must be taught appropriate behaviors, how to interact with others, etc.; and preschools devote much attention to the socialization process.

2. *Teachers as smorgasbord providers:* independent students take responsibility for their own decisions; faculty provide programs, courses, information, a productive environment, meaningful feedback, and serve as facilitators to help students learn effectively and efficiently. Students act as responsible, mature, and intelligent adults and consumers who select rationally from the program and course menus provided by faculty according to personal student goals. Within a course, students take responsibility to engage with the new material presented, understand new perspectives, analyze material critically, integrate ideas, and generally take best advantage of this stimulating learning environment.

3. *Teachers as servers, perhaps servants:* teachers facilitate learning following student concerns, demands, needs, interests, and even whims. Students control content, learning environment, pace and organization of material, nature of assignments and/or tests, etc. Students also can dictate changes in content, environment, pace, organization, assignments, due dates, etc. While this model may generate images of me-

dieval kingdoms, those monarchs likely controlled the teachers rather than the student children.

These models are extreme, and many readers will identify difficulties with each – at least for teaching computer science. However, in recent years, I have had numerous conversations with teachers, particularly new teachers, who seem to believe that they shoulder the entire responsibility for student learning and must accommodate every student circumstance. In this view, apparently, students need do nothing, any lack of learning represents a failure solely for teachers, and some teachers bow to most student requests; other actions yield low evaluations and a denial of contract renewal or tenure. While some of these faculty may consciously have chosen the *faculty-as-servant* model, these conversations raise questions regarding appropriate roles and actions for faculty.

More often, I expect schools and faculty seek some middle ground. Generally, schools for young children pay considerable attention to personal needs and include some parental roles. As students get older, schools expect students to take greater responsibility. In high school and college, faculty often have different expectations and take different roles for first-year students than for seniors three or four years later. Since most SIGCSE members teach computer science, we now consider expectations for schools, faculty, and courses after the primary years.

In any model, some expectations seem common. Faculty understand the discipline and take primary responsibility for curricula. Consistent with *in loco parentis*, faculty identify key elements, motivating degree requirements for graduation and course requirements for a computer science major. Electives within a degree or major allow students to select responsibly from specified alternatives, based on their interests and goals. "Topics" courses allow flexible study according to the short-term interests of student or instructor. Note, however, that schools vary significantly regarding the scope of requirements, the role of electives, and the inclusion of variable-content courses. Some schools have extensive requirements for graduation and a major, while other schools rely upon active advising with minimal formal requirements. Some encourage student planning by minimizing variable-content courses, while others promote these courses as providing exceptional flexibility.

Within a curricular framework, faculty teach courses and interact with students. A faculty member has a professional responsibility to cover essential topics that fit within the larger curriculum. This includes setup before the class starts, on-going preparation throughout the course, activity to convey understanding effectively, and creation of a supportive learning environment.

Beyond such generalities, many specifics depend upon the operative model of roles adopted by the school and its teachers and students. Here are some examples.

- Faculty can assign readings, but to what extent should students be expected to take the initiative to read the material?
- Faculty can assign exercises to expand background and clarify material, but how far should a faculty member go to make students do the exercises?
- Faculty can organize and present material, but what responsibilities do students have to study, think, analyze, and engage in the course?
- Faculty can present material in lectures or foster class discussions, but how much effort should be exerted to make students engage with the material during a class session?

Answers to these questions vary from one environment to another, and new faculty should consult colleagues regarding expectations and practices. Faculty and administrators also might collaborate to formulate institutional policies, establish practices when questions and complaints arise, and identify appropriate responses to common circumstances. Here are some examples.

- *Class Attendance:* Students often have the responsibility to attend classes and tackle assignments. However, sometimes students become ill, time conflicts arise, students address problems of over commitment by skipping classes, or students may not feel like attending. If faculty serve at the pleasure of students, faculty should accommodate these circumstances. If faculty present a smorgasbord, students have full responsibility to attend class and should not expect faculty to duplicate classes or make other accommodations If faculty serve as local parents, more analysis may be needed. When absences can be anticipated, students might be expected to take some responsibility, checking in advance about what will be missed, working to make up missed work, getting class notes, and reading material covered in textbooks. However, when students have been ill or missed a class due to circumstances beyond their control, other adjustments may seem appropriate. Personally, I have little sympathy for a student who misses a class without good justification, but I make great efforts to help a student who experiences unforeseeable circumstances (as verified by the Office of Student Affairs). If student indicate they missed my class because they were working on an assignment for another instructor, I tell them catching up is their problem; they need to take responsibility for their choice. Only after they have class notes and have done the reading can they come to office hours to ask specific questions. Not only does this reinforce students understandings regarding choices, it also may help them grow and mature as they learn to take responsibility for their actions.

- *Extensions for Assignments:* Students may ask for extensions on assignments or request that a test be taken at a later time. Again, the teacher's response may depend upon perceived roles. Depending upon the model, it may be appropriate to distinguish situations of high stress, illness, family emergency from other cases. In this regard, it also may be useful for the instructor to consider what long-term messages are being sent to the student. For example, if students consistently request and receive extensions, they may not learn to budget time and take responsibility. In one case, I know of a student who failed to turn in a polished thesis before the institution's deadline for graduation. While this school's policies were well publicized and known to be beyond the discretion of faculty, the student then blamed the instructor for not enforcing assignment deadlines earlier.

- *Missing a Test:* Some instructors try to telephone a student's room if the student is absent when a test starts. This approach fits with *in loco parentis* and can solve the problem of a student sleeping through an alarm clock. Further, some instructors prefer this type of effort to handling later debates if the student later wants another chance. Alternatively, responsibility could lie completely with the student, yielding a grade of zero. Allowing a student to make up a test may be a middle ground, although this approach raises many questions of fairness and test security.

- *Poor Student Performance:* If a student performs poorly in class, what actions should an instructor take? First, suppose the student is not trying, does not attend class, ignores offers for tutoring, etc., Then a viewpoint emphasizing student responsibility suggests an instructor need take little action, while an *in loco parentis* perspective might investigate the root causes for this non-productive behavior. If, however, the student is trying, the instructor may be willing to work with the student, perhaps at length, to help overcome obstacles and provide guidance. The extent and nature of this help, however, may relate to the roles of both the instructor and student.

Space limitations prevent an extensive discussion of these circumstances here, although a later column could incorporate reader feedback. I hope this discussion will stimulate readers to consider how the view of instructor and student roles can clarify faculty responsibilities and guide instructor responses to common circumstances.

Acknowledgment

Many thanks to Professor Pamela Ferguson and to the Science Teaching and Learning Group at Grinnell College for helping me clarify and refine my thoughts on this topic.

39.2 SUBSEQUENT REFLECTIONS

The roles of faculty continue to be discussed on many campuses. The models identified in the original column illustrate some extremes, and both schools and individual faculty are deciding what makes sense for them.

Beyond the general consideration of roles, three commonly discussed topics involve support services, the structuring of assignments, and "late days".

Support Services

As schools seek to improve student success and graduation rates, much focus has been placed on student services. New programs have come into being, and some existing programs have expanded. Common support services involve such resources as:

- physicians and counsellors for physical and mental health,
- opportunities for individual or group tutoring,
- advisors and/or workshops for academic advising, student life, time management, study skills,
- programs to address challenges faced by students with disabilities,
- etc.

On many campuses, such lists of support services may be extensive. Since faculty members focus on course content, pedagogy, and related academic matters, all of these resource areas are likely beyond the expertise of teachers; faculty members rarely are trained to address physical, mental, and personal matters.

However, faculty might be expected to learn about what resources are available. Faculty typically will not know answers to various personal issues, but faculty might be able to refer students to those specialists and counsellors who are. Further, since faculty may interact with students on a regular basis, students might seek out a faculty member when difficulties arise. In such situations, faculty might be sympathetic, and then an instructor may take the initiative to help a student obtain help. (Additional notes on the topic may be found in the previous chapter, Chapter 38.)

Structuring Assignments

Many discussions in recent years have explored approaches to break large assignments or projects into relatively small and manageable pieces, with each step building on work done previously. Sometimes called *scaffolding*, one common motivation is to assist students with time management. In principle, students might be expected to organize their time, but in practice may components of life (e.g., family responsibilities, work, study for other courses, etc.) interfere. Students may have good intentions in organizing their work, but short-term demands often get in the way.

For example, if a capstone or project-based course is to conclude with a 30-40 page paper, based on independent scholarship, one approach might specify a final deadline for the project and leave scheduling to the student. Although my experience on large projects is limited, work with students on small projects indicates such flexibility is rarely successful. Rather, final results seem best when milestones are set throughout a semester (e.g., weekly). Individual deadlines might be adjusted with discussion and negotiation, but a week-by-week outline of work helps keep students on task and progressing.

Similarly, a computing course may involve some programming projects that extend over 5-10 weeks. To guide students, various papers at computing-education conferences have described experiences dividing one large project into smaller activities with deadlines every week or two.

Also, in my own experience with weekly labs or problem sets, I have found results are substantially improved if the write up is divided into two parts (with due dates 3-4 days apart), rather than having a single weekly deadline.

With such divided assignments, a critic might argue that an instructor is adopting the perspective of *in loco parentis*; the instructor controls time management rather than placing responsibilities on the student. In some ways, the critic seems justified; the notion of scaffolding fits with this perspective. However, scaffolding also acknowledges practical needs for finding priorities within contemporary society, and a search in the ACM Digital Library under "scaffolding" identifies dozens of articles that report experiences with this approach.

"Late Days"

Another approach to balancing student's responsibility for time management with the demands of modern life involves the notion of "late days" for assignments. The idea is that each assignment in a course has a due date, but as demands on students arise, students are allowed to turn in a few assignments late—within specific bounds.

As an example, Thomas Murtaugh at Williams College credits colleague, Jeannie Albrecht, with the idea of late days for assignments in his CS1 course:

> You may use a maximum of four free late days during the course of the semester. A late day permits you to hand in a program or a written assignment up to 24 hours late, without penalty. To use a late day, simply email your instructor and let them know that you are using a late day. You should use no more than two late days for a single assignment. [134]

When late days are exhausted, some faculty may no longer accept late homework, others may impose a point or percentage penalty, etc.

Altogether, "late days" provide modest flexibility, but students also must assume some responsibility and decide when a late day might be best used.

CHAPTER 40

Thoughts about lecturing

This column first appeared in the *SIGCSE Bulletin*, Volume 38, Number 2, June 2006, pages 19–21[220] Subsequent reflections on this subject appear later in this chapter.

40.1 ORIGINAL ARTICLE

LECTURES have a negative image these days. Yet many conference sessions follow a lecture format, and anecdotal evidence suggests that many courses utilize lectures extensively. This column seeks to encourage discussions of pedagogy by identifying advantages and disadvantages of lecturing and presenting variations of the lecture format that address some of its criticisms. Please contact the author with your feedback and arguments in rebuttal!

Advantages

Even with its negative press, lecturing has some endearing qualities:

- Lectures provide a framework for the presentation of considerable information efficiently.
- A lecturer largely controls timing, format, approach, and content.
- Lecturing allows rehearsal, the refinement of structure and presentation, and the polishing of materials.
- Handouts and slides can supplement the oral presentation (e.g., adding details and references).
- If distributed in advance, listeners can annotate handouts during a talk.
- Utilizing presentations and written materials, lecturing can fit with learning styles that involve both visual and aural components.

Efficiency and control issues may explain the common use of lecture at conferences. For courses, lectures may be attractive in at least two settings. First, in large classes, student-instructor interaction is inherently limited. For example, if a course meets for 50 minutes three times a week and has 75 students, then a 2-minute question/answer for each student requires a full week of class time.

Second, lecture may seem attractive when teaching a course for the first time. A new course typically comes with few handouts, group assignments, and labs to support active-learning formats. Also, with little experience, it is sometimes difficult to anticipate what troubles students might encounter, and lecturing provides a framework to control the pace and content of a course. As an instructor gets new insights in subsequent course offerings, group and individual activities can be added, but these require time, experience, and ongoing creativity to develop.

Disadvantages

Many articles compare passive- and active-learning formats:

- Lectures allow listeners to be passive and thus less than fully engaged in the material.

Other potential disadvantages may not be as widely publicized:

- Lectures may be rather inflexible, especially when based on slides or published outlines; it may be difficult to respond to questions by listeners.
- If a lecture covers the same material found in reading, listeners may have little incentive to read a textbook or required articles.
- Slides or other media may distract both lecturers and listeners, so unnecessary attention is paid to glitziness—encouraging entertainment rather than content and effectiveness.
- Although an introduction may outline the structure of a talk, listeners may have difficulty keeping track of how subsequent details fit within that structure.

Some of these points are illustrated in a particularly nice way in the "PowerPoint Presentation of the Gettysburg Address" by Peter Novig. Readers unfamiliar with this piece may view the slides at http://www.norvig.com/Gettysburg/ [138].

Overall, a lecture format may provide an efficient way to present much material in a constrained amount of time and in a controlled way, but several issues may limit a lecture's effectiveness to promote learning.

Possible Variations

Some adjustments to a lecturing format can help address some of the difficulties identified earlier, In some cases, these thoughts regarding pedagogy also may apply to other class formats as well.

Slide Format

Years ago, lecturers often wrote an outline of a talk on a blackboard during the introduction and then checked off each topic when it was covered. This provided a framework for the overall presentation and tied various pieces together. Anecdotal observation suggests that some talks using computer software for projection return to the same outline periodically. However, such repetition seems inconsistent.

As an alternative, consider the following screen shot (Figure 40.1) for a talk on "Getting Published at CS Education Conferences: Factors for Acceptance, Approaches, Reflections" that I gave at a CCSC Midwest Conference in 2003.

FIGURE 40.1 Screen shot for a 2003 talk at a CCSC Midwest Conference, showing slide on right, outline at left, and highlight of slide in outline

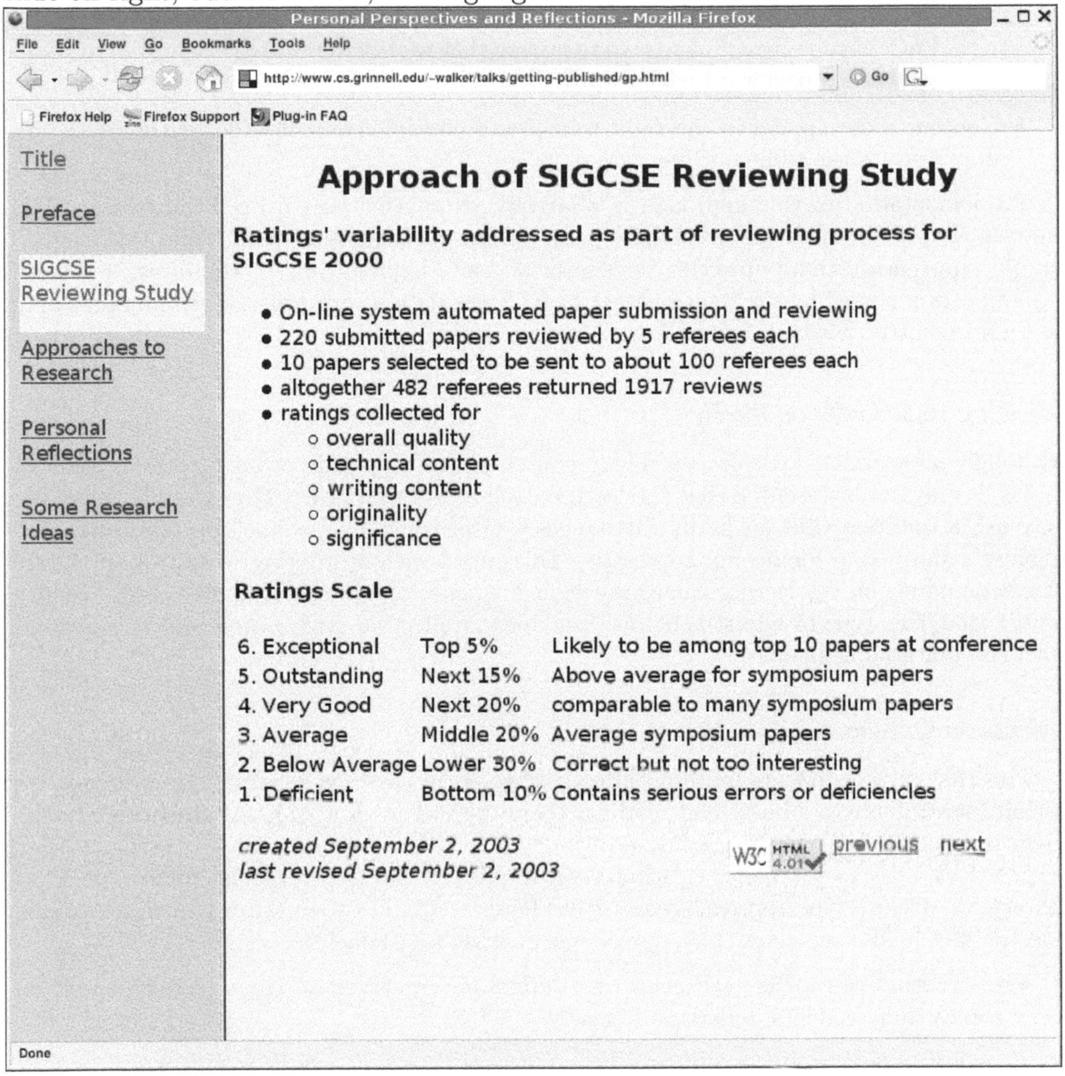

These slides have several useful characteristics:

- They are written in W3C standard html.
  - ◇ Most browsers render standard html in a reasonable way. (Going outside the standards may allow lovely effects in some browsers, but may not render well in others.)
  - ◇ Slides may be viewed through any browser (at the talk or by the user afterwards).
  - ◇ Font size may be adjusted within a browser to accommodate the projection system and conditions of the talk and the eyesight of viewers.
  - ◇ Highlighting the W3C standards can help inform listeners of the existence and value of these standards.

- An outline appears in a left frame throughout the talk
  - ◇ The current outline topic is highlighted to show progress through the presentation.
  - ◇ Each outline item links to the start of that section, enabling a lecturer to jump forward or backward easily through a talk.
- Content slides appear in the right frame, and multiple content slides within a section may refer to the same outline title.

Implementation of this approach is relatively straightforward using html frames. The indexes with different highlighted lines may be done as separate html pages, although a simpler approach uses a php script to present and add highlighting for the index, based on a query string passed by the right content page. Some details, examples, and templates may be found at http://www.cs.grinnell.edu/~walker/talks/

Adjusting Topic Order on the Fly

Although preparation for a lecture likely contemplates a specific order of presentation of topics, it may not be essential that the lecture itself follow that order. For example, a listener may ask a question that naturally introduces a later topic, or feedback on one topic may suggest a jump to a forthcoming example. Of course, such reordering of topics on the fly places demands on the lecturer and may reduce a sense of control. However, many faculty report that this type of adjustment also provides intellectual stimulation and engagement with the material at had.

Discussion Questions

A class that simply repeats material already covered in assigned reading can reduce motivation for students to do assigned reading; there may be no downside for students to come to class unprepared.

If a lecture is to supplement rather than duplicate, the lecturer must determine what material to discuss. One approach requires students to submit two questions on each reading, due the day before the class. This device has at least four benefits:

- instructors can focus each class on troublesome areas; little time need be spent on topics that students understand clearly.
- students must do the reading in order to prepare their questions, so they come to class with a reasonable level of preparation.
- students strengthen their abilities to read technical materials – a skill that is vital for lifelong learning.
- instructors can identify and correct misunderstandings at a relatively early stage.

Some questions indicate that students understand the main elements of the reading, and a class may focus on enrichment or discussion. Other questions indicate particularly difficult passages that require expansion and clarification. And other questions indicate student misconceptions that must be corrected before moving on to the main ideas for the reading itself.

In my experience, giving some minimal credit (e.g., 5%–10% of the semester grade) for the submission of reasonable questions provides adequate incentive for students to do this work.

Questions/Feedback at the End of a Session

Rather than give discussion questions ahead of time, instructors may encourage active listener involvement by asking a question to be completed in the last 2–3 minutes of a class. Answers suggest student understanding and provide feedback for the focus of the next class. An alternative approach asks listeners to write out at the end of each class what they believe were the 2–3 main points of the session.

Conclusion

A lecture format has some advantages, especially related to presentation efficiency and control. However, lectures also have potential difficulties. Rethinking the structure of slides and finding ways to address student questions without duplicating reading may help address some of these problems.

40.2 SUBSEQUENT REFLECTIONS

The original column identified several advantages of a lecture format, including efficiency for presenting material, instructor control, and coordination of in-class presentation with slides and handouts. However, the original column also indicated several disadvantages, with one primary issue being passive- versus active-learning: "Lectures allow listeners to be passive and thus less than fully engaged in the material."

Interestingly, in the 11+ years since the original column appeared, many discussions at computing-education conferences have highlighted two contrasting themes: the practice of utilizing lecture format for courses (particularly at the introductory level), and the need to engage students in class. From my observations, the widespread use of lecture continues for at least five reasons:

- *Inertia:* Traditionally, many classes followed a lecture model. Experienced faculty may have lectured for many years, courses progressed in a known way, and habit may be difficult to break. Some new faculty may have experienced lecture as a primary pedagogy during their schooling; perhaps lecture worked well for this select group. For experienced or new faculty who are accustomed to a lecture format, there may be little appreciation of modern pedagogy and correspondingly little incentive to change.

- *Instructor time:* In my experience, the preparation of group activities, individual exercises, questions for student in-class response, etc. requires considerable time. When first teaching a course, an instructor may have few resources, and lecturing may require substantially less time than developing numerous materials to support active learning. Over several semesters, resources may evolve for in-class activities, but time constraints may limit what can be done initially.

- *Student reactions:* Some students may want to be passive. If students lack motivation (e.g., in a course taken only to satisfy a requirement), students may want to be spoon fed and not want to be challenged.

- *Response to increasing enrollments:* High course enrollments and limited faculty may combine to require large class sizes. Various student-faculty interactions and small-group activities may fit well within small classes (e.g., classes with fewer than 40 students), but substantial creativity may be needed when managing large classes (e.g., over 50 and especially in the hundreds). (See Chapters 26 and 29 and the notes below for some ideas for handling large-enrollment classes.) i

- *Large introductory classes allow small upper-level courses:* When staffing is limited, a dean or a department may promote a strategy of providing relatively small classes at the upper level, with the tradeoff that introductory classes are large. In at least a few cases, deans at small colleges that advertise close student-faculty interactions have proposed this approach as a mechanism to encourage faculty support for student research projects and/or capstone experiences.

Whatever the motivation, many studies have demonstrated that the traditional lecture promotes little student engagement and rather modest learning when compared to pedagogy that engages students directly. For example, physicists examined several pedagogical approaches in the 1990s, and in 1998 Hake published "A six-thousand-student survey of mechanics test data for introductory physics courses." Between 1992 and 1996, Hake collected data from 6542 students from 62 introductory physics courses, comparing results of traditional lecture with "interactive engagement (IE)" (now called "active learning"). He concluded, "48 courses ... which made substantial use of IE methods achieved an average gain ... almost two standard deviations ... above that of the traditional courses." [80, p. 64]

When using lecture, therefore, an instructor faces the substantial challenge of how to actively engage students. With a large group, student engagement cannot require individual student-faculty interaction; if a course met three times a week for 50 minutes a class, and if 150 students were enrolled, then a single, one-minute question or response by each student would consume the entire week of classes—clearly not a feasible approach. Altogether, mechanisms are needed for student input, discussed or received individually, but tabulated collectively.

One relatively high-tech approach to this challenge involves the use of peer instruction and clickers.[149] A low-tech approach utilizes cards that students can display in response to an instructor's questions. Card format can be reasonably straight forward.

- One side of the card should allow a student to display an answer to a question.
  - ◇ If all questions from an instructor will require a yes/no binary answer, then the card might have red on the top and green on the bottom. (To accommodate color-blind students, these colors might also be marked with a black label.)
  - ◇ If all questions from an instructor might require up to four options, then each edge of a card might display a different, contrasting color (with labels or numbers 1, 2, 3, 4 for color-blind students).
- The other side of the card should be plain and uncolored.

With this arrangement, an instructor may ask a binary question or a question with up to four options. Following the style of clickers, each student decides upon the proper answer and raises a card with the correct color on top. Although exact tallies may be difficult, an instructor can review the displayed colors to obtain an approximate distribution of responses with each answer. Further, since the back of each card has a neutral color, students in back cannot be influenced by answers in front of them.

With such an arrangement of cards or clickers, students are strongly encouraged to respond regularly to questions during a lecture format. Peer instruction or other small-group activities may further encourage active learning—even within a large-class context [149]. Additional ideas for handling large-enrollment classes may be found in Chapters 26, 29, and 32. i

In summary, lecture seems to continue as a widely-used pedagogy, particularly in environments with high student demand and large class sizes. The on-going challenge is to engage students actively in the material within this lecture-based framework.

CHAPTER 41

Teaching and a sense of the dramatic

This column represents the first of a two-part series, with this piece first appearing in the *SIGCSE Bulletin*, Volume 33, Number 4, December 2001, pages 16–17[212]. Shortly after this original column appeared, Max Hailperin, Gustavus-Adolphus College, emailed me with several additional ideas for theatrical activities within the classroom, and several of his notes appear at the end of this chapter.

Chapter 42 in this book comprises the second part in the series, having first appeared in the *SIGCSE Bulletin*, Volume 34, Number 4, December 2002, pages 18–19[213]. Since these columns form a two-part series, most of my reflections on both this topic appear at the end of Chapter 42, the next chapter in this book.

41.1 ORIGINAL ARTICLE

IN RECENT CONVERSATIONS, several computer science teachers noted they became teachers, at least in part, because they wanted to act; they wanted an outlet for some form of career involving acting. While teaching seeks first to facilitate student learning, this article begins a two-part series exploring the idea that elements of theater (e.g., dramatics, stage effects, entertainment) might be employed to promote learning.

The Classroom as Theater

While teaching certainly is not equivalent to entertainment, class periods need not be dull and boring. Periodically, I hear about mathematicians at various institutions who reinterpret the Monotone Lecture Theorem at every class. Clearly even lecture-oriented sessions can be better.

In a good stage presentation, a playwright organizes a story or idea into a framework that connects with an audience. Beyond the basics of diction and articulation, an actor pays attention to pacing, emphasis, tone of voice, and delivery. Both playwright and actor observe audience reactions and make adjustments to obtain desired responses. Over years of experimentation, playwrights and actors learn how to achieve maximum impact by varying their delivery, their stage presence, their positions on the stage, and their interactions with each other and with the audience.

Of course, many of these same qualities apply in analogous ways in teaching; and as a beginning teacher, I found this parallel helpful in providing a context for planning lectures.

As a teacher, I spend much time considering how to organize and present topics. Within the classroom, I highlight main points through repetition, tone of voice, and the use of colored chalk or marker when writing. I consciously pause to make a point, and I consider what props might be most helpful in demonstrations. Perhaps most importantly, I constantly pay attention to student reactions; I try to read faces to determine to what extent my "audience" is with me and how they are reacting. And, naturally, I adjust my presentations based on what approaches seem to work most effectively – ceasing practices that produce little or no reactions and expanding behaviors that connect with my students. Classroom presentations always involve a vital spontaneous element, but advanced planning provides guidance.

The following two examples provide some ideas to introduce theatrical elements into classes that are either lecture- or discussion-based. I plan to include additional examples in my next column; readers' suggestions are particularly welcome.

A Destructive Binary Search

The first example combining a sense of the dramatic with active student learning and class presentation involves a "destructive" binary search, which I first saw visiting a class led by my Grinnell colleague, Samuel Rebelsky. Using an old student directory as a prop, Sam asked students how to locate the entry for one of the students in the class. After some discussion and guidance, a first step emerged: look for the name on a middle page of the directory. Following the traditional binary search algorithm, the desired name was determined to come in the second half of directory, at which time Sam tore the directory in half – discarding the first half of the directory (perhaps in a nearby recycling bin?) and retaining the first half. Applying the algorithm to the segment of the directory remaining, a middle page again was consulted, the directory segment was torn in half, part was discarded, and part retained. This process continued, until the relevant page of the directory was identified, with the physical tearing of a directory adding a sense of the dramatic and highlighting the main strategy.

Once the relevant page is found, the active demonstration can continue within the page. One can compare the desired name with a name in the middle of the page, tear the page in half, and discard the appropriate half. While this tends to be somewhat messier than the original tearing of the binding, the sense of theater continues, and students come away from the class with a particularly vivid understanding of the binary search. (I still have a clear image of Sam's class after 2+ years.)

"Do Re Mi" as an Example of Object-Oriented Problem Solving

Last semester, my beginning course introduced both object-oriented problem solving and Java. After a few weeks, students seemed comfortable with various basics of classes, objects, and syntax; but they lacked a broader problem-solving perspective. Thus, I worked with the class to consider how to develop a program to sing "Do Re Mi" from "The Sound of Music". (While only a third of the class knew the song by title, another third remembered it from the association with the movie, and virtually everyone remembered it after we sang the first part. Surprisingly, a higher percentage of international students seemed aware of the song than American students.)

Program development can follow the movie script. The movie scene begins with Maria teaching the children to sing using tones. The song starts, "Sol - do - la - fa - mi - do - re - . . . ".

To organize this within a simple object-oriented program, one might define a class for each tone and an object for each note. This yields the following Java declarations (which I write explicitly on the board):

```
Sol note1;
Do note2;
La note3;
. . .
```

While this gives the appropriate objects, interactions with the class discover that these declarations would not generate sound, leading to the use of a "sing" method in a program segment:

```
note1.sing();
note2.sing();
note3.sing();
. . .
```

While this handles the first part of the song, the children in the movie then remark that the song "does not mean anything" and Maria responds, "So we put in words, one word for each note." Clearly this specifies a "text" method (or "text" parameter in the constructor) and yields the code:

```
note1.text("When");
note2.text("you");
note3.text("know");
. . .
```

Additional design issues might involve the number of objects required for each class and whether these classes might inherit properties from a super "Note" class.

As with the class meeting, I conclude this column inviting you to join in the singing.

41.2 SUBSEQUENT REFLECTIONS

Within a few weeks of the appearance of this column in December, 2001, Max Hailperin, Gustavus-Adolphus College, suggested several additional activities that support kinesthetic learning. In subsequent email, Hailperin noted that some of these ideas may not be original with him, and his purpose was to pass them along to future readers. In what follows, I include his edited notes—with his permission.

- *Hardware components:* Given a simple program, a group of students might simulate its execution at a component level, with each student playing the role of a single functional unit. For a more detailed view, separate students might play the roles of an instruction register, a program counter, a general register, etc.

- *Pipelining:* After describing pipelining and sequential processing, corresponding algorithms might be developed for addition (or another basic operation). Within this framework, two teams might race each other to obtain the answer, with one team utilizing pipelining and the other not.

- *Halting Problem:* Following the approach for proving the Halting Problem is unsolvable, an instructor might ask a student to correctly predict whether the instructor will dismiss the class immediately or "babble on forever." In formulating this class activity, the instructor should stress that the instructor was going to prove that the student's was wrong.) (In email, Hailperin credited Gerry Sussman for the original idea of this activity.)

- *Parity:* With the instructor out of the room, students write a sequence of 0's and 1's on the board. When done, a student assistant adds a 0 or 1 to the sequence to obtain even parity. With this accomplished, one student covers one of the bits, the instructor re-enters the room, reviews the visible bits, and correctly identifies the value of the covered bit.

Additional reflections on Teaching and a Sense of the Dramatic appear at the end of the next chapter.

CHAPTER 42

Teaching and a sense of the dramatic, act ii

This column is the second in a two-part series that first appeared in the *SIGCSE Bulletin*, Volume 34, Number 4, December 2002, pages 18–19[213]. The first column in the series may be found as Chapter 41 in this book, first having appeared in the *SIGCSE Bulletin*, Volume 33, Number 4, December 2001, pages 16–17[212].

Subsequent reflections on both columns appear later in this chapter.

42.1 ORIGINAL ARTICLE

MY DECEMBER 2001 COLUMN began a series of articles exploring the idea of promoting learning through elements of theater (e.g., dramatics, stage effects, entertainment). That column noted how such elements as pacing, emphasis, tone of voice, and delivery are vital to both teaching and theater. Examples included a 'Destructive Binary Search' (suggested by colleague Samuel Rebelsky) and the 'song "Do Re Mi" as an example of object-oriented problem solving'. This column offers additional examples and touches on the use of literature to promote learning. Many of these examples were suggested by others, sometimes in response to my earlier column. Additional suggestions are solicited to continue this series.

Precision and Completeness of Specifications, Design, and Algorithms

While the idea of writing specifications or algorithms often seems straightforward, students (particularly at introductory levels) sometimes do not appreciate the need for precision, completeness, and attention to detail. Role-playing can help illustrate this issue, where one party carries out what another group writes.

- Colleagues, Janet Davis and Samuel Rebelsky, prepare a peanut butter and jelly sandwich, based on instructions from the class. At the start, a loaf of bread, jars of ingredients, and utensils are on a table. Dramatics arise when Sam tries to do exactly what is suggested - with very few assumptions. ("Oh, you wanted me to use a <u>knife</u> to spread the jelly on the top of the bread.") [56]
- At their SIGCSE 2002 Workshop on Extreme Programming, leaders Joe Bergin and Laurie Williams divided the overall group into teams of 6-8, and teams were further

subdivided into specialized groups. On a task of developing (a drawing of) a transportation device, one group of 2-3 identified desired features, while another group drew a picture of a vehicle that would meet exactly the specifications given - not anticipating future developments. ("But you didn't say the vehicle had to be steered.")

Algorithms

To highlight algorithms, classroom simulations may focus on various elements, such as data, structures, or functional units. Here are three examples:

- To illustrate sorting algorithms, Jeffrey McConnell (Canisius College) and Samuel Rebelsky describe several variations. For example, data values may be written on paper, with students playing the role of positions within an array. As the algorithm progresses, data on paper are passed from one array element or variable to another. Alternatively, students may represent the data themselves, and students move from one location to another according to the algorithm.
- Max Hailperin (Gustavus Adolphus College) describes "simulating the execution of a simple program by a simple computer, with one student per functional unit."
- To illustrate pipelining to introductory students, I have assigned successive tasks in the pipeline to students, organized within rows in the classroom. At first, each student verbalizes her or his work for a specific step. As the work progresses, multiple people work (and perhaps talk) in parallel. At a somewhat more advanced level, Max Hailperin describes "having a race between two teams of students doing addition problems, one use a pipelined approach, the other not."

Use of Music and Literature

Branching out somewhat from pure acting, music and literature provide wonderful examples of ideas, algorithms, and structures. Thus, as a high school student, I thoroughly enjoyed Tom Lehrer's classic, "New Math," as it explicitly applied arithmetic properties to subtraction. More subtly, the Eagles' "Take It To The Limit" might be considered as calculus mood music.

Similarly, teachers can draw upon familiar stories from literature. Three wonderful examples based on stories by Dr. Seuss come from Judith Hromcik (Computer Science Teacher, Arlington High School; Arlington, Texas). In the following narrative by Ms. Hromcik, she notes that not only did Dr. Seuss teach tolerance and responsibility, but also basic data structures.

Look even closer, and you will see that Dr. Seuss was a computer science teacher. I offer three examples as evidence:

- "Too Many Daves" from the book *The Sneetches and Other Stories* introduces the concept of arrays. This story is about Mrs. McCave, a woman who named all of her 23 sons, Dave. This caused her a problem, because if she wanted just one son and called out "Dave," all 23 sons came running. The story continues by telling the reader 23 different names that Mrs. McCave wished she had named her sons. I read this story to my students and then ask them if they can come up with a better solution to

Mrs. McCave's problem. Someone always suggests numbering (indexing) the Daves - Dave-1, Dave-2, etc. Now Mrs. McCave is the envy of every mother. She can call out one name, "Dave!!!" and have all of her children respond, or call out for the individual Dave-n, and have one specific Dave respond. Dr. Seuss was quite clever in writing such a cute story to introduce the concept of arrays.

- The second piece of evidence that I put forth is "Yertle the Turtle" from *Yertle the Turtle and Other Stories.* This is a story about a turtle king who wishes to be the king of all he that he sees. In order to see more and thereby be king of all he sees, he stacks up his turtle subjects and then **pushes** himself on top of the stack. Each time he wishes to increase his kingdom, he **pops** himself off of the top the stack, **pushes** on more loyal turtle subjects, and then **pushes** himself back on top. The story ends with the bottom of the stack sneezing, the stack collapsing, and Yertle the Turtle dumped into the mud. One could see this as stack overflow. I personally prefer to view this as trying to access the bottom of the stack - if you do that, you end up in the mud.

- The last piece of evidence is the story, "King Looie Katz" from *I Can Lick Thirty Tigers Today and Other Stories.* In this story King Looie Katz is a fanatic about keeping his tail clean and not letting it drag on the ground. He assigns the job of holding his tail up to another cat (linked list). Well, the second cat doesn't want his tail to drag on the ground, so he finds another cat to carry his tail. This process continues until the last cat in town joins the line and finds that his tail can never be held up, because there is no cat left to follow behind him and carry his tail. Dr. Seuss' solution is to have this small cat throw down the tail he is carrying, and then the next to the last cat throw down the tail, and so on (tail recursion?). I ask my students to come up with an alternate solution. One solution is to have the king carry the last cat's tail (circularly linked list). Another is to chop off the last cat's tail (null pointer the hard way). I then suggest something less traumatic for the last cat - a belt loop on which to hang his tail until a new cat can join the back of the line (null pointer the kinder way).

I believe the evidence supports my original conclusion. Dr. Seuss was a computer science teacher.

42.2 SUBSEQUENT REFLECTIONS

Several examples in these columns on "Teaching and a sense of the dramatic" involve kinesthetic learning; students learn by participating in physical activities rather than by listening to lectures, participating in class discussions, or watching demonstrations or the activities of others. McConnell describes several other kinesthetic-learning activities in [120, section 3] and [122, section 3]. Additional ideas for activities involving kinesthetic learning were emailed to me by Max Hailperin, Gustavus-Adolphus College, shortly after the December, 2001, article appeared, and several of those suggestions appear at the end of the previous chapter.

Other types of dramatic examples might utilize films/videos, demonstrations, labs, programming exercises, and projects. The following notes illustrate dramatic experiences with each of these types of activities. In each case, student observation, experimentation, or the writing of programs can make a substantial impact.

More Kinesthetic-Learning Activities

Films and Videos

The classic film, "Sorting Out Sorting" [46], describes nine sorting algorithms, provides a brief Big-O analysis of each, and uses graphical simulations to compare actual run times. Although the graphics and sound track are dated by today's standards, the film continues to have a strong impact—at least for many students. Viewers watch sorting algorithms in action and experience comparative run times. In comparing Tree Selection, Heapsort, and Straight Selection, for example, the narrator notes, "If you are getting bored, let this be a lesson about n^2 sorts." Toward the end of the film, all nine sorting algorithms are run in parallel on 2,500 identical data sets. As expected, the several $n \log n$ sorts finish rather quickly, and there is time to show the film credits as the n^2 sorts continue. Then, the narrator comments, "we won't show you the whole race, because it would take another 54 minutes for the Bubble Sort to finish. Instead why not see the whole film again." The film then does a quick review of all nine algorithms—still waiting for the Bubble Sort to finish.

Altogether, the film combines visual effects, selected narrative, simulation, and a clever story line to translate algorithms, Big-O analysis, and run-time experiments into a memorable viewer experience.

As a second example, Web search for "sorting algorithm folk dancing" produces videos that utilize elements of folk dancing to act out the steps of individual sorting algorithms. In some cases, dancers may wear costumes depicting various cultural traditions. With such inspiration applied to a local environment, computing faculty might initiate conversations with a local dance troupe to develop a suite of dances illustrating several algorithms.

Demonstrations

The "Sorting Out Sorting" film provides a sense of contrast and drama by combining graphics, sound, narration, and simulation. Of course, such elements of presentation need not be restricted to film or video, and in-class demonstrations can take advantage of the capabilities of modern desktop and laptop computers for sound, the display of data, etc.

With the film as motivation, an instructor might focus on a few favorite algorithms for a demonstration. In planning, a first step might focus upon what type of visual display or sound sequence might engage students in a class—what might make the demonstration memorable or dramatic? With this vision, program development might be tackled either as an instructor's individual project or as a software project within an upper-level course.

Yet another approach for demonstrations might entail searching the Web for visualizations of relevant algorithms. For example, in writing this subsection, a Web search for "sorting algorithm visualization" returned about 175,000 results. Review of such queries will likely yield several tools for class demonstrations; some might seem reasonably pedestrian and uninteresting, but some likely could be used for classroom demonstrations with substantial impact.

Turning to a different context, a demonstration can highlight both advantages and tradeoffs of common languages features. As an example, in an upper-level algorithms course, I often ask students to write code for six sorting algorithms and then to time the algorithms on data sets of 40,000 integers (in ascending, descending, and random order). Students are welcome to utilize code from the Web, as long as they believe the code to be of high quality.

Each time I have given this comparative assignment, radix sort is included, and one or more students elect to download Java code that relies heavily on auto-boxing and auto-unboxing. Although a Java-based radix sort can work very well, many versions on the Web convert between `int` values and `Integer` objects on almost every line. The result is that the Java code looks lovely, but may require 500,000 or more system calls to allocate memory. Students naturally conclude that the radix sort is by far the worst sorting algorithm (perhaps even worse than bubble sort).

With this background, a natural demonstration involves running two versions of a Java-based radix sort: one using auto-boxing and the other being careful to avoid auto-boxing whenever possible. In timing such algorithms, the differences in efficiency are substantial—often an order of magnitude different. With an appropriate narrative, the demonstration can have dramatic impact on students (without film or video or even fancy graphics).

Labs or Programming Exercises

Sometimes a hands-on programming activity can highlight an underlying principle in a dramatic way. For example, consider the following C code:

```
#include <stdio.h>

int main ()
{
  int a [8] = {2, 4, 6, 8, 10, 12, 14, 16};
  int i;

  printf ("addresses:␣␣a:␣␣%u,␣i:␣␣%u\n", a, &i);

  for (i = -8; i < 16; i++)
    {
      printf ("a[%2d]␣=␣%d\n", i, a[i]);
      a[i] = i - 6;
    }

  return 0;
}
```

This program is supposed to print values in memory in and near an array, but the program also changes values stored in those locations. Output from this program may vary somewhat, depending upon a compiler's layout of variables in memory. (For example, I obtained slightly different results when running the program on Linux and a Mac.) When run on one machine, the first part of the output generated by this program follows.

```
addresses:  a:  1480735424, i:  1480735416
a[-8] = 1480735472
a[-7] = 32767
a[-6] = 125685484
a[-5] = 1
a[-4] = 10
a[-3] = 43
a[-2] = -2
a[-7] = 32767
a[-6] = 125685484
a[-5] = 1
a[-4] = 10
```

```
a[-3] = -9
a[-2] = -2
a[-7] = 32767
a[-6] = 125685484
a[-5] = 1
a[-4] = 10
a[-3] = -9
a[-2] = -2
a[-7] = 32767
```

The point here is that the local variable `i` is stored a few memory locations before the array `a` (at least on some machines). In particular, with this compiler, the variable `i` is stored two `int` locations before the `a` array; i.e., at the location `a[-2]`. Thus, when `a[-2]` is reset to -8, the value of `i` is changed as well. The loop then increments `i`, and the loop starts again at -7. Altogether, this is an infinite loop, because a memory access (at `a[-2]`) changes a loop control variable.

Altogether, this type of example can provide a dramatic illustration of issues related to buffer overflow.

Yet another attention-grabbing lab or programming exercise involves a permutation sort, in which an array is sorted by generating all permutations of the original array until the ordered permutation is found. In a lab or assignment, one challenge involves students determining how to generate (and perhaps print) all permutations. Of course, for an array of n elements, the number of permutations is $n!$. Combining this with checking that a permutation is ordered yields an $O(n * n!)$ sorting algorithm. Timing this algorithm on small arrays (even size 8 or 9 or 10) can yield dramatic results and demonstrate clearly issues of a combinatorial explosion.

Projects

Sometimes the product of a project can be included in an activity that contains an element of excitement and drama. As an example, a CS2 course at Grinnell College involves the control of robots using C. At the end of the course, colleague, Peter-Michael Osera, developed a project, in which a program reads commands from a file and issues corresponding commands to a robot. When several programmers or groups have completed the same project, one can align the robots in a line, and have each robot follow the same commands. Numerous variations are possible in getting started, but the overall effect can be to create a "Robot parade".

When I tried this in my own CS2 class with robots, the robots followed each other down a classroom aisle, progressed out the door, and then moved down the hall—beeping and dancing as they went. Although students felt challenged in completing the project, all groups were successful, and the students were wildly enthusiastic about the parade itself. All were engaged in the parade itself, and some even took pictures!

Although a robot parade is specific to one type of class, one could imagine projects in which programs played music or displayed images or otherwise generated lively products. When run together, such programs may yield a festival atmosphere!

CHAPTER 43

Thoughts on student feedback to help teaching

This column first appeared in the *SIGCSE Bulletin*, Volume 38, Number 4, December 2006, pages 13–14[221] Subsequent reflections on this subject appear later in this chapter.

43.1 ORIGINAL ARTICLE

MOST TEACHERS have great interest in improving their courses, their interactions with students, and other components of the teaching and learning process. Although some student feedback can come from school-wide end-of-course evaluations, these typically focus on assessment for contracting, merit pay, and promotion, rather than on improvement. Thus, many school-wide evaluations contain general feedback that may provide little real insight. In some cases, the questions seem remarkably generic:

The instructor is breathing:

a all of the time,
b some of the time,
c none of the time.

John David Stone, a colleague of mine, notes that even the best of multiple choice questions (with up to 8 possible answers) yields only 3 bits of data – a very small amount of information.

This level of feedback may yield useful information about some mechanics, such as audibility and clarity of the speaker, legibility of writing, amount of eye contact, basic course organization, work load. However, teaching involves much more than such basics, and more insightful and extensive feedback is vital.

In any case, since institutions, not individual faculty, typically control college-wide evaluations, the rest of this column ignores them, focusing instead upon mechanisms that provide extensive, useful feedback to individual faculty.

Indirect Feedback through Observation

Perhaps the most basic observation is that faculty desiring feedback must ask, either indirectly or directly. Within a class, a faculty member can observe student reactions to presentations and in-class activities. If student's eyes glaze over, something about the class is likely not proceeding optimally. If students are focused and engaged, then learning may be progressing reasonably well. An observant teacher can tell the difference, but it may take considerable experimentation to determine just what elements produce various outcomes. In some ways, this process seems related to how an actor/actress tries different deliveries on stage and gauges audience reaction.

Directly Asking Students

Another effective mechanism for receiving feedback simply involves asking, either in class or outside. When soliciting feedback, a possible reaction from some students seems to be questioning a teacher's competence. However, if the instructor is careful not to appear timid or defensive, my experience suggests that discussions of a course can help student opinion by creating a partnership rather than undermining credibility.

Personally, I hesitate to devote much class time for such feedback, since courses contain much content already. On the other hand, when a student comes to my office with a question, it is easy to ask for the student's reactions. Informal questions near the end of a session allow students to give personal viewpoints, without feeling they must conform to a perceived norm. I ask for individual reactions, noting I will ask others about their individual thoughts as well.

Customized Evaluation Forms

After teaching a few years, I realized that I needed specific information to help me improve my teaching. General comments might indicate a course went well or less well, but generalities gave little insight in how to improve. Thus, I now distribute my own questionnaire to students, in addition to any mandated by the institution. To be more precise, I typically distribute my end-of-course evaluation at the same time the college-wide one is given. College policy requires students to complete the all-campus evaluation while in class, and these are returned to a central office. The instructor receives these evaluations after semester grades are turned in.

In contrast, I tell students to think for awhile about my evaluation form. Thus, I distribute the forms during the last week of classes, to be handed in at the start of the final exam. (Since students receive their exam only after turning in the evaluation, this approach yields a high response rate.) I promise that I will not read any evaluations until I have submitted semester grades.

My evaluation form has evolved over time, as I have sought to focus on one part of teaching or another. Consistently, each question is specific, so that I might get concrete suggestions.

As an example, the following form highlights common points. (In practice, I add space between questions, so the overall evaluation covers two sides of a single page.)

— Start of Questionnaire————————————

Individual End-of-Course Evaluation for Course

Please express your opinions about this class by commenting upon each of the following aspects of the course. Also, please answer the questions on the reverse side of this page.

These evaluations are due at the start of the exam.
Note: I will hand you an exam when you turn in a completed evaluation. If you forget your evaluation, then you will have the opportunity to complete a new evaluation before receiving a copy of the exam.

The evaluations will not be read until after grades are submitted to the Registrar's Office – so your evaluations here will not have any impact upon the grading.

Comments on Class Discussions and Lectures
Should more class time been devoted to presentation of material?

Comments on the Pace of the Course and Scheduling

Comments on Written Homework Assignments
Should these problems be continued another year?

Comments on the Programming Problems

Comments on Laboratory Exercises
What advantages and disadvantages do you see in having in-class laboratory work play a central role in the course? Were the labs effective in presenting the material and getting you to think about it? Do you have specific suggestions for improving the labs?

Comments on the Working Environment in the Lab
(E.g., did you find the Lab a good place to work? How might the atmosphere be improved?)
—— Page 2
Specific Questions

1. Which three or four laboratory exercises did you find the most useful?
2. Are there some laboratory exercises which you think should be revised?
   If so, please identify which of these lab exercises should be given highest priority for revision, and indicate any modifications you would recommend.
3. Were the labs the appropriate length and difficulty?
4. Did this course have the right balance among written homework, programming exercises, and laboratory exercises?
5. In retrospect, would you have preferred more tests in place of some of the other course activities? Please explain.
6. Did this course meet your expectations and needs?
7. How would you describe the general campus attitude and view of this course?
   (E.g., does the course have the reputation of being too hard or very easy, is the course considered irrelevant/very helpful for other disciplines?)
8. Was this course consistent with what you had heard about it?
   (I.e., do you believe the general campus view of the course is accurate?)
9. Would you recommend this course to other students? (Why or why not?)
10. Are you planning to take — [the next course in the sequence]? What factors influenced your decision?

— End of Questionnaire ————————————————

As a final thought, note that the above type of questionnaire can also be used halfway through a course to provide information about how a class is going. In that context, I emphasize that some elements of the course have already been determined and are unlikely to change. However, some small adjustments might be possible, based on student feedback, and larger changes might be possible the next time I teach the course.

43.2 SUBSEQUENT REFLECTIONS

The original column mentions that college-wide end-of-course evaluations may or may not provide much helpful feedback, but the column then highlights several mechanisms for collecting student feedback for a course:

- Taking advantage of instructor observations of students during in-class activities
- Asking students for feedback, either in class or during office hours
- Preparing a customized and detailed evaluation form

To be concrete, the column provides a sample questionnaire for use at the end of a semester. In addition, the column mentions the possible use of a short survey at a mid-semester point within a course. In recent years, I have used mid-semester surveys with some frequency, particularly when I am teaching a new course or when a course draws from a student population that is new to me (e.g., at an institution I have just joined, with non-majors from departments I have not encountered previously, or for a course that now satisfies a general education requirement).

For mid-semester surveys, I want to cover high points, determine the effectiveness of course components, and receive feedback regarding course balance, pace, and work load. For efficiency, I prepare a slide with a few numbered questions (no more than 10), I ask students to number their answers to correspond to the questions, and I try to collect student responses within the first 10 minutes of class.

As an example, I recently taught a non-majors course that satisfies a general education course at another institution. Although course details are not particularly relevant here, the course included four primary components:

- readings/class discussions on computing and its applications
- laboratory exercises to highlight elements of algorithmic thinking, supported by C programming
- research exercises to develop search and analysis skills (as described in [279])
- practice of technically-oriented writing (e.g., in writing two papers)

With background, the following 10-minute survey was administered just before the week-long mid-semester break:

— Start of Midterm Questions————————————————
This course has tried to integrate three activities:

- Provide background on computing and applications
- Provide experience with algorithmic thinking, logical reasoning, and imperative problem solving
- Provide practice in honing research and writing skills
    - ◇ Research Exercises (identify opportunities and raise questions regarding sources)
    - ◇ Papers on topics related to computing

Questions

1. To what extent do you think the readings and class discussions have helped provide background on computing and applications?
2. What are your thoughts regarding C programming, use of the robots, and the in-class labs in improving experience with problem solving?
3. Now that we have completed the research exercises, what are your thoughts about those exercises? (e.g., were they helpful?)
4. What are your thoughts about writing for this course (e.g, research exercises, paper 1)?
5. Class time has tried to balance work on research exercises, book discussion, and labs. What are your thoughts on this balance?
6. How would you describe the pace of this course?
7. How would you describe the workload for this course, compared with others? (Include the number of hours you spend on this course outside class.)
8. Through the semester to this point, labs have been done in pairs, and students have chosen their own pairings. After the break, would you prefer that students continued to chose their own partners, or should I pick the pairs (and change them every 1–2 weeks?

— End of Midterm Questions————————————————

As this example illustrates, the entire activity is short and focused on specific details of the course. In summary, student responses indicated satisfaction with the course, but also expressed a clear preference for more time for in-class labs and less in-class discussion. Overall, it was interesting that non-majors wanted an increased emphasis on labs, and also reported that the book was sufficiently clear that extensive time in classroom discussion was of secondary importance.

To conclude, short, detailed, and targeted feedback can suggest worthwhile course refinements that help tailor a course to current student interests. Further, distribution and reporting of a mid-course survey can foster strong positive student reaction—the instructor cares about what students think and makes modest adjustments in response. (With many course elements determined as a semester starts, logistics may not allow substantial changes. However, students may appreciate even small refinements in response to their feedback.)

CHAPTER 44

Selected/annotated references for the role of teachers in the classroom

AN INSTRUCTOR ASSUMES MANY ROLES, including coach, mentor, listener, presenter, manager, actor, and assessor. Each chapter of this Part IX has discussed elements of these roles, and interested readers may wish to explore these topics further. The following notes identify some starting places for additional reading, organized into these categories:

- the imposter phenomenon
- the Dunning-Kruger Effect
- approaches for classroom drama and active learning
- films

The Imposter Phenomenon

As described in Chapter 38, the imposter phenomenon describes a circumstance in which an individual may have a strong record of achievement and success, but may believe that this is by luck or accident rather than intelligence, ability, or insight.

- Pauline Rose Clance and Suzanne A. Imes, "The imposter phenomenon in high achieving women: dynamics and therapeutic intervention." *Psychotherapy: Theory, Research and Practice*, 1978 [35].
  This article presents the ground-breaking research that led to the identification of this phenomenon.

- Akira Miyake et al, "Reducing the Gender Achievement Gap in College Science: A Classroom Study of Values Affirmation", *Science*, 2010 [128].
  and
  Akira Miyake et al, "Supporting Online Material" *Science*, 2010[129]
  The primary article, together with its supporting materials, outlines an intervention that may help partially address the imposter phenomenon. This approach uses two 15-minute, in-class exercises that highlight an individual's values and perspectives and

that encourage self reflection. The supporting materials include an intermediate level of detail regarding these exercises.

- Maria Klawe, "Impostoritis: A Lifelong, but Treatable, Condition", *slate.com* [99]. Harvey Mudd President, Maria Klawe, describes her on-going experiences with the imposter phenomenon, and she describes several activities that can help students (and others) in overcoming the corresponding feelings of anxiety and insecurity.

The Dunning-Kruger Effect

Studies by Dunning and Kruger demonstrate that unskilled and low-ability individuals may have little awareness that they are incompetent. Within the classroom, this Dunning-Kruger Effect may explain some students whose tests or assignments are failing or near-failing, but who still may believe they are getting B's (or better).

- Justin Kruger and David Dunning, "How difficulties in recognizing one's own incompetence lead to inflated self-assessments", *Journal of Personality and Social Psychology*, 1999 [101].
A description of the original research that identified this effect

- Joyce Ehrlinger, Kerri Johnson, Matthew Banner, David Dunning, and Justin Kruger, "Why the unskilled are unaware: Further explorations of (absent) self-insight among the incompetent", *Organizational Behavior and Human Decision Processes*, 2008 [63].
This article briefly reviews the studies covered in the 1999 article and provides additional research results and commentary.

- David Dunning, "We Are All Confident Idiots". *Pacific Standard*, 2014 [62].
This column provides an update on current research, with modern examples—all written for a popular audience.

Approaches for Classroom Drama and Active Learning

Jeffrey McConnell wrote a wonderful series of four columns on active and cooperative learning, some of which present ideas for demonstrations and/or the dramatic.

- McConnell, Jeffrey J., "Active and cooperative learning: tips and tricks (Part I)" SIGCSE Bulletin, Volume 37 Issue 2, June 2005 [120].
This article provides background and gives several examples involving drama or kinesthetic learning in the classroom

- McConnell, Jeffrey J., "Active and cooperative learning: tips and tricks (Part II)" SIGCSE Bulletin, Volume 37 Issue 4, December 2005 [119].
In-class activities always involve a level of risk: will demonstrations or group interactions proceed in a way that is hoped? This article considers different levels of risk and mechanisms to minimize it.

- McConnell, Jeffrey J., "Active and cooperative learning: tips and tricks (Part 3)" SIGCSE Bulletin, Volume 38 Issue 2, June 2006 [121].
Many activities related to active and cooperative learning require the formation of groups. This column considers what characteristics might be considered in forming and evaluating effective groups.

- McConnell, Jeffrey J., "Active and cooperative learning: tips and tricks (Part IV)" SIGCSE Bulletin, Volume 38 Issue 4, December 2006 [122].
  This article discusses the design and assessment of activities related to active and cooperative learning.

Some articles describe specific in-class activities that can have a substantial impact.

- Janet Davis and Samuel A. Rebelsky, "Food-First Computer Science: Starting the First Course Right with PB&J", SIGCSE 2007 Proceedings [56].
  This charming article not only describes an entertaining demonstration, but also provides tips and more general ideas for in-class activities for introductory courses.
- Suthikshn Kumar, "A skit-based approach to teaching web and networking protocols", *ACM Inroads*, Volume 2, Number 4, December 2011, pages 30-32 [103].
  This article suggests several approaches and examples for using skits to help students learn Web and networking protocols.
- ITiCSE, SIGCSE's annual conference on Information Technology in Computer Science Education, regularly contains a session on "Tips and Techniques" for teaching and classroom activities. Dating back until at least 2002, each of these sessions contains 8-12 ideas for different aspects of a classroom environment. See the ACM Digital Library for ITiCSE Proceedings, and search the Table of Contents for "Tips and Techniques."

Film

Some computing-related films have become classics for their coverage of algorithms or other elements of the computing discipline.

- Computer Systems Research Group, "Sorting out Sorting", Dynamic Graphics Project, University of Toronto, 1981 [46].
  This film describes, simulates, analyzes, and compares nine sorting algorithms with a commentary that is both illuminating and entertaining.

IX

Exercises and Assignments

SOME STEM FACULTY DISTINGUISH between talking about science and doing science. A faculty member can reasonably talk about the scientific method, about the formulation of mathematical axioms, about the process of identifying principles for software development, etc. However, some STEM faculty flippantly categorize this area of study as "sociology" or "anthropology" or "social science." Although this categorization may not be fair to either science or these other disciplines, the underlying thought highlights the need for scientists to do more than talking about their disciplines.

In particular, STEM disciplines engage the scientific method, mathematical axioms and proofs, and software development principles to create new knowledge or to apply that knowledge to new application domains. One common buzz phrase is that "Scientists *do* science." For example, discussion of the problem-solving process may provide insights, but ultimately scientists and science students need to discover full solutions. Similarly, within computing, practitioners may find value in identifying common patterns for approaching various types of problems, but practitioners also must be able to apply those patterns.

As an example from mathematics, in a course on differential topology that I took in graduate school, Professor James Munkres' lectures presented a wonderful, high-level view of the discipline, and his proofs of results were truly elegant. After a class session the topic seemed clear, ideas were nicely integrated, and a student could not imagine how anyone could have trouble writing their own proofs. And then, the student tried the first homework exercise. Suddenly the ideas seemed complex, and the sequencing of arguments was far from clear in working toward a result. Only when working exercises did it become clear how polished a lecture's presentation might have been and how the class presentation avoided numerous pitfalls; working through an exercise and developing one's own proofs provided new levels of understanding. Listening provided a reasonable start, but true understanding came with doing.

In computing, full solutions often start with a high-level vision and general sense of an approach. However, one aspect of computing that I particularly value is the need to move from an initial idea through high-level specifications and design and down to detailed algorithms and software development. For me, each stage of the problem-solving process requires integrating ideas and creatively considering approaches; this engagement of ideas during the development of solutions draws upon many levels of insights and experiences through a creative and exciting process that leads to worthwhile solutions.

Within the context of a course, exercises and assignments provide an opportunity for students to *do computing.* These activities highlight course objectives, sharpen problem-solving skills, and encourage practice with many techniques and structures. In addition, exercises and assignments create opportunities to suggest broad applications, provide motivation, and send messages regarding the nature of the computing discipline.

CHAPTER 45

Homework assignments and Internet sources

This column first appeared in ACM *Inroads*, Volume 4, Number 4, December 2013, pages 16–17[248] Subsequent reflections on this subject appear later in this chapter.

45.1 ORIGINAL COLUMN

IN COURSES THROUGHOUT THE CURRICULUM, faculty members develop assignments for students to complete for homework.

However, the widespread availability of Internet resources provides students with extensive opportunities for consultation and exploration. This column suggests several principles and approaches that may help guide the development and use of homework assignments in this Internet age. Some of these principles involve the development and use of exercises and analysis that have curricular implications.

Principle 1: Don't Create Temptation or Encourage Dishonesty.

In today's culture, it seems naive to assume that students will be unaware of the numerous resources available online. As an example, a full decade ago, I had occasion to participate in an MAA summer workshop on discrete mathematics. One session discussed several standard textbooks. During the break that followed, I picked up the textbook on the top of the stack, opened to a random section, select- ed an exercise at the end of the section, and typed the author's name, textbook title, section, and exercise number into a search engine. Instantly, I received links to several solutions to the problem. In the intervening 10 years, solutions to textbook problems are widespread, and some Web sites allow anyone (e.g., students) to purchase answer books for a wide range of textbooks for a modest fee.

Within this context, I think it inappropriate to prohibit students to use the Internet for their homework.

- Students having difficulty will be tempted to break the rules and cheat.
- Students following the rules may be at a disadvantage in not obtaining help received by classmates.

Principle 2: Creative, New Assignments can Mitigate Copying from Online Sources, but only Partially.

One natural approach to resolve online solutions to problems is to create new assignments each semester. When problem assignments are truly new and unrelated to any past assignments—created by the instructor or anyone else in the world, solutions may not be available online. However, even this approach has at least three difficulties:

- New problems often take substantial time to develop, and past problems likely cannot be reused. Thus, the creation of new homework assignments may require extensive faculty time.
- New assignments may be variations of past problems assigned somewhere, so related solutions may be available.
- Once created, it may take substantial instructor time to search the Internet to determine if the problem is substantially different from previously created exercises. With thousands of instructors creating new exercises on a continual basis, checking new problems with those available elsewhere can require a significant and ongoing commitment of time and effort.

Principle 3: Require Careful Citation of all Usage of Sources (Internet and Others, including Course Textbooks).

Academic work in all disciplines expects that the sources of ideas will be acknowledged, words of others will be specified (e.g., with quotation marks or block quotes), and sources will be collected in a bibliography. I see no reason why computer science should be different and consistency among disciplines sends a uniform and clear message to students regarding expectations. Thus, my syllabi typically contain a section on Academic Honesty. For example, the following passage comes from a recent CS2-level course:

> All work in this course is governed by the rules of the college regarding academic honesty. In summary, standard practice requires that you must acknowledge all ideas from others.

- When working on homework, either individually or in a group, you may use any written source. However, the normal rules of citation must be followed, as described in the Student Handbook.
    - ◇ You must cite statements in the textbook, if you use them to guide your work.
    - ◇ Although the Web can be useful for reference, you are advised that much material on the Web is of poor quality.
    - ◇ You are responsible for the quality of what you turn in, regardless of the source of the material.

Principle 4: Although Students Often are Adept at Finding Material on the Internet, Students may have Neither the Experience nor the Skill to Analyze the Quality of that Material.

In 2003, Graham and Metaxas, conducted a survey/study of students at Wellesley College regarding the use of Internet sources. Their conclusions included:

- Students are relying more and more on the Internet as their primary source of information.
- Students have trouble recognizing trustworthy sources.
- Students are not consistently able to differentiate between advertising claims and fact.
- Very few students double-check information found on the Internet.
- Older students with stronger traditional research skills performed no better than other students.
- Students place greater emphasis on finding information than evaluating it. [75]

A couple years ago, I taught an upper-level algorithms course, in which one segment reviewed multiple approaches to sorting. In one assignment, students were to write comparable code for about six sorting algorithms and time the performance of the algorithms on various data sets. Several students used code for a radix sort from an online source (with attribution) and concluded the radix sort performed far worse than any of the other algorithms. Upon close inspection, it turned out that the code called the operating system to allocate space about 120,000 times (with malloc in C) for a data set with 40,000 items.

Overall, examples abound in which students locate information on the Web, but the information is obviously biased, incomplete, misleading, or at wrong.

Principle 5: Provide Students with Hands-on Experience Identifying and Evaluating Online Sources.

Since students typically use the Internet extensively to find information, my experience suggests that these students often believe they are adept at locating and using reliable information. With this perspective, I have found lectures have remarkably little impact regarding new search strategies or the need to analyze sources.

Rather, I suggest a type of shock therapy. Give students a problem to explore, have them report their findings, and then analyze their results. (Such an approach is described in some detail in [279].) Questions that seem straightforward often do not have straightforward answers. Thus, selecting questions for student exploration can highlight shortcomings in students' practices, which in turn can lead to productive conversations about the analysis of sources. Over the years, I have found collaboration with a librarian very helpful; personally, I work with Grinnell's Science Librarian or a librarian liaison with my department.

Principle 6: Hold Students Accountable for the Quality of the Work in the Sources they Utilize.

As my syllabus indicates, students are welcome to use Internet or written sources in my assignments, but I advertise that my grading scale gives 0 points for junk—regardless of its source. Further, when a student places her or his name on a paper, the student is taking full responsibility for the quality of the work—even if sections come from other sources with attribution. In my experience, giving 0 points to a radix sort with 120,000 calls to malloc captures students' attention and encourages substantial analysis in subsequent assignments.

Acknowledgment

Thanks to Marge Coahran for her reading of a draft of this column.

45.2 SUBSEQUENT REFLECTIONS

The original column suggested six principles regarding connections between homework assignments and the Internet. Two additional considerations include

- the posting of answers by students and
- instructor interest in comparing assignments from one course offering and the next.

Student Posting of Answers

In the past, I developed policies for students using Web resources as part of completing homework assignments. As noted in the original column, the Web contains a wealth of information, and students naturally explore the Web to gather ideas and potential solutions to current homework problems.

In the past couple of years, however, I have become aware of at least one instance (school to remain anonymous) for which a student posted on the Web answers to homework in that course. If an assignment forbids discussion among students, posting of a student solution might be considered as a mechanism that allows one student to help another without face-to-face conversation.

With this potential loophole regarding student collaboration, I now include the following paragraph within instructions for exercises that must be completed individually, without consultation.

> Posting of solutions for this exercise in any form (e.g., on the Web, on a bulletin board, via email) is expressly forbidden. Also, check that all files on your account with solutions have permissions that prevent access by others. Any posting of solutions or the existence of any exercise-related files made public on your account will yield an automatic COURSE GRADE of "F".

Years ago, general rules of a college suggested (at least to me) that one student should not copy the work of another when doing their homework. With the current ease of posting solutions on the Web, it now seems that one must be explicit about disallowing such a practice.

Comparing Assignments from one Offering to the Next

In my original column, Principle 2 stated, "Creative, new assignments can mitigate copying from online sources, but only partially." Interestingly, discussion on the SIGCSE.MEMBERS listserv in December 2016 highlighted a different perspective: When past assignments are reused, an instructor can gain insight about whether a new teaching technique is effective or how students in the class compare with those in the past. Altogether, reuse of an assignment allows comparisons from one year to the next, and comparisons may provide valuable feedback.

On the negative side, reusing an old assignment may have several drawbacks:

- Reuse of an assignment may increase the risk of copying from past student solutions. At some schools, student groups (e.g., in a dorm, fraternity, or sorority) may collect assignments and student work from semester to semester, in which case old solutions may be readily available. In such cases, student work in subsequent semesters may reflect, in part, the degree to which files of past work are available, rather than the relative success of changed pedagogy or student quality.

- When reusing assignments, some faculty may modify an assignment, perhaps changing its wording or making small adjustments in specifications. Such changes may help combat plagiarism from past student work and may complicate students' searching for solutions online. However, as assignments change, comparisons of student work from one course offering to the next become more difficult. As an example, Advanced Placement (AP) Examinations in Computer Science include some multiple-choice questions on multiple exams, as a mechanism to compare student performance from one year to another. In order to be reliable, such questions, called "equaters", not only must have identical wording, but also must appear at the same position on the test. That is, question 4 on one test must also be question 4 on the other, so students will be equally fresh and focused in answering both questions. From the perspective of AP, any change of even a single word on an equater is considered grounds for undermining the value of the question in any statistical comparison of student performance on two exams. From this perspective, changing a question from one course offering to another may be a useful approach for minimizing plagiarism, but substantial change in wording likely means that the comparison of student performance will require considerable interpretation and caution.

Even with these difficulties, the reuse of questions on a final examination may be useful in one specific context under certain circumstances. If students turn in both the test questions and their answers for a final examination, and if the exam is never returned to students, then the test questions might be considered reasonably secure for reuse. In today's society, students still might use a cell phone or other device to record questions, but such risk may be mitigated if cell phones and hand-held devices are forbidden during an exam. With tight controls, one might consider whether one or more final exam questions might be reused to evaluate new pedagogy or compare student populations. Such an approach still may have some challenges, but I have found the reuse of questions on a final exam to provide some useful data for assessment.

Altogether, Internet sources and assignment files from student organizations present several challenges:

- Creation of new assignments may be useful, but can be time consuming.
- Reusing old questions may be tempting, but also troublesome.
    - ◇ Use of the same problems may invite copying from past student work.
    - ◇ Changing old problems may limit comparisons of student performance from one year to the next.
- Repeating final examination questions might provide a modest mechanism for comparing student performance, if
    - ◇ students must turn in both the questions and their answers,
    - ◇ neither the examination questions nor student answers are returned, and
    - ◇ students taking an exam one semester do not write down what they remember of the questions, in anticipation of students taking the course in a later semester.

CHAPTER 46

A racquetball-volleyball simulation

This article launched my service as a columnist for the *SIGCSE Bulletin* and later ACM *Inroads*, first appearing in the *SIGCSE Bulletin*, Volume 29, Number 4, December 1997, pages 22–23 [204]. Subsequent reflections on this subject appear later in this chapter.

46.1 ORIGINAL COLUMN

THIS NEW COLUMN will address assignments, class projects, research ideas, pedagogical approaches, and other issues related to effective classroom teaching. Each article will introduce a new topic, giving relevant background, possible materials for classroom use, and a discussion of teaching considerations. As appropriate, comments will indicate in which course(s) this material might fit.

Please note that contributions are actively solicited for this column. Feedback about ways to improve this column also would be welcomed. Send ideas, full articles, and comments to me at walker@math.grin.edu.

The Game Simulation

The following exercise may be given reasonably early within a CS 1 course; the required background involves only I/O, Boolean expressions, conditional statements, conditional loops or recursion, and random numbers. A racquetball version is given first; modifications for a volleyball version follow.

Racquetball Version

Racquetball is a game played by two players on an indoor, enclosed court. Scoring proceeds as follows:

The score starts at 0 - 0.

Player A starts serving.

When Player A wins a volley, she scores a point and is allowed to serve again.

When Player A loses a volley, she loses the serve but no points are scored.

Player B starts serving.

When Player B wins a volley, she scores a point and is allowed to serve again.

When Player B loses a volley, she loses the serve but no points are scored.

A player can only score points while she has the serve. A player loses the serve when she loses a volley, but no points are scored on the change of serve. Play continues until either the score is 11-0 (a shut-out) or one player scores 21 points. (A player need not win by two points.)

Write a program that reads the probability of Player A winning a volley and then simulates the playing of 1000 games with Player A having the first serve on each game. Record the number of wins (including shut-outs) for each player and the percentage of wins. Also record the number and percentage of shut-outs for each player.

Volleyball Version

Volleyball rules are similar, except play continues until one team scores 15 points, and a team must win by at least two points (if the score is 15-14, play must continue until one team leads by 2 points). There is no special rule for ending a game due to a shut-out.

Solution Notes

While details of a solution may be handled in many ways, students typically view game play in one of two ways. Ignoring initialization and follow-up, outlines of each approach follow:

Approach 1

Perspective: A serves then B serves

1. Continue until someone wins:
   (a) A serves
   until A wins game or loses serve
   (b) If A has not won,
   B serves until B wins game or loses serve.

Approach 2

Perspective: a game is a sequence of serves

1. Continue until someone wins:
   (a) Server begins
      i. If server wins, server gets point
      ii. If server loses, serve goes to the other person

Either of these approaches may be implemented iteratively or recursively

Assignment Variations

Virtually all introductory students find the flow of activity in this problem to be rather complex, and few can solve it immediately. Thus, this exercise can motivate a discussion of problem-solving design principles. For example, the assignment might be divided into several steps:

1. Students might write an outline describing the main and secondary steps for the simulation.
2. After outlines are reviewed and returned with comments, the instructor might distribute several correct outlines. [Multiple outlines illustrate that many problems may be solved in various ways.]
3. Students base their programs either on their own (corrected) outlines or on the instructor's.

Pedagogically, an instructor can emphasize that main procedures should follow the main outline elements, with subprograms handling details.

As another variation, in a course introducing OOP and object design, students might be asked to describe the appropriate objects for a solution.

Simulation Assumptions

As stated, the exercise specifies the same probability of A winning a volley, regardless of who is serving. (E.g., A wins 40% of the serves, when either A or B serves.) Students commonly ask whether such assumptions are realistic, and this provides a motivation to discuss general questions about assumptions of modeling.

Testing Considerations

As Table 46.1 illustrates, typical results from this simulation are very sensitive to the initial probabilities. For example, if A has a 40% chance of winning a volley, then A will win almost no games. Since these results may not be intuitive, student questions provide a natural framework for raising issues of testing. How does one know if a program is correct? What approaches to testing might be helpful? At a more general level, one can raise questions of testing simulations when answers may not be known in advance. This can motivate discussions of the appropriate use of computer models.

> *Technical Aside:* The results shown may be explained mathematically by noting that, to score a point, a player must win two volleys in a row. Thus, if A's chance of winning a volley is 0.4, A's chance of winning a point is $0.4 \times 0.4 = 0.16$. B's chance of winning a point in this case is $0.6 \times 0.6 = 0.36$, which is more than double A's chance of scoring a single point.

Conclusions

Altogether, this exercise can provide a helpful framework for motivating considerations of algorithm design, program structure, and testing as well as general issues of the appropriate usage of simulation and computer modeling.

Table 46.1: Simulated Racquetball Results					
Probabilities for winning volley		Percentage of wins		Percentage of shutouts	
A	B	A	B	A	B
0.40	0.60	3.0	97.0	0.0	4.3
0.41	0.59	6.2	93.8	0.0	4.4
0.42	0.58	6.9	93.1	0.2	3.4
0.43	0.57	11.6	88.4	0.2	1.8
0.44	0.56	15.1	84.9	0.3	1.8
0.45	0.55	21.3	78.7	0.5	2.0
0.46	0.54	28.4	71.6	0.2	0.9
0.47	0.53	29.1	70.9	0.4	1.4
0.48	0.52	37.8	62.2	0.9	1.0
0.49	0.51	45.8	54.2	1.3	1.1
0.50	0.50	54.5	45.5	2.1	0.3
0.51	0.49	61.8	38.2	1.7	0.7
0.52	0.48	67.1	32.9	1.7	0.3
0.53	0.47	72.0	28.0	1.8	0.5
0.54	0.46	79.8	20.2	2.2	0.2
0.55	0.45	82.7	17.3	2.6	0.3
0.56	0.44	86.1	13.9	3.2	0.0
0.57	0.43	91.3	8.7	4.1	0.2
0.58	0.42	93.8	6.2	5.3	0.2
0.59	0.41	95.6	4.4	7.6	0.1
0.60	0.40	97.2	2.8	8.3	0.1

46.2 SUBSEQUENT REFLECTIONS

Considerable time has passed since the original column first appeared in December 1997. At that time, many introductory computing courses focused on imperative problem solving, and iteration was often emphasized for repeating operations. In this context, the racquetball-volleyball simulation yielded a lovely programming assignment that required only modest background: I/O, Boolean expressions, conditional statements, conditional loops, and random numbers.

Iterative Solutions

Initially, I envisioned the flow of the game as a way to identify steps in a simulation, where basic steps corresponded to updating variables, conditional statements, and iterative loops. From this perspective, each step translated naturally to one (or a few) lines of code. Also, for clarity, some groups of steps (e.g., a volley or a game) mapped nicely to a function or procedure, giving the overall program a natural structure.

In implementing the algorithm, iteration served as a natural mechanism to move from one volley or one serve or one game to the next, and I wrote iterative code to implement both Approach 1 and Approach 2 in the original column.

Recursive Solutions

Upon further reflection, the outlines in the original column required repetition, but nothing in either outline required iteration. For example, consider the outline for Approach 1:

Perspective: A serves then B serves

1. Continue until someone wins:
   (a) A serves
   until A wins game or loses serve
   (b) If A has not won,
   B serves until B wins game or loses serve.

Within this outline, the key activity involves serving, and such processing maps easily to a recursive algorithm. The following Java code outlines a **playUntilWin** method that draws upon supporting methods to determine if the server has won the game:

```
/**
 * Play one game of racquetball or volleyball to conclusion
 * @parms   server and receiver indicate the team "A" or "B"
 *          probWinVolley specifies the likelihood the server
 *                wins a volley
 *          serverScore, recScore contain current score of
 *               server and receiver
 * @returns winner of game: either "A" or "B"
 */
public String playUntilWin (String server, String receiver,
                            double probWinVolley,
                            int serverScore, int recScore)
{   // serve
    if (Math.random() < probWinVolley)
    {   // score point
        serverScore++;
        // if win, return winner
        if (serverWins (serverScore, recScore))
           return server;
        // if not win, serve again
        else
           return playUntilWin (server, receiver, probWinVolley,
                                serverScore, recScore);
     }
   else
     {  // other side wins; other player serves
        return playUntilWin (receiver, server, 1.0-probWinVolley,
                             recScore, serverScore);
    }
}
```

With this serve-by-serve approach, a full game is accomplished with the initial call:

```
playUntilWin ("A", "B", probWinVolley, 0, 0);
```

As with many recursive algorithms, this approach is clear and clean—highlighting the main outline in Approach 1. (An analogous recursive solution works well for Approach 2 as well.)

Object-Oriented Solutions

Moving from an imperative perspective to object-oriented problem solving, initial work requires consideration of several design questions. Here are several considerations.

- What objects are appropriate? One approach is to consider a `Game` class (perhaps an Abstract class) with subclasses `Racquetball` and `Volleyball` that clarify the rules to win.

- How might a program be structured? One approach is to have one method to simulate a single game and a second method to tabulate the results for 500 or 1000 games. Also, methods/variables might be appropriate for considering when a player must win by 2, what probability should be associated with player A winning a volley, when a game is won.

- What game parameters are needed? Some choices include the number of games to be simulated and whether a particular game should be racquetball or volleyball.

- Should game parameters be defined within a class, specified as static variables, entered by a user during run time, and included as command-line parameters?

Overall, such issues and choices can provide a rich context for the discussion of many object-oriented concepts. Materials to support classroom activities include readings, discussions, and labs, such as are found in [241].

CHAPTER 47

The balance between programming and other assignments

This column first appeared in the SIGCSE *Bulletin*, Volume 30, Number 4, December 1998, pages 23a–25a[205] Subsequent reflections on this subject appear later in this chapter.

47.1 ORIGINAL COLUMN

AMONG THE MANY REMARKS I have heard over the years regarding both principles and details for various assignments and test questions, I find myself mulling over a few comments repeatedly. Two such remarks are:

1. If computer science is not just programming, why are so many assignments in computer science courses programming problems?
2. In today's world, even beginning computer users work with applications that feature interesting graphics, fancy fonts, and sophisticated user interfaces. However, to focus on fundamental algorithms, structures, and control elements, CS1/2 assignments often feature quart-to-liter conversions and other equally dull problems. How can introductory computer science courses compete effectively with the flashy applications in other areas, and still focus on basic concepts and techniques?

This column beings a discussion of the first of these remarks. Subsequent columns will continue this discussion and consider the second question.
Note: Reader comments, suggestions, ideas, etc. are sought to continue and expand this discussion. Insights regarding the second remark are especially welcomed.

Goals of Assignments

While one practical value of assignments may be to generate grades, one hopes that assignments will promote learning. Six components of understanding are identified in Benjamin Bloom's Taxonomy of Education Objectives in the Cognitive Domain [25]. Altogether, these six areas cover a wide range of knowledge and skill which students should master:

1. Knowledge of specifics (e.g., terminology, dates, events, persons, places, specific techniques, ideas)
2. Comprehension (e.g., understanding information, making use of communications received)
3. Application (e.g., using "abstractions in particular and concrete situations")
4. Analysis (e.g., "the breakdown of a communication into its constituent elements or parts")
5. Synthesis (e.g., "putting together of elements and parts so as to form a whole")
6. Evaluation (e.g., "judgments about the value of material and methods for given purposes.")

As part of learning, assignments should help students master information at one level and help push them to the next levels of understanding. Thus, at a practical level, assignments may have such goals as:

- clarifying an idea, algorithm, etc.,
- encouraging a student to understand some reading,
- providing practice in working through an idea, algorithm, etc.,
- tying new material to topics mastered previously,
- encouraging the integration of ideas,
- encouraging generalization or specification, or
- anticipating new work by raising issues, highlighting possible problems, or identifying special cases.

The Role of Programming

While one can imagine programming assignments that foster learning at many levels, one might speculate that many programming assignments focus on relatively low-level details, specific algorithms and structures, syntactic constructs, and machine- or language- dependent elements. In addition, one can hope that a nicely written program requires thoughtful practice in problem solving, careful design, the use of appropriate structures or classes, etc.

All of this suggests that programming assignments can touch upon many elements of computer science and foster learning at many levels. For some courses, therefore, it may be appropriate for students to spend much time with assignments that emphasize programming.

Other Types of Assignments

However, Bloom's Taxonomy also highlights the need to foster understanding at many levels, and one might seek a variety of assignment types to encourage broad understanding. The following list suggests some of the variety of assignments possible.

1. Identify what is printed by a given program or code segment, given relevant input.
2. Describe data structures, classes, and algorithms which would be appropriate to a given problem.
3. Compare two solutions, structures, or algorithms.
4. Explain why a given construct is used (or not used) in a given context.
5. Analyze the efficiency or predict times for a given algorithm with specified data.
6. Solve a problem, where an elegant or concise solution produces particularly short code.
7. Identify the social impact, ethical issues, or potential consequences of a potential decision or program
8. Identify n advantages and n disadvantages of a specific algorithm or technique or structure (where n is a number specified in the problem).
9. Identify which of several algorithms or structures you would use in a given situation.
10. Identify the loop invariants or assertions at a given point in code.
11. Write a loop or procedure, given a loop invariant or pre- and post-conditions.
12. Identify and correct an error in a program that produces a specified (incorrect) result.
13. Given a program, identify elements requiring change to add a feature or change an implementation.

Some Sample Alternative Assignments

The following illustrate some of these types of assignments. As already noted, *readers suggestions are solicited for more examples and for the identification of other types of assignments.*

Problems Requiring Much Thought, But Relatively Little Programming:

Babysitting [200, Modified from problem 5, p.105]: A baby sitter earns $1.75 per hour until 9:00 pm (while the children are awake and active), $1.00 per hour between 9:00 pm and midnight (after bedtimes and the baby sitter can study), and $1.25 per hour after midnight (when babysitting cuts into prime sleeping time). Write a program that reads the starting and ending times (in hours and minutes) for a babysitter and computes the babysitter's fee. Assume all times are between 6:00 pm and 6:00 am.

Parameter Passage: Consider the following parameter passage methods: reference, value, value-result, name. Also consider the dynamic versus static binding of variables.

Write a program (using Pascal-, C-, or Ada-like syntax) that contains no more than two (non-nested) procedures (in addition to a main procedure) and that generates the following output for each type of parameter passage and each nesting strategy:

Type of Binding	Static Binding	Dynamic Binding
Value	1	5
Value Result	2	6
Reference	3	7
Name	4	8

Throughout, assume a temporary variable or location is generated for expressions, if needed. (E.g., if passage by reference or value-result is used with an actual parameter $n+1$, assume a temporary variable is created and assigned the value $n+1$, and this new address is passed. Of course, the value would be discarded when the procedure returns.)

Other Problems: Gries' text [76] includes many wonderful problems of this type, particularly in latter chapters.

A Problem to Investigate an Algorithm:

Tree-Shape Problem: Draw a binary search tree with the nodes containing each letter of the alphabet (ignoring case). Then describe an ordering of the letters, so that your tree results using simple insertion of the ordered letters into an initially nil tree. Extend your answer to describe an algorithm which will order the letters to produce any given binary trees containing each letter exactly once.

A Design Problem to Encourage Synthesis [201, Modified from exercise 20.12, p. 677.]:

A druggist must keep records of all prescriptions she has filled, including information on the drug, patient's name, and date filled; and she must be able to retrieve these records by patient's name (for processing refill), patient's family name (for tax purposes), date (for doctor's inquiries), and month (for billing).

Design a data structure to store and retrieve this information. Your structure should be practical and efficient; the druggist fills many prescriptions each day, and new customers arrive in town frequently. Also, your design should assume that there is too much data to rely solely upon storage in main memory, so disk storage is needed.

In discussing your data structure, specify the needed declarations and draw a picture showing any pointers that might connect pieces of this structure. Also, identify the algorithms needed for storing new prescription information, searching by patient's name or date, and printing by family name or date. Finally, indicate how your structure might be modified if retrieval also was required by doctor's name and/or by drug name.

47.2 SUBSEQUENT REFLECTIONS

In the original column in December, 1998, my thinking focused generally on introductory computing courses designed for students who might become majors. As background, ACM/IEEE *Computing Curricula 1991* discussed two basic alternatives for introductory computer science: a breadth-first view and a more traditional, depth-first approach. [8]

- The envisioned breadth-first perspective provided a high-level overview of such topics as computer organization, operating systems, security and privacy, theoretical underpinnings, proofs, algorithms, and problem solving. Typically, programming was included as part of problem solving and algorithms, but its scope often was modest.
- The traditional, depth-first approach covered such basic problem-solving elements as control structures, recursion, iteration, data structures (e.g., arrays and/or records/structs), simple searching, and simple sorting.

Although the breadth-first approach was appealing in its goal of providing a wide-ranging perspective of the discipline, the coverage of various topics often was insufficient to allow students to satisfy prerequisites for later courses. Overall, a breadth-first course might add another required course for the major, without yielding depth at an upper level. In practice, several computing faculty experimented with the breadth-first approach, but after a few years, most schools continued with the traditional, depth-first approach.

As this historical note suggests, the balance between programming and other topics has been a long-standing and controversial topic among computing educators for many years. In this discussion, some educational leaders make a distinction between courses intended for non-majors and courses for [potential] computing majors or general audiences. The following notes review directions for both of these audiences and courses.

Courses for General Audiences

The March 2015 issue of ACM *Inroads* included a "Special Section on The Role of Programming in a Non-Major, CS Course," and the "Welcome" to that issue provides a worthwhile overview of the appropriate balance between programming and other topics within the context of a course for non-majors. [96]

- As described by Kick and Trees, the newly-developed Advanced Placement Course for Computer Science Principles (CSP) targets a broad audience and works to motivate students to engage ideas of computing. To illustrate the range of areas touched by computing, a CSP course highlights seven "Big Ideas" (creativity, abstraction data, algorithms, programming, Internet, impact) and six "computational thinking practices" (connecting, creating, abstracting, analyzing, communicating, and collaborating). As Kick and Trees observe, "Programming is one component of AP CSP and one of the seven big ideas. But, more than a unit of study, AP CSP integrates programming concepts throughout the curriculum. " [98, p. 43]

  In practice, several instructors developed prototype courses, as CSP was being formulated. Although programming was included in each prototype, the amount of programming varied substantially.

  With an initial breadth-based experience, the CSP effort hopes some students will want to study computing further, perhaps taking a more traditional course later.

- Although programming is part of CS Principles, does it follow that programming should be part of every introductory computer science course, designed for non-majors? Answers to this question encompass several contrasting perspectives.

  - ◇ Goldweber argues, "Programming should not be part of a CS course for non-majors," not because programming is harmful or inappropriate, but because other topics are more important. Some vital topics include algorithmic problem solving and strategies (e.g., "brute-force, greedy, and divide and conquer"), a top-down methodology, abstraction, the need to handle complexity, and a basic understanding of what is and is not computationally possible. Unfortunately in this view, introductory programming often degenerates into extensive attention to syntax, semantics, compiler errors, and similar distractions. Altogether, Goldweber identifies several important learning outcomes, and discusses how mechanics of a programming language (even at a simple level) can interfere with achieving those outcomes. [71, p. 56]
  - ◇ As a contrast, Cooper and Dann describe the need for a non-majors course to provide experience with computational thinking that includes "problem solving via decomposition, data representation and modeling. In our experience, we have found that using a tool, such as Alice, Scratch, or a robot, lends a consistency

of context—which makes it easier to integrate data representation and modeling into a managed curriculum." Further, in several studies, they found that the Alice environment can help students focus on programming concepts while avoiding some distractions of syntax and compiler messages. [47, pp. 51-52] Overall, such environments can engage students, introduce fundamental concepts, and provide a reasonably smooth path to later experiences with Java or other languages.

◇ In an article complementing Goldweber's perspective, I consider a range of "Priorities for the non-majors, CS course" and note "programming may not make the cut". When developing courses for diverse audiences, an instructor must identify relevant themes for the target audience. Some possible themes might include "how computers work,", "computers in daily life," "computers in business and finance," "computing in entertainment," "social and ethical dimensions of computing," etc. Table 1 in the article suggests specific topics to support these themes. [258, pp. 46-47] In considering such possibilities, some courses might naturally include programming as a component, but others (e.g, the course described by Goldweber) might focus on other topics—repurposing time that might otherwise have been devoted to programming.

- As an additional perspective, if a non-majors course includes practice with computational thinking as a goal, then an instructor must consider how students will communicate their ideas for algorithms and how algorithms might be compared. The discussion of solutions to problems requires a precision of language—natural language (e.g., English) is not sufficiently clear, unambiguous, and consistent. With this in mind, my second article argues "Computational thinking in a non-majors CS course requires a programming component." [257]

As with all course, content for computing courses targeting non-majors must depend upon desired learning outcomes. For some courses, programming may be an essential component—particularly when combined with topics that support other themes. Other courses may prioritize other themes and topics, and programming have little or no role.

Courses for Computing Majors and Mixed Audiences

An introductory CS1 or CS2 course typically contains a substantial programming component. As with all components of a course, inclusion of programming should support clearly-identified learning outcomes. Here are just a few of the possibilities:

- *Precise communication:* Students should be able to communicate algorithms precisely, using a specified programming language.
- *Problem solving:* Students should be able to demonstrate problem-solving ability by solving a problem with a program that will compile and run.
- *Preparation for later work:* Students should demonstrate mastery of programming techniques and tools, so that they have the needed background for a later course.
- *Full or substantial coverage of a language:* Students should be able to explain and use all syntactical elements and features of a language, so they have a deep knowledge of problem solving capabilities, concepts, and techniques.
- *Career preparation:* Students should demonstrate sufficient mastery of specific skills needed in the workplace (e.g., for an internship or job).

Articulation of desired programming outcomes seems an important step in clarifying the role of programming within a course and in the construction of assignments.

However, the original column also suggests that desired outcomes for a course should extend beyond simple programming. Here are just a few of the many possible options for a CS1 or CS2 course:

- *Problem decomposition:* Students should be able to approach a problem using a top-down methodology.
- *Abstraction:* Students should not only be able to explain how abstraction can aid the problem-solving process, but also when abstraction can add layers of complexity with little gain.
- *Applications:* Students should be able to articulate a range of applications that computational problem solving might help address.
- *Impact of computing:* When developing solutions, students should be able to anticipate the possible social, ethical, and cultural impacts of alternative solutions.
- *Limits of computing:* When considering a problem, students should be able to discuss when a computing-related solution might be helpful, but also when a solution may not be feasible or possible.

Naturally, one or two courses (e.g., CS1 and/or CS2) cannot cover all conceivable topics. However, when identifying outcomes, the original article suggests that both the outcomes and the accompanying assignments should seek a balance of topics—not just programming.

Finding interesting examples and assignments for CS1 and CS2

This column first appeared in the *SIGCSE Bulletin*, Volume 31, Number 2, June 1999, pages 24–25[208] with the title, "Finding interesting examples and assignments for CS1/2 assignments". This revised title is somewhat more descriptive and concise. Since it appeared, some cited references are no longer available, and the column is slightly edited to remove non-accessible references and to update some others.

Subsequent reflections on this subject appear later in this chapter.

48.1 ORIGINAL COLUMN

MY PREVIOUS COLUMN contained two comments, which I find myself pondering repeatedly. This column begins a discussion of the second of these statements:

> In today's world, even beginning computer users work with applications that feature interesting graphics, fancy fonts, and sophisticated user interfaces. However, to focus on fundamental algorithms, structures, and control elements, CS1/2 assignments often feature quart-to-liter conversions and other equally dull problems. How can introductory CS courses compete effectively with the flashy applications in other areas, and still focus on basic concepts and techniques? [205]

As a simple example, in 1967 I was excited to write a program to use Newton's Method to approximate the nth root of a number, using a simple loop. Today's calculators make such computations seem routine.

While one might react to student demands by adjusting large parts of a course to add pizzazz, a more conservative approach seeks topics which appeal to a wide audience, with relatively little change in core coverage. Specifically, one might look for examples and assignments with considerable appeal that can be introduced in a CS l-level course with no more than a class or two preparation. Some of these areas include:

- simple simulations,
- restating problems in new contexts,

- identifying problems from many domains,
- using real data,
- providing hints of real applications or research,
- taking advantage of a browser interface, and
- placing a problem within a Web-CGI context.

The first five of these address student expectations by finding particularly interesting or challenging topics for examples and assignments, while the last two areas consider the human-computer interface directly. This column discusses the first five of these for courses at the CS1/2 level. I leave the last two topics, together with readers' contributions, to another column.

Simple Simulations

Simple simulations are a particularly fertile area for interesting examples in beginning courses. Initially, one typically must spend a class or two on random number generators and their use a topic that may provide an opportunity to introduce basic security concepts (e.g., what makes a good password?). Thereafter, simulations may be used to address many types of problems, such as the following:

> A gambler starts an evening with \$20; he bets \$2 on each game and stops when he either runs out of money or has a total of \$50. For one game, the gambler bets \$2. If he loses, then \$2 is deducted from his account. If the gambler wins the bet, the \$2 is not deducted and a specified payoff is added. For example, if the payoff is \$5 and the gambler starts with \$20, then his new balance would be \$18 if he loses the bet and \$25 if he wins. Write a program which asks the user to enter the likelihood of the gambler winning one bet and the payoff from the bet and then simulates the gambler's successes after 1000 nights of gambling.

Other simple examples can include simulating young children's games (e.g., Hi-Ho Cherrio), to determine how many turns are needed on average until someone wins. The racquetball/volleyball simulation discussed in an earlier column [204] and in Chapter 46 is a somewhat more ambitious example of this type of assignment.

Note: While such simulations can provide much practice with simple loops and conditionals, care is needed so that the simulations will have a broad appeal. In particular, various war games or sporting events may appeal largely to one gender.

Restating Problems

Sometimes, a change in the apparent context of a problem can make it seem more interesting, while providing an opportunity to discuss broader issues. For example, consider the following simple example of conditional iteration and simulation:

> Simulate tossing a coin. In 1000 experiments, what is the average number of tosses to obtain both a head and a tail, and what is the longest number of tosses in your 1000 experiments?

The following problem [202, pp. 200-203] requires the same programming skills, but the context may appeal to a wider audience. Further, it provides an opportunity to discuss assumptions made in simulations. Students with strong math backgrounds also could try solving the problem using infinite series.

> A couple decides that they want to raise at least one boy and one girl. They decide to keep having children until they have one child of each gender, and then stop having children. Assume that having a boy or a girl is equally likely, and assume the gender of one child has no influence on the gender of the next. Run a simulation for 1000 couples to determine the average number of children each couple will have. Also record the maximum number.

New Problem Domains

Computations from many disciplines are within the capability of CS1/CS2 students. The challenge for CS faculty may be to find appropriate problems in other domains which might be accessible to beginning CS students. As a side benefit, a diversity of examples can help introductory students understand the need to read and analyze in order to formulate appropriate requirements, designs, and algorithms.

For example, in his CS1 course last spring, John Stone included exercises on oxygen transfer between mother and fetus, the Consumer Price Index (CPI), a new gambling game, numbers in the Aleut language, finding specific holidays (e.g., Thanksgiving) in various years, differences among various versions of a John Keats' poem, population trends among Iowa cities in 1980 and 1990, and the use of stated preferences to assign students to tutorials [4]. Such a range of exercises provides practice with many standard algorithms, approaches, and problem-solving skills, while appealing to many intellectual interests.

Real Data

The Web not only greatly simplifies the acquisition of real data, but also enables problems to ask interesting and realistic questions. For example, a simple file problem for CS1/2 might ask to sort members of the U.S. Senate by state or to retrieve all members of a House-specific committee. Data to support such applications are available from [190, 191]. Similarly, an exercise might ask for state representatives (often available on line) to be sorted by zip code for efficient mailing.

The Census Department [189] is another wonderful source of data, with easily useable tables for the population by age and gender. Other areas include anticipated survival rates, number of households, poverty levels, and income levels. Much of this information is available by state, giving rise to numerous assignments that stimulate interest, but still use fundamental file- access and processing techniques.

Hints of Real Applications or Research

While one cannot expect to include the full complexity of real applications or CS research in introductory courses, a course can provide examples of such work. For example, a CS 1 lab [207] might outline an expert system to place incoming students in math and CS courses. Other CS research or development projects might give rise to similar case studies. While students may not follow all of the details of such applications, they can gain insights on how basic ideas, algorithms, and structures may be extended to handle real problems.

Conclusions

The ideas given here just begin a discussion of what might be done to enliven CS1/CS2 classes. Each idea can be introduced in a class or two, each can provide opportunities for discussing computing in a broad context, and each appeals to a diverse audience. Now that these examples are in print, I hope readers will send in their own favorites. My next column will include any suggestions received from readers, as well as ideas of using the Web or CGI programming at the introductory level without interfering greatly with more traditional CS1/CS2 themes and topics.

48.2 SUBSEQUENT REFLECTIONS

The original column identifies a challenge: finding interesting examples and assignments for introductory courses. As noted in the column, students in classes change over time, as do their experiences and expectations. Applications that might have excited students several decades ago may not resonate with current students. Overall, the identification of examples and assignments remains as much of a challenge now as it did when the original column appeared, although cultural tastes and parameters may have evolved.

Three approaches for meeting this challenge include identifying application themes for courses, utilizing"nifty assignments", and basing problems on public databases. Each of these options can provide a wealth of materials.

This section concludes with the statement of a problem that requires relatively modest background (e.g., conditionals, loops, 2-dimensional arrays, characters). As with several problems identified in this part of the book, the problem's solution can be reasonably short and simple when time is devoted to developing a careful design, but coding can be long and complex without a well-understood design.

Application Themes for Courses

In recent years, several course-development projects have evolved to integrate course content with interesting examples and assignments. With this approach, for example, an introductory course might cover program structure, conditionals, control structures, iteration and recursion, some common algorithms, and basic data structures. However, examples and exercises that illustrate these ideas may be drawn from a common application theme, which can add interest, illustrate common applications of computing, connect computing with other disciplines, and allow scaffolding of assignments that lead to an motivating result by the end of the course.

Two well-known efforts at the introductory level involve image processing and the use of robots within introductory courses.

- *Image processing or media scripting:* The creation and manipulation of images provides a visual and artistic environment for introducing many basic programming constructs. A course combining image manipulation with computing can appeal to students with interests in the fine arts, furthering outreach to groups beyond traditional STEM fields.
  - ◇ Mark Guzdial and Barbara Ericson's text combining multimedia and computing is widely used and now in its fourth edition [78]. As students learn new computing techniques, they can experiment with applications to image processing, where they can see the direct impact of code on images. Versions of the book support both Python and Java programming environments, following a general imperative or object-oriented perspective.
  - ◇ Rebelsky, Davis, and Weinman utilize image processing within the framework of functional problem solving. A Scheme-based environment interacts directly with

the GIMP image processing package, allowing students to program naturally while using a high-powered image manipulation tool behind the scenes. [156].

- *Control of robots:* Scribbler 2 robots provide a cost-effective environment, with which students gain experience with many basic elements of introductory programming. Control of robot behavior may include sequential execution, conditionals, iteration, recursion, and elementary data structures. Since robots respond immediately to commands, students can receive immediate feedback from robot actions.

  ◇ Deepak Kumar has a wonderful, introductory book, *Learning Computing with Robots*, available through the Institute for Personal Robots in Education (IPRE) [102]. Although this book might be used in several contexts, the primary target audience includes general audiences—perhaps non-majors, and the course assumes a Python-based environment.

  ◇ Since summer 2011, I have developed full course materials for an introductory course on *Imperative problem solving and data structures.* Course materials, available online, include readings, labs, examples, projects, programming assignments, and a course infrastructure. The target is general audiences who might want to take later computing courses; that is, this course might be considered a traditional CS1 course. Course materials are updated each time the course is offered; [267] identifies the Fall 2016 version. Also, the underlying C-based environment is named MyroC (the C version of the Myro–my robot–project from IPRE), and the MyroC infrastructure is available from the MyroC home page [276].

"Nifty Assignments"

Several computing conferences include on-going sessions involving "nifty assignments." In one session, organizers select 6-12 assignments submitted by contributors, and each author presents their assignment idea within a 5-10 minute window. Although each presentation is short, session attendees hear a lovely range of interesting ideas within a single conference session.

In many cases, since the session is a formal part of an organized conference, abstracts of each "nifty assignment" are available within the ACM Digital Library. For a current listing of these assignments, the reader should use the Digital Library's URL: http://www.acm.org/dl, and search under the term "nifty assignment."

Public Data Sets

Similarly, the Web contains a wide range of data sets available for big data analytics. Such resources greatly expand the databases identified in the original column: [189, 190, 191].

In many cases, extensive data sets are available without charge, although an instructor or researcher may have to register for an account. To access many of these data sets, an instructor may use a favorite Web-based search engine, using the query "public data sets." Here are three possibilities, out of the many options returned by such a search.

- *Data.gov:* "The home of the U.S. Government's open data Here you will find data, tools, and resources to conduct research, develop web and mobile applications, design data visualizations, and more." [55]

- *Amazon Web Services:* "AWS hosts a variety of public datasets that anyone can access for free." [164].
- *Google BigQuery Public Datasets:* Google BigQuery hosts several public data sets. "Google pays for the storage of these data sets and provides public access to the data via BigQuery. You pay only for the queries that you perform on the data (the first 1 TB per month is free, subject to query pricing details)." [73]

Although instructors may need to extract data from large databases in order to shape an appropriate problem for an introductory course, the availability of numerous data sets provides a base for many creative examples and assignments.

A Simple Route Cipher

I have used the following problem several times within a CS1 level course. Prerequisite material includes character data, conditionals, nested loops and simple I/O, so this problem can fit reasonably within the middle or end of an introductory course. I like this problem, because several simple, but different, solutions (60-90 lines, with comments, blank lines for clarity, etc.) are possible—all require initial thought for a clear and simple design. However, students who jump directly into coding often have difficulties with muddled logic and unnecessary complexity.

Introduction: When sending a message from one place to another, it is common for the sender to encode the message before it is sent with the understanding that the receiver would know how to decode the message. With this encoding process, anyone intercepting the message in transit would not be able read the text.

For encoding, one approach is a *substitution cipher*, in which each letter in original message is replaced by another letter. (For example, each "a" in the message might be replaced by "d" and each "t" might be replaced by "w". This type of cipher is commonly used in many word puzzles in newspapers and puzzle books.

A second approach for encoding is called *transposition*, in which the characters of the original message are rearranged in a different order. This problem implements a simple type of transition cipher, called a route cipher. (Historically, the Union forces in the American Civil War used a variation of a route cipher, called the Union Route Cipher.)

Encoding: In a simple route cipher, letters of a message are placed into a rectangular table. As an example, suppose the cipher is based on a table of 5 rows and 9 columns, and suppose we wish to encode the text "this short example illustrates a route cipher". The first step of a route cipher is to insert the message row-by-row into the table, on character at a time.

t	h	i	s		s	h	o	r
t		e	x	a	m	p	l	e
	i	l	l	u	s	t	r	a
t	e	s		a		r	o	u
t	e		c	i	p	h	e	r

With this arrangement, the encoded message is obtained by retrieving the letters according a designated path or route from the rectangle. For this problem, we will retrieve the letters from the table column by column. For example, reading column-by-column from the above table, we obtain the coded message "tt tth icciels sxl c auaisms phptrholroereaur".

Decoding: Given an encoded message, the receiver places the text character-by-character into the rectangle according the prescribed path (e.g., column by column). With the letters in the rectangle, the original message can be restored by reading the rectangle row-by-row.

Extensions: In the basic encoding approach, the original message is placed in a rectangle of a designated size. If the rectangle has r rows and c columns, this approach works well if the message has length r*c, the size of the rectangle. Extensions are needed if the original message has length other than r*c characters.

- If the original message has less than r*c characters, additional characters might be added to get the needed number. In particular, for this problem, one should add letters of the alphabet a, b, c, d, e, ... at the end of message as needed to fill the rectangle.
- If the original message has more than r*c characters, the message is divided into blocks of r*c characters, and each block is encoded separately.

As another example, suppose the rectangle is specified with 3 rows and 4 columns, and suppose we want to encode the message "this extended example shows the full algorithm".

Encoding follows these steps:

1. Divide the message into blocks of 3*4 = 12 characters. The last block would have only 10 characters, so 'a' and 'b' have been added to complete the block.

```
t h i s   e x t e n d e
d   e x a m p l e   s h
o w s   t h e   f u l l
  a l g o r i t h m a b
```

2. Place each block into a rectangle, row-by-row:

```
t h i s   d   e x   o w s       a l g
  e x t   a m p l   t h e     o r i t
e n d e   e   s h   f u l l   h m a b
```

3. Read characters from each block, column-by-column: "t ehenixdste" "dae m epsxlh" "otfwhusel l" " oharmliagtb"

Combining the encoded blocks gives: "t ehenixdstedae m epsxlhotfwhusel l oharmliagtb"

Problem:

- Write a program that reads the rectangle size (a row and a column) and the text of a message and prints the encoded message.
- Explain how the above program can also be used for decoding and illustrate your explanation with an example.

Programming Notes:

- Although conceptually the first encoding step reads the entire message and then divides it into pieces, in practice, the program should read and process one block at a time:
  1. Read rectangle size and create rectangle(s) of the appropriate dimensions to process a single block.
  2. Continue until all input is processed

(a) Read one block
(b) Process characters for the one block
(c) Print encoded characters for that block

C, Java, and other languages allow arrays to be declared of a length that is determined during run time.

Background Reading:

A nice treatment of transposition ciphers may be found in *Elementary Cryptanalysis: A Mathematical Approach* by Abraham Sinkov [172, Chapter 5]. A revised edition of the book appeared in 2009 [173].

CHAPTER 49

Academic honesty in the classroom

This column first appeared in the *SIGCSE Bulletin*, Volume 36, Number 4, December 2004, pages 18–19[216] Subsequent reflections on this subject appear later in this chapter.

49.1 ORIGINAL COLUMN

DISCUSSIONS OF ACADEMIC HONESTY typically focus on student responsibilities:

- What are effective approaches to teach students to follow basic rules of citation, quotation, and paraphrasing?
- What mechanisms should teachers employ to identify potential cases of plagiarism?
- What procedures should classroom or college establish to investigate suspected cases of academic dishonesty, maintaining the integrity of academic work while protecting students' rights for due process?

Over the years, several SIGCSE papers have explored these issues, as has an ITiCSE 2002 Working Group report. This article explores parallel opportunities and responsibilities for faculty while teaching, specifically considering three settings:

- use of textbook ancillary materials, such as PowerPoint slides;
- discussion of correct solutions to problems; and
- test development.

In each case, some observations follow a real-life anecdote.

Use of Textbook Ancillary Materials

An Anecdote: During the spring semester 2004, several of my students worked through a standard textbook on computer networking. As with many published books, the authors provided a publicly-available Web site of supporting materials, although one section was password protected for instructors only. This latter section included PowerPoint slides for each chapter.

During my guided reading course, I happened to use a Web search engine (e.g., Google) using the authors' names and chapter title as keywords. As expected, this search identified the authors' slides. The index of those slides contained a prominent statement encouraging faculty use, with the request that the copyright notice be displayed and the authors be given credit.

However, my Web search also yielded many other listings. Examining the first six carefully, I found all were either exact copies of the authors' materials or lightly edited versions. None included the authors' copyright notice, and none mentioned the authors in any way. Additional searching of the Web sites for these courses yielded the same results. Each course site contained slides, but I could not find any course sites that credited the authors for their supporting materials.

Reflection: Before the advent of widespread electronic file sharing, authors often prepared Instructors' Manuals for textbooks. These often contained transparencies or masters for handouts, and copyright notices commonly appeared at the bottom of each page. At that time, such notices were required for protection under the copyright laws. In such a context, simply copying materials provided credit to the authors for their work.

Now, U.S. laws have changed, all materials saved in a permanent form are protected, explicit copyright notices are not needed (although an explicit notice aids legal actions). Since a blanket notice on an index page may suffice for court challenges, copyright notices may not be visible on individual materials. Even ignoring such legal issues, however, academic honesty would dictate that credit be given for all materials prepared by others.

Altogether, it seemed surprising that PowerPoint slides from textbook authors were widely available on course Web sites without acknowledgment.

Solutions to Problems

An Anecdote: In the summer 2003, I participated in an MAA Prep Workshop on Discrete Mathematics, organized by Bill Marion at Valparaiso State University. One session described various problems that might be appropriate for class use; another discussed popular textbooks. In a moment of curiosity during a break, I selected the nearest textbook from a table, opened the book to a random section, and performed a Web search on a nearby workstation. Keywords were the author's name, "exercise", and the section number. Immediately, the search returned full answers to all problems in the section. This led to searches for other textbooks and other sections, with similar results.

In most cases, instructors apparently had posted solutions to assigned problems, after their assignments were due, to provide feedback to their students. Students could check their answers, learning what they did wrong and how to correct their work. Of course, this material also gives correct answers to students around the world.

Reflection: With world-wide access to materials on the Web, faculty might consider how they make solutions available to students. Before the age of the Internet, instructors commonly posted answers to solutions on bulletin boards, and many faculty and students found this approach useful. In a global setting, however, such local benefits may have global implications. And, since materials can be scanned easily, distribution of solutions in paper form still may yield solution sets posted on the Web.

Practically, faculty today may need to assume that solutions to all textbook exercises are available through the World Wide Web. This would seem to have broad implications for the nature of assignments and mechanisms for grading.

Test Development

An Anecdote: Some years ago (not at Grinnell), I knew an instructor who used a generally-available self-study package for part of a course. The package, designed for home use by individuals, included audio, handouts, exercises, and self-assessment tests. Since the materials were not designed for a formal classroom setting, anyone could purchase the entire package. Due to the high quality of the materials, the instructor duplicated and distributed them to students in class and then used the self-assessment tests for quizzes and exams.
Reflection: Although the duplication of handouts raises copyright issues, let's focus just on the tests here. In this case, the instructor copied and administered tests without attribution. Note, however, that citation would have other practical implications. If the instructor cites the source of the tests, then students would know what self-study package to purchase. Secrecy is essential so that students will not have complete prior knowledge of all tests. On the other hand, instructors' use of others' material without citation raises questions of academic honesty.

While this case seems extreme, consider the related question of what citation is appropriate when using tests supplied specifically to faculty by publishers. If a publisher provides test banks to instructors who adopt books, when and how should the use of these materials be cited? Lack of citation would seem contrary to principles of academic honesty, but citation could encourage students to try to track down test bank data and answer books. (I regularly receive requests by students throughout the world for answer books and instructors' manuals for my textbooks. Citing test banks might encourage such attempts further.)

In summary, academic honesty is a well established principle, and most institutions have clear policies to enforce such concepts in student work. Similarly, faculty traditionally follow principles of citation, quotation, and paraphrasing in writing papers, giving talks, and other scholarly endeavors. However, some common practices raise questions regarding the extent to which principles of academic honesty are appropriate and observed within the realm of teaching.

49.2 SUBSEQUENT REFLECTIONS

Conversations with colleagues around the country and with conference attendees suggest that challenges with academic honesty have continued since the original column appeared. At some schools, cases of academic dishonesty apparently seem modest and isolated. At other schools, some faculty report that 20% to 30% of all academic dishonesty cases at a school involve computing courses.

The following notes consider four matters related to academic honesty and dishonesty:

- School rules and procedures
- Plagiarism-detection software
- Course rules and practices
- Multiple versions of tests

School Rules and Procedures

Many [most, all?] schools have rules regarding academic honesty. Students must report lab data accurately, give credit to others for their ideas and words, etc. Further, if an instructor suspects possible academic dishonesty has occurred, the schools have well-established procedures for reporting and investigating suspicions.

With increased use of the Internet and social media, schools may need to review their rules and procedures to accommodate emerging circumstances.

- The original column observed that solutions to many textbook exercises were easily obtained on the Internet. Today, the availability of such solutions seems to even more widespread. For example, in some cases, I have found literally thousands of solutions on the Internet to an individual exercise of a popular textbook. Overall, when exercises come directly from a textbook, most students can find solutions online, if they wish.
- When a student solves a problem (e.g., from a homework assignment or a take-home exam), the student may post the solution on a Web page or on social media. Perhaps the student wants to show off, or the student hopes to help peers. Since the solution is posted and available to others, there may be no first-hand or direct contact among students, so this scenario may or may not be covered by traditional rules regarding "collaboration" or "discussion".
- As a textbook author and a course instructor, I periodically receive inquiries asking for solutions to problems in my books or assignments. Also, sometimes upper-level computing majors ask for solutions—out of a "sense of curiosity." In such cases, I may be willing to help other instructors (if the request comes on departmental letterhead), but such requests also may be attempts to obtain answers to course assignments.
- Sometimes a policy specifies that submitted work must be the student's own, not someone else's. However, in recent years, I have become aware of cases when a student responds, "Of course that is my work; I paid for it, so it's mine."

Although some faculty might consider such situations dishonest, school policies may or may not address them. Further, if an upper-level student posts a solution or otherwise helps a student in a lower-level course, questions arise regarding the consequences, if any, to the upper-level student. They are not enrolled in the course, so their grades cannot be lowered. With new scenarios for possible dishonesty emerging, schools may need to review policies.

Plagiarism-Detection Software

In recent years, software has emerged that compares current student submissions (e.g., papers, computer programs) with previously-written material (e.g., published articles, books, Web materials, work submitted by students in the past).

Often, a software vendor maintains an extensive database of previously-written material. At some schools, students submit their work for some or all courses electronically, and the submission process automatically forwards the work to a software vendor for review. The vendor then compares the new material to all entries in the database. In some cases, the submitted material may receive a score of the percentage of material that has appeared previously. Sometimes individual passages may be identified as repeating work elsewhere.

As an example, as of this writing, turnitin.com reports that its database includes 62 billion indexed Web pages, 734 million student papers, and 165 journal articles, periodicals and books [188]. Similarly, Modiba, Pieterse, and Haskins identify at least 11 software tools that test code similarity [130].

One system that seems particularly popular for comparing computer programs is Moss (Measure of Software Similarity), a widely-used and freely-available system developed at Stanford University by Alex Aiken and collaborators.

> In response to a query the Moss server produces HTML pages listing pairs of programs with similar code. Moss also highlights individual passages in programs that appear the same, making it easy to quickly compare the files. Finally, Moss can automatically eliminate matches to code that one expects to be shared (e.g., libraries or instructor-supplied code), thereby eliminating false positives that arise from legitimate sharing of code.

Interestingly, reactions to such automated tools vary among educators.

- Some faculty embrace automated systems as particularly helpful mechanisms for detecting and combating plagiarism.
- Some faculty observe that the automated systems require the uploading of student work, and such storage undermines intellectual property rights; students cannot control the distribution of their work—-at least in this context.

Course Practices and Rules

Colleges and universities typically have school-wide policies and procedures regarding plagiarism, individual work, collaboration, and similar matters related to academic honesty. However, such policy statements may be general in nature, and clarification may be needed for work within specific departments or courses. For example, when students work in pairs to collect data in a lab or when students collaborate on a computing lab, what part(s) of a write-up must be done individually, and what part(s) can be written collaboratively.

To address such issues, some computing departments develop their own policies and practices regarding academic honesty. Typically, these are consistent with school-wide statements, but departmental rules may be specific about what is and is not acceptable regarding the development of programs. For example,

- If students collaborate on part of a lab and then use common code for individual programs that come later, how should common code segments be identified and cited?
- If some assignments are to be done individually and some allow collaboration, how will students know the rules for an individual assignment?
- If one student solves a problem, to what extent can that solution be posted on a Web site, on a bulletin board, or on social media? And, if the solution is posted, what consequences might be expected?

Similarly, instructors of individual courses need to be clear regarding the rules for individual or collaborative work. For example, in my courses, I often include an explicit statement on each assignment about whether collaboration is allowed or not.

Further, I have found it useful to require the use of an "academic honesty certification" for each exercise, program, etc. A common wording, for programs written in a language similar to C or Java, appears in Figure 49.1.

Although such a certification does not completely eliminate issues of plagiarism, requiring this for every program submitted has at least three advantages:

- Before turning in a program or other assignment, a student must think about sources (even if the time spent on reflection is brief). Further, I require a handwritten signature (a typed signature is not adequate), so the student must consciously confirm the certification is complete.

- In grading, if I wonder what sources might have been used, the certification indicates places where I might start a search.
- If an issue of academic dishonesty arises, the student has identified a complete list of sources. In presenting circumstances of a case, the student has no recourse if their work was copied from someone else without citation.

In practice, I have found the requirement of an academic honesty certification has had a substantial impact on code turned in. I still may encounter plagiarized code from time to time, but such issues arise much less than in the past.

FIGURE 49.1 A Sample Academic Honesty Certification

```
/*****************************************************************
  Academic honesty certification:
    Written/online sources used:
      [include textbook(s), course materials;
        complete citations for Web or other written sources
       write "none" if no sources used]
    Help obtained
      [indicate names of instructor, students, or others
       consulted; write "none" if none of these sources used]
    My/our signature(s) below confirms that the above list of
       sources is complete AND that I/we have not consulted
       anyone else about the solution to this problem
    Signature:
 ****************************************************************/
```

Multiple Versions of a Test

One mechanism to discourage copying utilizes multiple versions of an assignment or test. For example, I have developed multiple versions of tests for introductory courses since the mid 1980s, and I interleave the tests before I distribute them in class. In passing out tests, students usually take the test on top and pass the rest onward, ensuring that adjacent students have different versions. The instructor also can hand out tests one-by-one, going down a row to similar effect.

When developing multiple test versions, a challenge can arise in writing problems of similar difficulty. Here are some approaches that are sometimes helpful:

- When writing a problem to build a data structure (e.g., a binary search tree),
    - ◇ one data set might be obtained from another by adding a common value to each element; the new data structure will have the same structure as the original, but all data values will be different.
    - ◇ when one algorithm produces a specific data structure with given data, an instructor might write out a similar data structure with the location of elements switched left-to-right. By changing data values, the same algorithm can be used for an alternative structure, where insertion to the left in one structure is mirrored by insertion to the right in the second.

- After writing one problem (e.g., that uses a loop in a specified way), one can consider an alternative problem, in which the problem-solving approach is identical, but the narrative appears different.

Although writing corresponding problems may seem reasonably straight forward in many contexts, instructors sometimes may misjudge the difficulty of a problem. Often an instructor may accurately predict the difficulty of a problem, but sometimes a prediction may be off. For example, when making up multiple versions of a test, I usually am reasonably good at writing problems that cover similar material but look different, and usually the parallel problems turn out to have similar statistics. Also, the distribution of scores on one test often is quite similar to the distribution on another version. However, sometimes a review of statistics on two versions of a test show substantial differences.

In my experience, different grade distributions may indicate one version of a test might be harder than another, but sometimes the distribution of students may not have been as random as I might have expected. In one notable example for a particular semester, scores on four different versions of a test suggested that three versions were of similar difficulty, but the fourth was more difficult. That is, the distribution of scores on three versions seemed statistically similar, but the distribution of the fourth was lower. Later in the semester, however, the students who scored poorly on the fourth test also performed poorly on subsequent homework assignments, programming tasks, labs, etc. Overall, use of multiple versions of a test may help address concerns over plagiarism, but analysis of results on multiple tests requires care: when are differences in performance statistics due to differences in the tests themselves, and when are differences due to the populations of students taking each version?

CHAPTER 50

Exercise solutions: motivations, messages sent, and possible distribution

This column first appeared in ACM *Inroads*, Volume 4, Number 1, March 2013, pages 14–16[247] Subsequent reflections on this subject appear later in this chapter.

50.1 ORIGINAL COLUMN

IN MOST COMPUTING COURSES, students work on assignments, write programs, develop lab reports, and/or complete tests. Once these activities are completed, instructors may develop and/or distribute sample solutions. This column explores some relative merits and drawbacks in preparing and distributing these sample answers. The column suggests that distributing solutions may be helpful in some contexts, but may be counter-productive in others.

Some Possible Purposes for Preparing Solutions

Motivations for writing out solutions to problems might be organized into at least three (non-exclusive) categories: problem development, promoting learning, and facilitating grading.

Problem development: Often I try to write a solution to a problem as I am developing an assignment, test question, or lab. As I draft a problem or exercise, I try to think through what an answer might entail. Then I try to explain the problem as completely and clearly as I can. However, when I try to write a solution, it is not uncommon for additional details and surprises to emerge. For example, one part of a solution may require a special technique, or a code segment may depend upon an unintended procedure. When developing a test, I can time myself in supplying answers. (For Calculus I or CS 1, I have found that if an hour test for beginning students takes me more than 8 minutes, students likely will find the test too long.) Altogether parallel writing of questions and answers helps clarify both the problem statement and my desired answer (just as developing test cases and code together is part of various agile programming methodologies).
Promoting learning: After students have completed an exercise, they will have viewed a solution in one way. A sample solution can help confirm a student's approach, but it also

may provide insights about alternatives. Nicely documented answers can illustrate whether an approach is plausible, effective, and efficient. Further, sample solutions have the potential to identify possible pitfalls or subtleties. For example, from time to time, I may develop 4–6 alternative solutions to a problem to illustrate a range of design choices and problem-solving perspectives.

Facilitating grading: Development of sample solutions can help guide grading. A possible solution might be organized into parts, with points assigned to various pieces. Further, possible solutions can assist the reading of student answers, helping anticipate what approaches might be encountered. When an instructor works with one or more graders, sample solutions can aid communication about expectations. Of course, graders must recognize that students might attempt alternative approaches, but sample solutions can provide a base for the grading process.

Possible Motivations to Distribute Solutions

When I was a graduate student in mathematics, it was common to call a problem "trivial" if there was an obvious way to proceed, and if that way worked. I think factors regarding the distribution of solutions for learning may be different for "trivial" and "non-trivial" problems,

Distributing solutions to "trivial" problems: Currently, one of the courses I am teaching is calculus I. In this context, most problems from the textbook are remarkably standard—variations of the same problems may be found in many textbooks. Further, a publisher typically prepares answer books that show an expected solution to each problem. In many cases, these problems fit within the "trivial" category. Often there is a typical set up of a problem, the set up leads to an expected formula, differentiation or integration techniques apply in a standard way, and an answer is obtained by plugging numbers in at the end. A similar situation arises for simple exercises in beginning programming, in which one writes a simple code segment (e.g., a loop or a conditional) to perform a simple task.

For these "trivial" problems, sample solutions can help students check basic techniques. Can they set up a problem, can they use an appropriate formula or syntax, can they work through the details to obtain a desired answer?

Altogether, answer books and solutions to "trivial" problems can help reinforce basic levels of student understanding, and answers can help improve the efficiency of graders by showing anticipated solutions.

Distributing solutions to "non-trivial" problems: Following the above categorization for "trivial", a problem may be considered "non-trivial" if either there is not an obvious way to approach it or if the obvious approach fails.

When no obvious approach exists, the problem may require considerable analysis and insight, or there may be several alternative approaches to consider. In my experience, when I attack a problem and find my solution becoming progressively more complex, I usually discover that I have missed something. With insight, I can replace a convoluted and complex approach with a much simpler one. When a problem has this type of quality, distribution of a solution after students have done their work can help highlight qualities of the problem and insights about possible solutions. In these situations, I have found that simply handing out a solution, however, is inadequate. Either I need to discuss the solution in class, or I must include commentary in my answer, so that students can understand the process and characteristics that might have led to the elegant solution. (Personally, I find class discussion is most successful in this regard.)

When a problem might yield to multiple problem-solving approaches, students need to understand what approaches are possible and why some approaches might be chosen. For example, this situation is a common motivation for the case-study approach, in which an experienced author can guide a student through multiple steps and design choices on the way to a solution. Although such discussion can be very helpful, extended commentary may have at least two practical drawbacks:

- Development of extensive commentary may require an author to spend substantial time. Often the author must write two or more complete programs and commentaries on those programs.
- Reading of an extensive commentary may require students to spend substantial time in studying. Thus, commentary without a corresponding assignment may not be read.

When a problem suggests an obvious approach for solution, but that approach fails, a sample solution can help students identify multiple strategies, but again commentary may be needed. Why did one approach fail, what characteristics of the problem suggested an alternative approach, and how might one anticipate such difficulties?

Messages Sent When Solutions are Distributed

The distribution of sample solutions to students may convey several messages—sometimes intended and sometimes not. Some positive messages may help promote learning.

- Solutions to "trivial" problems can reinforce standard techniques.
- Solutions can help students gain confidence in their basic understanding of common constructions.
- Students can expand their repertoire of skills by reading new or alternative approaches.
- Solutions can help serve as a study guide for tests and later assignments.
- Case studies can describe design choices and alternative problem-solving techniques.

Other messages from sample solutions may not be as positive.

- Solutions to "non-trivial" problems may suggest that there is only one appropriate way to solve a complex problem. The distributed approach is good, but students may think that all others are bad.
- Distribution of a single solution may suggest that problem solving is primarily a challenge to happen upon the "right" way. Students do not understand that problem solving involves analysis or creativity, but rather they may try to stumble upon a hidden formula or answer. For example, this view may encourage random changing of code until it apparently works.
- Students may focus upon details of a sample solution rather than an overall approach. ("I was thinking of a while loop, but I guess I should have used a for loop.")

In my experience, distributing answers to "trivial" problems can work well, but I need to be much more careful with assignments that are relatively open ended or require creativity. For example, in my recent book, *The Tao of Computing (Second Edition)*, I consciously do not provide an answer book for the many discussion questions and research exercises throughout the book. In this setting, questions are supposed to encourage students to explore topics on their own, investigate recent advances in the field, consider their own perspectives, and come to their own conclusions. Student answers should be logically constructed, based on evidence, but many different discussions and conclusions are possible. Printed answers would tend to limit the range of possible answers and potentially discourage creativity.

Where to Distribute Solutions

Several years ago, at a summer workshop about discrete mathematics in the undergraduate curriculum, a session identified several commonly-used textbooks. At the break after the session, I picked up the textbook on top of a pile, opened to an arbitrary section, and picked a problem at random. I then used a browser to search the Web with the author's name and "exercise m.n" as the search term. Instantly, I obtained a full solution to the problem. Subsequent checking indicated that answers were available online to all exercises in the book.

Several decades ago, various groups (e.g., fraternities, sororities, and student clubs) on various campuses maintained files of exercises and tests, allowing their members to access answers to a wide range of activities. Although such archives had implications locally, students at other locations rarely had such access. In today's age of the Internet, however, answers abound for world-wide consumption. Even if an instructor makes an answer available only on a local server with access restricted to a specific campus, students commonly send answers to their friends or post materials on public sites. Also, as an author, I periodically receive requests for answer books for my books—clearly from students worldwide. Further, I have seen copies of some answer books posted on public (non-author) sites.

With the communication capabilities of the World Wide Web, instructors likely need to be particularly careful in how sample solutions will be distributed. In some cases, posting of case studies and annotated answer commentaries may encourage in-depth study and analysis. However, once answers are posted, students may be tempted to copy rather than think.

Conclusion

The development of sample solutions can have several positive motivations, but instructors should review the messages they are sending if/when sample solutions are distributed. Further, the World Wide Web has had a dramatic impact on the widespread availability of solutions to problems, and instructors may want to consider distribution before they post solutions—either publicly or on a private course site.

Acknowledgment

I am deeply indebted to Marge Coahran for providing me with feedback on an earlier idea for this column and then for helping me polish the material presented here.

50.2 SUBSEQUENT REFLECTIONS

Solutions to exercises still can be beneficial, as discussed in the original column. However, the availability of solutions to textbook problems and other commonly-used problems seems substantially increased from even a few years ago. Here are a few observations I noted while writing this subsection:

- Sipser's book on the *Theory of Computation* [174] is widely used for upper-level courses through North America (and perhaps more broadly), and one of its features is a wonderful collection of problems. Many exercises highlight important ideas, require insight, and help students engage with the material at hand. However, as a popular text, many instructors and other groups have posted solutions to most (perhaps all) of its problems. For example, a recent Web search, with the search terms "Sipser 1.17 solution," yielded 2200 hits. Although some of these links may not provide answers, hundreds of solutions are available to Exercise 1.17 in this book.

- In mathematics, Stewart's *Calculus* is used extensively in college courses. In this case, a recent Web search revealed numerous opportunities to obtain not only individual answers to specific questions, but also entire answer books. In some cases, the solutions within answer books seem available (section-by-section) online.

- In many subjects, for a fee, some companies advertise they will supply answers to book-related or course-related questions. With a subscription, an individual can access answer books and [sometimes] solutions to course-specific problems.

Altogether, when using an established textbook, an instructor likely is safe to assume that students can obtain answers to all questions online. A few years ago, one might have contacted faculty, asking them to take their solutions of textbook problems down from the Web. However, the practice seems sufficiently widespread that solutions to textbook exercises will be available to any student with modest Internet access.

In some cases, of course, Internet solutions may be terrible—either in terms of correctness, efficiency, structure, readability, or clarity. From time to time, I have encountered a range of solutions that I would be embarrassed to display!

Overall, this availability of textbook solutions yields several suggestions for instructors:

- A syllabus and/or an individual assignment should be clear regarding the rules of citation for homework exercises and assignments. At the least, solutions from the Web should be cited:
  - ◇ Without a citation, copied solutions likely represent academic dishonesty.
  - ◇ With a citation, copied solutions may or may not be eligible for partial or full credit.

- Students should be advised that they are responsible for the work they submit. Poor-quality work is worth little credit, even if the approach or the details come from a Web source.

- Faculty may try to rewrite problems to undermine Web searching. Sometimes changing phrasing in a problem can thwart many search strategies. On the other hand, the revision process may change a problem in a substantive way:
  - ◇ The problem may no longer addresses the same topics.
  - ◇ The new problem may be much easier or much harder.
  - ◇ Searches for phrases in a rewording may not retrieve the same sources as searches for the original, but the new resources still may be plentiful and provide extensive help.

As noted in the conclusion of the original column: "The development of sample solutions can have several positive motivations, but instructors should review the messages they are sending if/when sample solutions are distributed." Further, instructors now must consider how the widespread availability of solutions on the Web may impact course exercises and assignments.

X

Student Progress in Courses

WITHIN A COURSE SETTING, STUDENT LEARNING may be considered a collaborative enterprise involving both faculty and students. Faculty help provide a framework, introduce concepts and perspectives, identify assignments, develop laboratory sessions, provide feedback, mentor students, guide learning, etc. Students engage material, expand understanding through Bloom's six levels of mastery [25], investigate problems (e.g., from assignments), explore laboratory exercises, draft solutions and programs, etc.

As part of this faculty-student collaboration, faculty can use their knowledge of course content and pedagogy to organize topics, formulate an effective class format, and compile appropriate assignments, labs, and projects. However, faculty also may want to anticipate likely challenges and student issues. Several examples follow.

- Many students may have wonderful intentions, but they often lack time-management skills.
- Since social and cultural factors, as well as other academic demands, may constrain student time, courses (particularly at the introductory level) may need to help students structure their time and encourage productive use of that time. For example, an instructor may need to find creative ways to encourage students to be prepared for class (when other demands are pulling in other directions).
- Students from one semester often differ from one semester to another, so courses may need on-going adjustment to fit current students.
- At times, student performance on tests or other assessments may suggest a need to address immediate challenges as a semester progresses.

Altogether, many factors can have a substantial impact on student progress in a course, and an instructor may need to make refinements to meet circumstances as they arise. Each chapter of this Part explores aspects of each of these challenges.

Beyond the course environment itself, of course, many external factors can have dramatic impact on students' academic success.

- In today's society, many students may be overcommitted with demands from courses, jobs, family, etc.
- An atmosphere of inclusion or exclusion (either intentional or inadvertent) can have a substantial impact on students' feeling of community and their perceptions of success.
- Work schedules may have a dramatic impact on the ability of students to work in labs or to collaborate with partners on assignments and projects.

Although these factors may lie outside the realm of focused academic work, their impact can be substantial. The selected (annotated) references at the end of this part suggest an initial collection of readings for further study of student success within courses.

CHAPTER 51

Structuring student work

This column first appeared in ACM *Inroads*, Volume 5, Number 4, December 2014, pages 30–33[254]

51.1 ORIGINAL COLUMN

QUESTIONS REGULARLY ARISE when crafting a programming problem, written assignment, laboratory exercise, etc. concerning how much structure and guidance to include.

- Should instructions provide step-by-step guidance on how to achieve a specified task?
- Should the work for one activity serve as the basis for the next activity?
- Should instructions specify a sequence of deadlines for parts of the work, or should a deadline be given only for a final product?

After some general comments, this column explores three components of this topic: structure within an activity, structure among activities, and the specification of deadlines.

General Comments

Student background evolves over time. Students starting a program may have little idea of what words mean, what algorithms to consider, what data structures are relevant, or how to put pieces together. When students graduate, however, they should be able to tackle large, complex problems, from design through implementation, testing, and maintenance. Thus, a basic question for educational programs is, "How can courses help students progress from the beginning stages to more advanced levels of competence?"

For any assignment or laboratory exercise, at least three extreme positions seem possible:

- "Tell everything:" Provide extensive instructions, telling students every detail for each step.
  - ◇ *Example:* Books describing word processors, spreadsheets, or other packages may present sequences of specific instructions, with or without explanations.
  - ◇ *Advantages:* Students know what to do at a basic level and may be able to solve similar problems on routine assignments or tests.

  - ◇ *Disadvantages:* Students may not understand underlying concepts, so may be learning only at the most elementary levels of Bloom's Taxonomy [25].

- "Sink or swim:" Since students eventually will have to learn how to solve various problems, require students to develop problem-solving skills on their own from the start.

  - ◇ *Example:* An assignment or lab exercise might ask, "Experiment with this software to identify the types of information it can produce." (Beginning students may have little idea what "experiment" might mean in this context, but an assignment may expect students to determine what this instruction means.)
  - ◇ *Advantages:* Students become resourceful researchers and problem solvers from the very beginning and gain experience as they move toward graduation.
  - ◇ *Disadvantages:* Students may flounder with little idea of how to proceed and eventually drop out.

- "Education as research/development projects," sometimes called the "Moore Method" in mathematics: Give students the major results, and require them to discover their own proofs, algorithms, etc.

  - ◇ *Example:* A workbook in abstract algebra or topology may list a sequence of lemmas, theorems, and corollaries, without any accompanying text (e.g., proofs, discussion of concepts). Definitions may or may not be given.
  - ◇ *Advantages:* Students experience elements of research, in which they must brainstorm varying approaches and create their own techniques to obtain desired results.
  - ◇ *Disadvantages:* Students largely must re-invent the wheel for each topic, consuming substantial time to cover only a modest level of material.

Within an individual course, assignments and laboratory exercises may blend approaches, and the nature of activities may evolve from introductory to advanced courses. Faculty may want to smooth the learning process as students develop their skills, but students should not expect that all problems have quick and easy solutions. Overall problem solving requires experience with complex, multi-step problems; and activities should highlight concepts and foundational approaches, not just trivial problems. In addition, students need to develop the motivation, skill, and confidence to see solutions through—not stopping part way through.

The next sections discuss how various approaches might come together various activities and how deadlines might be developed. Throughout, much of my narrative focuses upon activities at an introductory level, but many ideas extend to advanced courses as well.

Structure within an Activity

When contemplating an assignment or laboratory exercise, I often find that Bloom's Taxonomy provides a helpful conceptual framework. As a reminder, in 1956, Benjamin Bloom identified six levels of learning from an elementary knowledge of definitions and details through the mastery of concepts and the ability to synthesize and evaluate [25, p. 201-207]. *Often in designing an activity, my goal is to help move students from whatever understanding they have to the next one or two levels of mastery.* Although this goal may seem clear, implementing it requires reaching a careful balance. I want steps within a structured activity to stretch student understanding—limiting routine or repetitive work. At the same time, I want activities to be sufficiently manageable that students have a good chance of success, at least much of the time—driving their motivation. Put together, steps in a structured activity should move a student up Bloom's levels of mastery, while promoting motivation and yielding considerable success.

For in-class activities, I typically want students to think and stretch their understanding. However, in the interests of time, I often want to provide some structure to guide them. For homework assignments, where time is less constrained, I often emphasize integration of topics and provide little guidance.

What follows are four practices, with variations, I often find helpful in designing activities:

1. Assume students have done the reading (if this assumption is false, they will have incentive to do the background work next time). This implies my assignments almost never ask for a simple rehashing of a basic algorithm, code, etc.

2. If students have a reading knowledge of an algorithm, data structure, etc. structure a laboratory exercise that actively engages students. For example,

    - before class, students may be required to answer structured questions or articulate questions they have encountered.
    - a sequence of steps might ask students to apply an approach to specific data (requiring understanding and a command of how processing proceeds).
    - additional steps may require students to explain not only what result occurs, but why (not only reinforcing understanding, but also sharpening communication skills).

3. If students have a basic understanding, ask students how one approach compares with another—either through analysis of the algorithm or through experimentation with data.

    - Students might determine which, if any, of several code attempts are correct, how to correct identified errors, or how to revise code to meet different requirements.
    - Students might develop test cases that will prove whether an approach works.
    - Students might compare the relative efficiency of several algorithms under varying circumstances.

4. If students have moderate mastery of a range of techniques, present a homework problem or programming assignment with little guidance.

    - If the assignment is given as part of a course module, students might expect elements of recent work might be relevant, providing possible implicit structure to the activity.
    - If the assignment is given as one of several independent activities (possibly as extra credit or as something to be completed sometime during a semester), then little structure encourages students to synthesize ideas and place approaches in context (largely in the "sink or swim" model).

Altogether, I rarely follow a "Tell everything" pedagogy, as that approach seems aimed at the most elementary levels of Bloom's Taxonomy. If basic comprehension needs checking, some faculty use an in-class quiz to test that students have read assigned materials. Alternatively, for beginning courses, I start classes with activities that assume students have done the reading (or be embarrassed at their lack of preparation when they collaborate with others). For advanced courses, I often ask students to submit discussion questions or answers to questions on daily reading. In both contexts, I structure in-class activities to move students to higher levels of mastery. Homework can have somewhat less structure to promote the integration and analysis of the material.

Structure among Activities

In structuring assignments and laboratory exercises through a semester, at least two approaches seem common. One approach divides a course into separate pieces, and independent activities provide practice with each piece. The second approach divides an overall project into multiple activities, and the activities build upon each other to yield a relatively extensive or complex result.

In my experience, the first approach has substantial advantages—at least for an instructor. When topics for a semester are organized into an overall plan, perhaps day-by-day or week-by-week, the schedule guides what material to emphasize during each time period. Effectively, the schedule provides a mechanism for managing course complexity by encapsulating small portions of a course. Each activity can focus on a reasonably small and manageable collection of topics, and the development of each activity can proceed independently. The overall class schedule ensures that pieces fit together, and details of one assignment or laboratory exercise need not relate to details of another. As a further advantage, refinement of a course often can focus on individual activities with little impact on other course components. One part of a course can be improved or refined without changing other pieces.

On the downside, although this approach provides experience with individual topics, students may gain little understanding of how the pieces fit together. Further, work on small pieces may provide students with only limited experience with the design and implementation of interacting components.

Scaffolding: The second main approach for structuring multiple course activities organizes a large project into component steps that are spread over multiple assignments and completed by individuals or teams of students. As a common example, in a course on compiler construction, a first activity might include writing the specification of a language (e.g., in BNF), a second might write a parser, a third might use the parser to construct a parse tree, ... If all goes well during the semester, students construct the substantial part of a compiler—giving students a strong sense of accomplishment and satisfaction. Students also gain insights into how various components of a course may be integrated, helping students master content at a high level within Bloom's Taxonomy.

Although this approach has great potential for student learning, challenges arise for the instructor (and students). When each step depends on earlier steps, special care is needed in the design of activities to ensure that the steps connect properly. If specification of a later step is slightly different than anticipated at an early step, rewriting may be required. Modest difficulties may help reinforce the need for careful specifications and the maintainability of code—highlighting worthwhile principles of software design. However, substantial difficulties may yield substantial student frustration, require extensive time as students correct earlier code, and possibly prevent successful completion of the final project.

Another difficulty arises if an instructor wants to refine some steps in preparation for the next offering of a course. Since the activities build upon each other, a change at one point may require adjustments both in earlier and in later activities.

Altogether, development of a series of activities requires considerable care and preparation. In my discussions with colleagues, faculty who make extensive use of scaffolding often suggest two aids that can help. First, to reduce frustrations and extensive work for students redoing early steps, an instructor may make solutions from one stage available to students for later steps. For example, once students complete an assignment or laboratory exercise, the instructor may distribute an alternative solution. In the next step, students may use either the solutions they have developed or the instructor's alternative. In a variation, once one step is completed, student solutions may be distributed, and students may use any of these versions as the base for their next step. Returning to the compiler example, once students develop the BNF grammar and parser, students might develop the next step (code to build a parse tree) based on any of the parsers already turned in.

A second aid may involve an extensive package of core code, perhaps constructed and distributed by the instructor. In this environment, successive activities may build primarily on the instructor's code—adding functionality and integrating library functions toward an overall goal. This type of aid may limit gains from students building an entire project by themselves, but it may reduce risks from one stage undermining success of the overall project.

Specification of Deadlines

In a perfect world, I might prefer to publish final deadlines for activities and expect students to schedule their activities, so that their work was completed on time. Students need to sharpen time management skills, so they can meet deadlines.

In practice, however, contemporary life is hectic. Even well-meaning plans can go astray, and outside demands may require students to address immediate needs—even if they had planned to spread a long-range activity over weeks or months. As a result, work may be left to the last minute, and quality may suffer. Of course, one response might be for an instructor to consider less-than-wonderful work (and the corresponding low grade) as part of the learning process.

However, three observations may encourage a more pro-active approach. First, although faculty may complain about students waiting until the last minute, the track record for faculty may not be much better than for students. As many ACM Inroads readers may know, deadlines for paper submissions for SIGCSE symposia are publicized months in advance. However, the consistent track record shows that authors submit at the last minute. For example, Figure 51.1 shows when the papers submitted to SIGCSE 2010 arrived. Of 305 papers submitted, only 41 (13%) were submitted a week before the deadline, 264 (86%) were submitted the last week, and 164 (54%) the last day. [278]Some observers suggest that students often leave work to the last minute, and faculty may represent the most successful of those students. Other observers hypothesize that some authors start early and then refine their drafts as long as possible before actual submission. (While some submissions may indeed represent extensive revision, a review of late submissions suggests that some authors likely started rather late.)

Second, some years ago, I taught a course that included one lab each Tuesday. Originally, full reports were due the following Monday, just before the next lab. In reviewing student submissions, both my students and I observed that many lab reports appeared to have been started late, on Sunday night. In response, I divided the labs into two parts, with the first due on Friday and the rest on Monday. In this division, the lab work did not change—only deadlines changed from once a week to twice a week. The result was dramatically improved student performance. In practice, the dual deadlines encouraged students to work on parts of each lab multiple times during a week—not just once a week. Effectively, these deadlines raised the time allocated to a report and improved qualtiy, as students reacted to short-term deadlines.

Third, when working with students on semester-long, independent projects, the work is consistently most successful when the students and I identify milestones for each week. Students may have good intentions to work steadily on a project, but other demands interfere. Having short-term deadlines helps them keep on track, the way they prefer.

Altogether, structuring activities and deadlines represents an on-going challenge. Presenting open-ended tasks and allowing flexibility in time management may promote desirable long-term skills, and a progression of activities may help students to plan and organize. However, providing guidance through structured steps may have substantial impact on learning, motivation, and success. Further, structuring deadlines may allow work for one course to compete in time management with other life demands. Finding the right balance for assignments and laboratory exercises requires balancing many competing factors.

FIGURE 51.1 Papers submitted to SIGCSE 2010, based on the days before the published deadline (prepared with permission of the SIGCSE 2010 leadership)

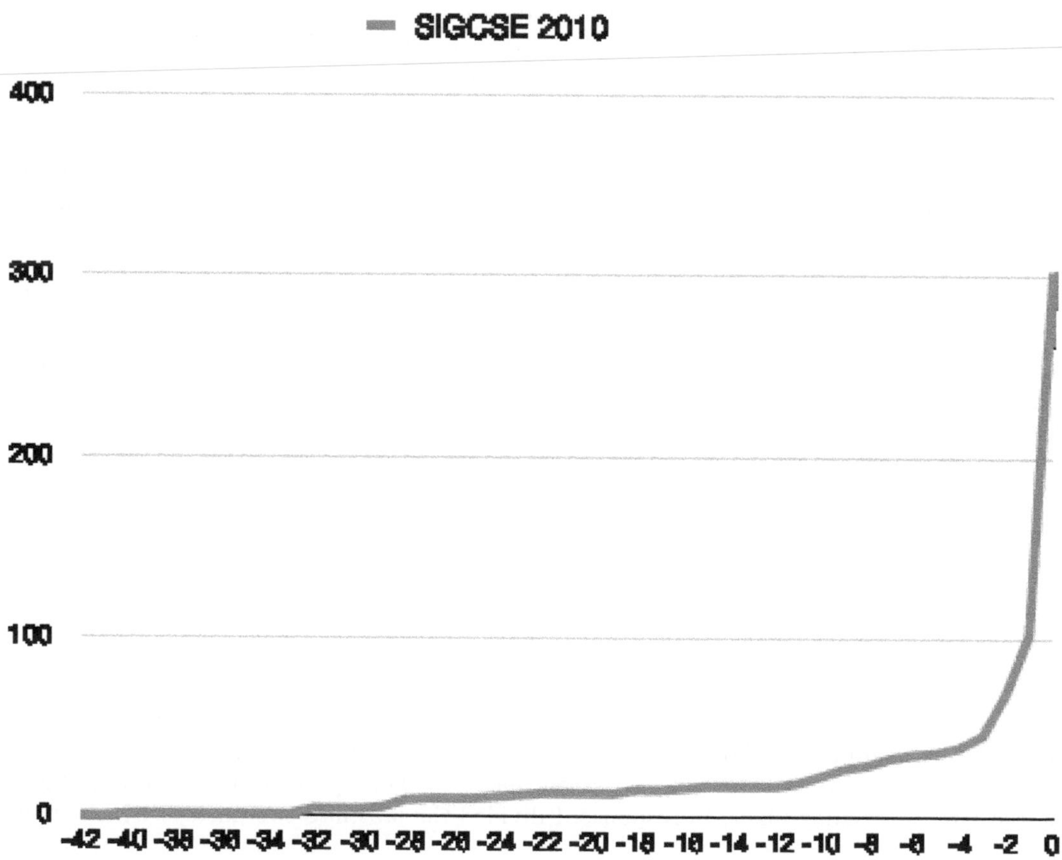

CHAPTER 52

Encouraging student preparation for class

This column first appeared in ACM *Inroads*, Volume 5, Number 1, March 2014, pages 24–25[252] Subsequent reflections on this subject appear later in this chapter.

52.1 ORIGINAL COLUMN

YEARS AGO, perhaps when I was in college, I heard comments about the "quiet, unfettered, contemplative life" of academia. Faculty could devote all their time to intellectual pursuits, and students had no outside involvements that might distract them from their studies. In this idealistic world, students likely would devote full time to their courses, they always would have time and interest for preparing for their next classes, they would participate actively in all class activities, and they would have all of their homework assignments completed on time.

Unfortunately, at least for me, reality bears little resemblance to this vision. An important consequence is that students must balance demands for a course with demands from other courses and with jobs, family, etc. Even if students are highly motivated to learn a subject, they must identify priorities in allocating their time. Preparing for class is one important element in this competition for a student's time.

With wide-ranging demands on student time, I want to organize my classes to provide incentives for students to come to class prepared. Overall, I find it useful to organize possible expectations and/or assignments into three categories, related to time.

- Activities due before a class session
- Activities due as a class starts
- Activities within a class session

The remainder of this column reviews each of these types of activities.

Activities Due Before a Class Session

Often, students are supposed to complete a reading or assignment in preparation for a class session. Several approaches can be helpful to encourage students to do the work and to help guide the class session itself.

- For many classes, I require students to respond to a reading in writing.
  - ◇ If the material seems clear, students write a summary (typically 5-8 sentences).
  - ◇ If the reading does not seem completely clear, they should write several questions for class discussion.
  - ◇ This feedback is due the day before class (e.g., 5:00 pm the previous afternoon), and I give up to 10 points per reading. Trivial or vague responses may receive few points; thoughtful statements receive most or all points.
  - ◇ Overall, student responses may count only 5%-10% of a final grade, but this often provides adequate incentive to encourage most students to read the assigned material.
  - ◇ I typically organize the responses into categories and then post the results on a Web page accessible only locally. Often the summaries and questions seem of better quality, if I post responses with attribution (who wrote what).
  - ◇ I plan each class session based on the questions submitted. If all students understand a topic, there is little reason to talk about it in class; if several questions relate to a common issue, class discussion seems important. Sometimes, summaries and/or questions suggest common misconceptions, and I build a class session to address the underlying difficulties.
- Although my summaries/questions are open ended, a colleague, Jerod Weinman, usually asks 3-5 questions on each reading.
  - ◇ Students answer questions ahead of time in an on-line journal that Jerod can review.
  - ◇ Questions may be general (e.g., what was most challenging, or most interesting, or most important), or questions may be focused (e.g., explain in words the meaning of a specific equation).

The main point here is that an instructor can ask students to respond in some way to a reading, and the student feedback may be helpful in guiding subsequent discussion.

Activities Due as a Class Starts

Assignments or computations due at the beginning of a class can encourage student preparation for material to be discussed, and also can provide a basis for in-class exercises.

- Students have special incentive to complete an assignment or computation, if some class activity explicitly references the homework.
- For one biology instructor whom I worked with for several years, preliminary computations (to be used in a laboratory exercise) served as an admission ticket to lab. If the student had the computations completed, the student could participate in lab; if not, the student could not enter the classroom, with a corresponding penalty for missing lab work.
- If a class activity will depend upon prepared work, the student may need to retain a copy of the homework or exercise for use in the class.
  - ◇ A student may be asked to bring two copies to class; one to hand in and one for class reference.

- ◇ Alternatively, if the instructor only checks the work was completed, the student or an assistant might ask to see the exercise as the student enters the room.
- ◇ To simplify grading or checking the work is completed, a standardized form can be helpful. A glance at the page might suggest whether the assignment was largely completed, even if a quick review would not provide a careful analysis of the correctness of all details
- ◇ If the student fails to bring the assignment to class, the student may receive a point penalty, or the student may have to work separately from the others to complete the assignment before being allowed to start the lab.
- ◇ If submitted work is graded, unprepared students might earn 0 points, and students submitting only cursory work may earn few points.

Connecting a homework assignment with in-class activities can provide a strong incentive for students to complete assignments. However, mechanics for checking that work was completed likely will depend upon the school and the personality of the faculty member.

Activities Within a Class Session

Work within a class may build upon required reading or homework. Connections between homework and in-class activities can provide strong incentives for students to prepare for class, even if students are not required to turn in materials. Building on connections with homework also can aid active student engagement within a class.

- Some faculty give 5-10 minute quizzes at the start of each class, or at random class periods, to check if students are prepared.
- Faculty may call upon students (perhaps at random) during class to explain something from the reading.
- If students are to bring discussion questions to class, an instructor periodically might ask a student to start the next section of discussion by offering one of their questions. (One approach is to write a simple program that randomly picks a student and displays the name on the screen. I suggest showing the program to the class to demonstrate a sense of fairness. Also, I suggest writing the program to choose names with replacement—once a student is chosen, that student still has to be ready the next time.)
- Faculty may organize students into small groups to discuss a topic, in which case a student's lack of preparation may be exposed to her/his peers.
  indexsoft skills—seealso reading

In my experience, peer pressure can be particularly helpful in encouraging students to come to class prepared. Students do not want to be embarrassed during class discussion, so the class format can provide encouragement. On the other hand, if faculty repeat much of a reading during a lecture, students quickly learn that they need not come prepared for class—students can wait for the summary by the instructor.

Acknowledgment

Many thanks to Marge Coahran for her suggestions and feedback on an early draft of this column.

Conclusions

Contemporary society often places significant demands on students' time, so students must balance class preparation with other pressures of life. Students may have fine intentions, but adding incentives to come to class prepared can help. Interestingly, in my experience, incentives need not be great, but some credit toward a final grade can be quite useful.

If readers have additional ideas for encouraging students to come to class prepared, please email me. I may include your thoughts in a later column.

52.2 SUBSEQUENT REFLECTIONS

In recent years, I have become progressively more aware of how student motivators differ from one campus to another.

- If collaboration is required in a lab setting, students in some environments come prepared because they do not want to be embarrassed in front of their peers. Students in other environments seem comfortable if they are unprepared when they start working with a partner. (E.g, "I did not do the reading, and I gather you did not either. Snicker, giggle. Let's start at the beginning, or we could take a long break.')
- On some campuses, points and grades seem to be a primary motivation; if an activity (e.g., required submission of commentary before class or an expected quiz at the start of class) will impact a course grade, students are reasonably likely to be prepared. On other campuses, grades seem less important than intellectual curiosity or relevance to potential jobs or opportunities to apply knowledge to help others.
- Sometimes students will read required materials before class, if they believe the reading will have a direct impact on class or lab work. In other contexts, students may devote hours to surfing the Web for answers, even after they observe that course readings provide concise descriptions for needed background.

Underlying all of these examples is the basic issue of what motivates individual students on a specific campus. Activities in the original column provide a range of approaches that work well in some environments. Ultimately, however, experimentation may be needed to discover what incentives, rewards, or penalties resonate with students in a specific course.

- If students respond primarily to points and grades, then creative ways may be needed to allocate points for course preparation.
- If students want to minimize time spent, then ways might be found to demonstrate how preparing for class now will have a payoff in reducing time for assignments later.
- If students seek positive relationships with their peers, then rewards might include constructive roles in class for those that are well prepared.

Once identified, course activities and assignments might be structured to incorporate those motivators in a substantive way.

CHAPTER 53

Mid-course corrections

This column first appeared in ACM *Inroads*, Volume 3, Number 1, March 2012, pages 20–21[245] Subsequent reflections on this subject appear later in this chapter.

53.1 ORIGINAL COLUMN

BEFORE EACH SEMESTER, a faculty member typically reviews course goals and objectives, identifies appropriate topics, chooses textbooks and/or other supporting materials, constructs a careful class schedule, selects assignments and labs, etc. By the middle of the term, dynamics within the classroom are largely established, students and the instructor have expectations regarding required activities and quality standards, and behavior patterns are generally determined.

With classwork underway, an instructor may wonder what parts of the course are going well and what refinements might be appropriate. Typically mid-course corrections involve two main steps: gathering data and making adjustments.

Gathering Data

In assessing how well a course is going, an instructor needs to obtain honest or objective feedback.

- *Formative Assessment:* Although formative assessment is often considered as a way to provide feedback to students, formative assessment also can help identify concepts or techniques that may be causing student difficulties. For example, programming assignments may indicate numerous students are having trouble applying programming constructs when solving problems.

- *Summative Assessment:* Summative assessment typically helps determine student competencies, often focusing on grades. However, these tools also may identify weak areas in a course that require additional attention.

- *Middle-of-Semester Survey:* A mid-semester survey may be administered in class, for homework, or with an on-line form. Possible questions include:

    - ◇ Do class sessions have the right balance of lecture, small group activities, labs, etc.?
    - ◇ Is the pace of the course too slow, too fast, or about right?
    - ◇ Is the amount of homework too much, too little, or about right?

- ◇ How do expectations for the homework in this course compare to other courses?
- ◇ Are readings interesting and at the right level?
- ◇ Do the labs have the appropriate length and difficulty?
- ◇ Should the course include more (or less) group activities?
- ◇ How might the course be improved to promote learning?

- *Mid-semester Class Discussion:* Rather than distribute written surveys, an instructor may devote some class time (perhaps 20 minutes) for discussion on positives and negatives for a course. What is going well, and what might be improved? Of course, if an instructor wants honest feedback, the instructor must not be defensive—rather, what do students really think?
- *Email:* Students might email the instructor with feedback about specific questions, with their identification of course elements that are working well, and with suggestions for what might help the course go even better.
- *Office-hour appointments:* If a course is not too large, the instructor might ask each student to schedule a 5-10 minute office appointment. In addition to inviting course feedback, the appointment might encourage students to ask questions about the material at hand. Alternatively, small-group appointments can serve as focus groups on elements of a course.

Altogether, surveys, email, and individual appointments provide feedback from individual students. Class discussions and group appointments encourage brainstorming, but also can allow a few students to dominate discussion and influence class attitudes. In some settings, instructors may ask for individual feedback (perhaps a survey), followed by a group or class discussion.

Once obtained, student feedback requires interpretation and analysis. Students can indicate their feelings, but students likely do not have perspective about content or pedagogy. For example, students may comment about the level of rigor without knowing what expectations are appropriate in a specific course. Also, students may identify symptoms of possible difficulties, but may not understand underlying causes. To avoid debate, some faculty prefer to avoid some types of questions, so students will not expect instructors to address certain complaints. Personally, I prefer to know what students are experiencing, and I often comment in class on the feedback I receive. My students consistently respond positively when I provide background on how the course fits in a broad context and why I use specific pedagogy. I also discuss possible responses to the student feedback.

Some Possible Adjustments

At the midpoint of a term, many course elements are reasonably fixed, such as class meeting times, textbooks, and promises in the syllabus (e.g., a schedule of tests). Even with these constraints, however, some adjustments are possible. The following list illustrates changes that have worked for me or for other faculty with whom I have talked.

- *Change balance of lecture, discussion, lab, small group, etc.:* Although class schedules specify when and where courses meet, the instructor normally determines what activities occur during those times. For example, time could be shifted from some lectures to collaborative activities. However, facilities in assigned rooms may dictate the extent to which lab activities might be changed or expanded.

- *Explain unstated assumptions:* At some schools, inadvertently every course may assume some topics are covered in that "other" course. When identified, an instructor may explain areas that somehow students have been expected to learn by osmosis.

- *Change topic schedule:* A preliminary schedule identifies hopes for topic coverage, but interactions with students indicate the extent to which learning is actually occurring. If students are mastering material quickly, the pace of the course might be increased. On the other hand, if students are having trouble, some topics might be dropped; touching on many topics might not be productive if students will remember few of them after the course is over.

- *Add student presentations:* Some topics, previously planned for lecture, might be assigned for student presentations. Alternatively, as extra credit, students might prepare oral presentations on enrichment topics to supplement coverage of regular topics.

- *Change assignments:* Assignments can be expanded, simplified, re-oriented, rethought, etc. More or fewer assignments might come from a textbook; more or fewer assignments might involve programming, proofs, computations, work on paper; some required assignments might be converted to extra credit, etc.

- *Change difficulty of assignments:* Student performance on assignments provide insight on progress in student learning. For example, impossibly-difficult assignments may undermine student morale. If assignments are too hard or too easy, make unnecessary or inappropriate assumptions, etc, then adjustments can allow assignments to better match student backgrounds, experiences, and abilities.

- *Change due dates of assignments:* Sometimes course syllabi prescribe when assignments are due. When syllabi permit, adjustments can help tailor assignments to student schedules. For example, several years ago, a class of mine had weekly labs, and the lab for one week was due the next week. Examination of student work indicated most students completed their write-ups the night before it was due. By changing deadlines, so half the lab was due mid-week and the second half due just before the next lab, students devoted considerably more time to their write-ups and quality was noticeably higher.

- *Clarify assignments:* Should student feedback or the analysis of student answers indicate miscommunication between instructor and students, instructions can be rewritten. For example, assumptions may be written out, rules for citation or collaboration made explicit, and expectations identified.

Altogether, teaching requires initial planning, followed by continual adjustments. Each class is different: topics and materials evolve, student personalities and backgrounds change, and pedagogy continues to develop. Within this environment, an instructor should pay attention to how the course is working—what is going well, and what might be refined. Mid-course corrections allow a perceptive instructor to build on those successful course elements and alter those that might be improved.

Acknowledgment

Many thanks to Marge Coahran, Dickinson College, for her suggestions in refining this column.

53.2 SUBSEQUENT REFLECTIONS

When teaching, I welcome feedback. Regardless of how well I think a course is progressing, I keep searching for ways to make it even better. Thus, I regularly utilize mid-course surveys as a mechanism to discover student perceptions. As noted in the original article, various constraints limit the adjustments possible within a semester, but sometimes even modest refinements can make a difference. In considering seeking feedback and making refinements, at least three issues can arise: a factor of risk, instructor control, and class size/context.

Risk

When an instructor asks for feedback, students may report what they really think. Students may not be able to identify specific difficulties, but they can report their perceptions.

Some faculty might consider such feedback a potential risk. If an instructor views education as a hierarchical institution, negative student comments or suggestions may be interpreted as disrespectful or challenge to a faculty member's position as the class head. In this model, a faculty member is supposed to have all of the answers, so feedback might be considered a threat. In past decades, an instructor might be viewed as "the sage on the stage", and the role of a course was to transmit information in one direction: from the instructor to students. Although this perspective might have been common in the past, it seems outdated now and contrary to modern educational models.

On the other hand, when viewing education as a faculty-student collaboration, feedback need not be threatening, identification of challenges can help both faculty and students, and ideas in the original column may apply.

Instructor Control

Related to the issue of risk, adjustments of course content and/or pedagogy suggest a course was not perfect. If an instructor is devoted to exactly one approach, any suggestion of adjustment might be considered as undermining the instructor's authority and control. If feedback suggests an instructor's approach is non-optimal, the instructor may feel threatened.

Personally, I think of myself as a guide and facilitator, and I rarely have the delusion that I am in total control. Students have their own perspectives, priorities, and understandings, and I cannot expect to control all that they think or do. I can try to help and offer directions, but I cannot make students do much. Further, as students change from one semester to another, I likely will need to make adjustments in response. With different students in class, I cannot expect to be in total control in interacting with each new group of students.

Class Size/Context

From my perspective, the course syllabus serves as a contract between a faculty member and students. For example, if a syllabus specifies three hour-long tests on specific dates, then students will expect the three tests and plan their work accordingly.

However, a syllabus may allow some adjustment in the topics covered, the pedagogy involved, and the pace of new material. For example, tests may be administered on three specified dates, but adjustments might impact the material assessed. Similarly, a course might shift the relative time spent on lectures, in-class discussions, and labs.

Also, logistics may allow relatively more flexibility for small classes than large ones. With several hundred students attending, modest adjustments might allow more use of clickers or small groups, but the basic class format likely must remain unchanged during a semester.

CHAPTER 54

Recovering from disappointing test results

This column first appeared in ACM *Inroads*, Volume 6, Number 3, September 2015, pages 38–39[259] Subsequent reflections on this subject appear later in this chapter.

54.1 ORIGINAL COLUMN

I UTILIZE IN-CLASS TESTS in many courses for summative assessment. In constructing these tests, I want to cover much of the relevant course material, in part to obtain a reading of how well the course has facilitated learning of the important ideas and student outcomes. Toward that end during test development, I outline concepts and techniques to be covered, and I construct problems that I believe will provide reasonable evidence regarding whether or not students understand the prescribed material. Since in-class tests must be completed within a specified time, I try to select problems which I expect students can complete reasonably in the allotted time. My hope is that problems will be sufficiently non-trivial to provide evidence of student mastery, but not so difficult that students will be overwhelmed within the time constraints. Further, I generally avoid problems that hinge in identifying a trick; I am much more interested in gathering evidence about mastery of material than in expecting students to happen upon a small, detailed key that unlocks a solution—at least for in-class tests.

Often, after administering these tests, the test results turn out roughly as I might have expected. Many students demonstrate mastery; problems are challenging, but manageable (at least for many students); and student scores seem generally consistent with student assessments from other sources.

However, from time to time, student scores may seem disappointingly low—certainly lower than I might have hoped or expected. This column considers some possible responses when students earn lower-than-anticipated scores on their tests. In what follows, I focus on in-class tests with free-response questions rather than multiple-choice or fill-in-the-blank questions. Within my usual teaching environment, tests usually require students to write program code, explain ideas with sentences and paragraphs, or identify output as they trace code, so this column focuses on free-response problems with text-based answers. I might hypothesize that similar comments might apply to other test formats, but those are largely beyond my direct experience.

Some Factors Influencing Student Scores

When grading tests, at least three factors likely influence student scores:

- Student mastery of the material,
- The difficulty of the test, and
- The grading rubric.

In principle, I would like to assign grades based on student mastery, but the other factors certainly have an impact. In some contexts, techniques, such as pre-testing and long-term statistical studies, may help gauge test difficulty and scoring distributions, but such approaches have limited applicability for day-to-day testing in small classes.

In practice, in spite of my best efforts, student test results sometimes seem disappointing:

- Individual students may have difficulties on certain problems.
- In rare instances, students generally might find a test much harder than I had anticipated.

In the long term, I might consider how I might adjust future offerings of a course to address possible shortcomings . However, in the short term, I need to consider an immediate response to results on student tests, encourage student mastery, and provide motivation.

Responses to Disappointing Student Tests

Since teasing issues of student mastery apart from test difficulty and rubric scales in practice seems speculative at best, my preference is to find ways to utilize these tests to promote learning. Since I use in-class tests as factors within a final course grade, I believe test scores (even low test scores) should count in a substantive way.

However, low scores also can provide students incentives to engage material further—if an appropriate context can be found. In considering options, my starting point is that a low score on a problem of a test likely indicates some level of misunderstandings or incomplete mastery—even if the problem was difficult or the rubric unforgiving. In my experience, if students are presented with a mechanism to partially make up for a lower-than-desired score, many students will take the initiative to talk to an instructor, teaching assistant, mentor, lab assistant, or other resource to dig into relevant material.

To provide an incentive, I often offer an optional activity (e.g., written assignment, or in-class test, or oral presentation) to focus a student's thinking about a topic. After grading, I then compute a weighted average of the original in-class test and the optional activity. Although this idea seems straightforward, some details may require some thought.

- Often I compute a simple average of the percentage received on the original test and the percentage on the optional activity.
- If class performance on a question suggests an issue in wording or other difficulty, a different weighting might seem more appropriate.
- Since I want to encourage students to dig into material, I tell students that the revision cannot lower their original recorded score; if they earn fewer points on the optional activity, I will not count it, but keep the original score.

Within this framework, what follows is a partial list of practical approaches that I know have been used in various class settings.

- *Test revisions, out-of-class:* I may allow students to revise their solutions to the original test as a homework assignment.
    - ◇ If a student misses few points on a problem (e.g., less than 20% of points), it seems there will be little educational value in the student revising the problem. Thus, I allow a student to turn in the original test for this problem and receive full credit.
    - ◇ If a student misses more than 20% on a problem, then the student likely is missing something regarding the material, and my revision policy states the student must revise the entire problem. (If the part has 3 parts, and the student has missed only a few points on each part, the student still must revise all parts.)
    - ◇ On some in-class tests, I allow choice (e.g., do 4 of 5 problems). On a revision in this setting, the student must do the additional problem that they omitted on the in-class test.
    - ◇ Since collaboration may be an issue on out-of-class activities, students must complete an "academic honesty certification" for each revised problem, citing any written sources used and certifying they have neither given nor received aid from any individual (except the instructor) for that problem.
- *Repeated, in-class tests—same questions:* Some faculty take class time for a second administration of the original test. In this context, students already know the test questions, and they have feedback on their original answers. Thus, they know what they got right and where they lost credit. Students have motivation to read, review labs, or talk to instructors (or others) to correct misunderstandings, but they must develop sufficient mastery by themselves to be able to write correct solutions on their own for the in-class retest.
- *Repeated, in-class tests—similar questions:* Rather than use the same questions on a second text, some faculty take class time for a second test with different—but similar—questions. Students can use their original test results as a guide to what topics they need to work on. However, rather than trying to memorize answers, students must apply their understanding to different questions on the revision.
- *Make-up, in-class tests—questions selected from long list:* As a variation of the previous two approaches, an instructor might identify a collection of possible questions for a make-up test. The in-class test would select a subset of these problems. As with the earlier alternatives, students can use results from an original test to help determine areas they need to explore further, and they can practice on the collection of possible questions. If the collection is relatively large, memorization of answers may not be practical, and students will need to think through solutions during the in-class test revision.
- *Repeated questions on subsequent tests:* As an alternative to taking class time for a repeated, in-class test, selected problems from one test might reappear on subsequent tests. To provide incentive for students to learn material that did not go well previously, an instructor might indicate one or more questions of previous tests will appear on the subsequent tests. In some cases, a previous problem might be edited somewhat.

If students do not know just what questions will reappear, there may be incentive to review all previous problems.

- *Oral, in-office tests:* Although time consuming for an instructor (and likely impractical for large classes), a retest—with the same questions or different ones—might take place as an oral exam within an instructor?s office or lab. Students would be asked questions, and they might give their answers orally—perhaps using a whiteboard or pad of paper to aid their presentation. This type of exercise has the advantage that students learn to think on their feet, formulate their ideas on the fly, and present their ideas orally—skills often not emphasized in many computing courses. On the downside, such tests may require substantial time for an instructor.

- *On-line retesting options:* Many of the options already presented here might be adapted for online testing environments. As with any online environment, issues of test security and user authentication must be addressed, but online testing may avoid use of class time to repeat the original test.

Many readers will recognize that each of the above approaches has possible drawbacks—time, opportunities for collaboration, additional test preparation, additional required grading, etc. However, these approaches can provide incentives for students to review difficult material and address misconceptions. Semester grades should count an original test in a meaningful way, so students take each assessment seriously. However, with some type of follow-up activity, students can actively engage material from a new perspective and work through challenges.

54.2 SUBSEQUENT REFLECTIONS

Although the original column presents several initial ideas regarding constructive responses to low test scores, a faculty member's first reactions may not be particularly positive. For example, an instructor might be inclined to blame students: Perhaps students are not working hard enough, doing the required reading or assigned exercises, or devoting adequate time outside class. An instructor likely believes that the class is properly structured, and class time has been effective, so any fault lies with students rather than the faculty member.

In some setting, this blaming of students might be quite appropriate, but an instructor also might consider other factors. For example,

- If students seem unprepared for the work at hand, brainstorming may help find ways to supply the needed background.

- If many students did not finish a test, the test might have been too long. (For introductory courses, I often find that if it takes me longer than 7-9 minutes to complete a test, then likely the test will be too long for an in-class exercise targeting one hour.)

- If incorrect student answers follow a pattern, errors may suggest an underlying misunderstanding of one or more concepts that should be reviewed.

- If students seem to understand various pieces of a topic, then a challenge may be how to help the students integrate those pieces into complete solutions.

- If students seem unfocused or poorly motivated, then some class discussion or discussion with faculty colleagues might help identify approaches that will help engage students effectively within the course.

Altogether, an instructor may be inclined to blame students for test results, but a more constructive approach may involve finding constructive ways to move forward. Test results may identify unresolved issues which might need to be addressed.

CHAPTER 55

Selected/annotated references for student progress in courses

MANY FACTORS CONTRIBUTE TO STUDENT ACADEMIC PROGRESS. As the organizer and leader of a course, an instructor can structure content, select appropriate pedagogy, help students with time management, provide encouragement, identify and suggest outside resources, and make mid-course adjustments. However, an instructor's influence largely falls within reasonable bounds of the course itself. The instructor may have only modest impact on student preparation and motivation. Further, an instructor likely has very little influence on broad social, financial, cultural, and personal matters.

Chapters within this Part X largely focus on a student's and instructor's activities within the classroom. For readers interested in further study, the first readings suggested below highlight these course-related elements that may contribute to student progress and success. The second collection of readings then consider outside factors and influences.

Course-Related Elements for Student Progress and Success

Numerous articles report a blend of refinements that can impact student progress in courses and enhance success.

- Sheryl Burgstahler, Richard E. Ladner, Scott Bellman, "Increasing the participation in computing of students with disabilities", *ACM Inroads*, Volume 3, Number 4, December 2012, pages 42-48 [31]
  A presentation of numerous insights about students with disabilities in computing classes, as part of an *Inroads* Special Issue on Broadening Participation.

- Aharon Yadin, "Reducing the dropout rate in an introductory programming course", *ACM Inroads*, Volume 2, Number 4, December, 2011 pages 71-76.[291]
  Discussion of a change in programming language, the use of a micro-world throughout the course, and "using individual assignments that enforced better learning habits" to reduce the fraction of students failing a CS1 course by 77%. [291, p. 71] Although individual course refinements can be useful, an integrated approach can have substantial effects!

- Aharon Yadin, "Using Unique Assignments for Reducing the Bimodal Grade Distribution", *ACM Inroads*, Volume 4, Number 1, March, 2013 pages 38-42.[292]
A background description for courses which show bimodal distributions for student grades, and a discussion of some approaches for addressing this type of teaching challenge.

Outside Factors and Influences

In recent years, much attention has been paid to learning environments in and around campuses. The literature is rich with numerous accounts and experiments; the following references may provide a start with some well-known accounts and approaches.

- Jane Margolis and Allan Fisher, *Unlocking the Clubhouse: Women in Computing*, MIT Press, November 2001. [114]
A widely-circulated account of experiences of women, as they interact with the computing culture at Carnegie-Mellon University. The book explores many factors, both in and outside the classroom, that may impact student success in classes—particularly among women and people from under-represented groups.

- Jane Margolis, *Stuck in the Shallow End, Updated Edition: Education, Race, and Computing*, MIT Press, March 2017. [115]
A follow-up/ground-breaking study of the experiences of students and teachers at three Los Angeles public high schools and issues that surround students from under-represented minorities, especially African American and Latino students.

- Amber Settle, John Lalor, and Theresa Steinbach, "A Computer Science Linked-courses Learning Community", *Proceedings of the 2015 ACM Conference on Innovation and Technology in Computer Science Education*, July 2015, pages 123-128. [165]
.
A description of learning communities for computing students, and an analysis of the effectiveness of these communities in contributing to student success in computing.

XI

Assessment and Grading

NEAR THE END OF SEVERAL RECENT SEMESTERS, colleagues and I have observed:

- Many faculty bemoan having to devote substantial time to developing tests.
- Students often complain about the time required as they prepare for tests.
- Students frequently lament the stress and pressure associated with tests.
- Faculty and many teaching assistants often complain about having to grade tests.

Conference attendees periodically express similar experiences.

With such negative commentary regarding the testing process, a natural question arises, "why bother with tests and assessment at all?"

As a partial response, faculty, students, and outside parties might keep in mind that assessment and grading have at least two components.

- *Formative assessment* provides feedback to students on their work—typically with the purpose of promoting learning.
- *Summative assessment* evaluates student mastery—typically with the goal of determining course grades or certifying a level of competence.

FIGURE 55.1 Instructors Word Cloud

As students engage with a subject, they may read and experience ideas and approaches. In this process, they form some initial perspectives, but these preliminary understandings often are untested and may be subject to misunderstandings, confused logic, or unexamined assumptions. Often within STEM fields, students sharpen their understandings and examine their perspectives by applying techniques to solve problems, conducting experiments, and synthesizing ideas into an overall framework. (See Chapter 47, for example, for Bloom's Taxonomy of Education Objectives in the Cognitive Domain [25].)

As students progress, feedback on their work can help them determine what understandings seem on track and when adjustments may be helpful in their thinking. Formative assessment provides this feedback.

Similarly, studying or working within a field does not necessarily indicate mastery. Further, some people may require more time or study than others to obtain designated levels of competence. Summative assessment provides one mechanism to determine and certify degrees of mastery.

Altogether, assessment has an important role, both within the learning process and for certification of achievement to outside audiences.

An important challenge within education involves considering how to supply feedback and certification fairly, consistently, effectively, and efficiently. Traditionally, faculty might review and notate student work in detail, providing extensive feedback, and such analysis often consumed considerable time. Unfortunately, such one-on-one commenting may not scale well to classes with large enrollments.

Ideally, assessment should address both high-level concepts and low-level details, but the mechanics of grading and the need to review the work of many students may lead to grading that focuses on some levels (e. g., often emphasizing low-level details) while providing little consideration of other levels (e.g., largely ignoring high-level concepts and approaches).

Discussion in this chapter considers some mechanisms to provide worthwhile feedback, while applying to a range of course settings.

CHAPTER 56

Notes on grading

This column first appeared in the *SIGCSE Bulletin*, Volume 32, Number 2, June 2000, pages 18–19[211] Subsequent reflections on this subject appear later in this chapter.

56.1 ORIGINAL COLUMN

IN THIS COLUMN, I HOPE TO BEGIN A DIALOG about how to grade as effectively and efficiently as possible. Please send me your ideas to continue this dialog. Also, as I plan to address the issue of plagiarism in a future column, I would like to hear how other faculty address this problem.

Grading Is Hard and Time Consuming

Virtually any CS faculty member knows by direct experience that grading in our discipline is difficult and time consuming. For example, in 1990, the MAA/ACM/IEEE Computer Science Task Force on the Teaching of Computer Science with Mathematics Departments cited grading rates of readers at Advanced Placement (AP) exams as evidence that grading CS takes longer than comparable assignments in mathematics or other sciences; CS grading rates appear more closely tied to foreign language than science. (See [1] and the Appendix for some relevant data.) However, while upper-level foreign-language courses often require only 4 or 5 short papers during a semester, many CS faculty believe that one assignment every 2 or 3 weeks provides students with inadequate practice. Thus, in CS, assignment frequency may parallel other sciences, while grading time parallels foreign language.

Two Purposes of Evaluation

As with faculty teaching evaluations, it is useful to remember that evaluation of student work typically has two purposes:

- *Formative assessment:* Providing students with feedback on their work, so they can improve their understanding.
- *Summative assessment:* Providing a mechanism to assign grades.

Often, grading of an assignment must satisfy both purposes. However, that need not always be the case, and it is worthwhile to keep in mind specific purposes when grading a specific exercise.

Some Ideas to Handle Grading Efficiency

One approach to efficiency is to anticipate grading when devising an assignment. For example, solutions involving diagrams, multiple-choice, and fill-in-the-blank may grade more quickly than code or paragraphs.

Another way to finish grading quickly is not to do it. While this is rarely an option for all parts of all assignments, some alternatives might be possible:

- Ask an assistant (either an upper-level student or a graduate student) to grade. (Note the assistant may need directions and a grading scale.)
- Grade only selected parts of an assignment or program (but be sure to announce this policy in advance).
- If an assignment has several phases, provide only incremental feedback on successive phases.
- If an assignment is divided into stages to keep students on schedule, provide only minimal grading for some stages (e.g., just check that an intermediate step has been received).
- Develop a checklist highlighting basic elements expected for an assignment. E.g., for a program, devise a form for ratings on documentation, readability, program structure, appropriate algorithm selection, testing, and correctness.
- Such a checklist can provide considerable feedback quickly and minimize the need for lengthy comments.

Some Thoughts on Grading Itself

Here are some comments I often make to my grading assistants. While many of these draw on my experiences grading AP math and CS exams for almost 20 years, what follows also has been guided by extensive discussions with colleagues.
ndexgrading! rubric

- First, determine the total number of points for the exercise, based on its importance and difficulty.
- Next, before starting to grade, prepare some sample solutions, and assign points to the various parts. (While I have been doing this for years on written assignments, last semester I found this was vital for oral exams as well.)
- Grade one problem at a time for everyone, so you can remember each problem as you go.
- Keep notes on how you handle specific cases.

Altogether, this approach allows you to apply a consistent standard without having to look back though all the papers.

- Be constructive, but terse, in your comments. (At times, the best short, constructive statement might be "please see me about this".)
- Do not give credit for junk. While you can reward great effort with encouraging comments, the score should reflect understanding.

- Use credit to emphasize what is important. For example, for several years, I complained on papers about documentation, readability, use of pre- and-post-conditions, and appropriate testing. However, assignments did not improve until I deducted points for incorrect grammar, poor documentation, and incomplete test plans.

Appendix: Reading Rates of AP Examinations

The following paragraph and table come from a 1990 report of the MAA/ACM/IEEE Computer Science Task Force on the Teaching of Computer Science with Mathematics Departments [83].

> The data summarized in Table 2 give the comparative rates for grading the free-response section of the Advanced Placement Examinations. For most subjects, students have 90 minutes to answer, so the table provides an objective measure of the relative time required to evaluate student work in the listed subjects. Columns 2 through 5 give the average number of 90-minute exam books read by a single grader in a 6-day period for the 1983 through 1986 examination years. It is important to note that these figures are not comparable in an absolute sense. Different subject areas differ in terms of test format, time a student has to take the test, percentage that the free-response counts towards the final AP grade, reader training, and grading strategies. For example, the French and Spanish exams include a 15- to 20-minute tape made by the student which must be listened to and graded.
>
> Calculus and Computer Science compare the most favorably on these variables and, in particular, are quite similar in terms of reader training and grading strategies. Yet the table implies that computer science is comparable, in grading demands, to a foreign language rather than to mathematics or to the natural sciences. One can grade approximately one and one-half [times] as many English examinations and almost twice as many calculus examinations in a fixed time period.

Table 2: Number of AP Examination Books Read by Discipline

Subject	1983	1984	1985	1986
French Language	147	145	144	128
Spanish Language	146	151	146	159
Computer Science	191	173	172	168
German Language	252	218	246	235
English Language	222	287	253	280
English Literature	262	256	257	268
Calculus (A, B, BC)	351	314	324	322
Biology	338	356	410	337
Chemistry	449	457	410	425
Physics	548	440	487	445

56.2 SUBSEQUENT REFLECTIONS

Much has happened within undergraduate computer science education since the original article appeared in June, 2000.

- Enrollments dropped in the late 2000s, but enrollments have since soared to new, high levels.

- At many schools, high enrollments, coupled with limited staffing, has resulted in large class sizes—particularly at the introductory level.
- At some schools, increased enrollments leads to an increase in the number of teaching assistants, graders, or other support staff.
- Various tools have emerged that can grade lab exercises, examples, and focused code segments, providing feedback to students.
- Some tools have emerged to support the grading of student work by teams of instructors and teaching assistants.

Even if a course remains the same over time with regard to the number of assignments, programs, projects, and tests, increased enrollments yield an increase in the number of student pieces that must be graded. When class sizes are modest, an instructor may be able to handle feedback and grading—perhaps with modest help from an assistant, and many of the comments in the original column focus on this type of personal grading. As class sizes increase, however, additional mechanisms become vital. Several comments and observations follow, with additional suggestions in Chapter 57 (for the grading process itself) and in Chapter 58 (for additional reading).

Formative and Summative Assessment

The original column noted that formative assessment has a different purpose than summative assessment. As a result, the mechanisms for providing assessment may differ substantially between these two evaluations, and the nature of feedback may be similarly different.

The goal of formative assessment is to provide feedback to the student.

- Fundamental issues include: is the work correct, does the student understand basic terms and principles, is there a better way to approach a problem, has the student mastered basic syntax and semantics (in programming), etc.?
- Particularly in introductory courses, assignments often involve relatively short exercises to allow students to solidify basic understandings and to gain initial practice.
- An emphasis of formative assessment is feedback—often quick feedback, so that students will understand when their understanding is on track and when correction is needed.
- If points are assigned at all, the purpose may be to encourage students to complete an assignment, and the total number of points may be small.
- Since feedback is a primary purpose, issues of academic honesty may have a secondary priority.

In contrast, the goal of summative assessment is to determine mastery and to provide a grade.

- Fundamental issues include: to what extent is mastery demonstrated, to what extent is a statement or solution correct, how does the quality of this work related to some type of standard or professional expectation?

- An overall grade might identify the degree of understanding, such as is identified in Bloom's Taxonomy (e.g., remembering, understanding, applying, analyzing, evaluating, and creating). [25]
- The emphasis of summative assessment is a number of points or an overall grade.
- Other than a grade, students may receive little or no feedback.
- Since grading is a central goal, issues of plagiarism can be a significant concern.

Since the purposes of formative and summative assessment are quite different, mechanisms for providing this evaluation also can be different. Certainly some approaches might be used for both, but separation of the two seems helpful in the following discussion.

Notes on Formative Assessment

When students are learning material, quick feedback can help clarify ideas and identify misconceptions at an early stage. Particularly in introductory courses, extensive, ongoing, and immediate feedback can guide student learning. Pragmatically, the amount of feedback an instructor can provide is limited, and large classes may require feedback opportunities beyond what one (or a few) individuals can provide.

Fortunately, in recent years, question banks and exercises have emerged from a range of vendors, and numerous tools can provide immediate, online feedback for many, relatively short exercises—particularly at the introductory level. In these settings, students typically are given a question (often involving the writing of a short piece of code). Students' answers then are compiled and run using a suite of test cases, and answers are compared to instructor-created and certified programs. The result is almost-instant feedback.

Depending on the tool, students may or may not be given feedback regarding algorithmic design, program decomposition, readability, style, etc.

As an example, Barr and Trytten describe their use of Turing's Craft CodeLab and compare this tutoring tool with CodingBat, Problets, zyBooks, and Python Tutor [22]. These tools typically present a wide range of basic assignments, and students write short code segments in response. The software then provides immediate feedback regarding whether the student answer solves the problem given.

Notes for Summative Assessment

Summative assessment focuses upon mastery and grading. Typically this type of assessment occurs only a few times in a semester (e.g., with 2-4 tests, a few projects, and/or a final examination), but the scope of each evaluation may be substantial. Also, with grades and certification of mastery at stake, substantial concerns regarding plagiarism or other academic dishonesty may arise.

Traditionally, one instructor handles grading tests for a relatively small class, while a team (e.g., an instructor with several assistants) may organize a "grading party" to address materials submitted within a large class. In principle, this approach can scale with sufficient time and resources; the grading of tens-of-thousands of Advanced Placement Computer Science A exams with a few hundred readers working for a week or more provides an example. However, relatively few schools and instructors are likely to enthusiastically embrace this approach when confronted with hundreds of tests to evaluate in a short time.

To address issues of grading within large classes, at least three types of automated approaches have evolved in recent years:

1. *Tools to support grading teams:* One type of automated tool, such as Gradescope [69], provides an environment that provides a mechanism to develop and modify rubrics in real time, displays different parts of scanned tests to various graders, and allows individuals within a grading team to record scores by checking boxes within a rubric. With such systems, graders still must review each question of each test, but mechanics are streamlined: "No paper shuffling, collaborative not-necessarily-in-person grading, dynamic rubric creation with change-anytime points and categorization, no data entry, immediate grade returns, and no post-grading cheating!" [69, p. 83]

2. *Tools for automated grading:* Other types of software, such as Coderunner [111], support assignments, labs, programs, and a variety of types of test questions (e.g., multiple choice, short-answer, matching, programming problems). To begin, Coderunner provides templates for some common types of questions, such as writing a function, program, or class, and additional templates may be defined for other types of questions. Each template specifies a program that will run behind the scenes to evaluate a student answer (e.g., does the student program produce correct results on test data), but Coderunner also utilizes a pre-processor that can check a submission against style sheets or other constraints. Altogether, this type of automated tool may be able to handle a full range of grading tasks for many types of problems, but in some cases an instructor may need to spend substantial time refining a template that can analyze student answers completely and correctly.

3. *Checkers for detecting plagiarism and/or program similarities among submissions:* The "Subsequent Reflections" section of Chapter 49 on *Academic Honesty in the Classroom* describes some plagiarism-detection software, particularly the popular Moss system that was developed at Stanford University by Alex Aiken and collaborators [13]. Such software is freely available and provides listings of programs with similar code. Further, this type of software may be able to identify differences in software due to easy editing, such as the change of variable names and text within comments. Other, more general systems, such as [188], examine similarity of papers and other types of materials. Additional references and systems may be found in Chapter 49.

Some additional thoughts on grading and allocating points may be found in the next chapter, with suggestions for further reading in Chapter 58.

CHAPTER 57

Grading and the allocation of points

This column first appeared in the *SIGCSE Bulletin*, Volume 41, Number 4, December 2009, pages 14–16[229] Subsequent reflections on this subject appear later in this chapter.

57.1 ORIGINAL COLUMN

WHEN CONSIDERING HOW TO GRADE, at least four principles come to mind; grading should

- be fair and consistent,
- reinforce high-level course goals and themes,
- provide students with feedback, and
- be reasonably fast for the instructor.

This column considers several ideas that may help with the first two of these principles. Due to space limitations, comments on the last two principles are deferred to a later column.

Fair and Consistent Grading

Of course, assigning every paper the same score (e.g., 100%) efficiently provides great consistency. Using a random number generator for determining grades also is remarkably efficient, but consistency, course goals, and feedback can be problematic.

To be more serious, experience indicates that different seasoned teachers may assign remarkably varying scores to the same student work. Upon reviewing an assignment, one teacher may conclude,

> This student clearly has the main ideas. Some details are off, but the student understands the important goals and themes.
> Conclusion: 80%

Another teacher may think:

> How could anyone who really understands the important goals and themes of this course make these types of errors.
> Conclusion: 20%

Although such differences seem inevitable, the development of a road map or rubric for each exercise can help an instructor match detailed point allocations with high-level themes. A rubric breaks an exercise into pieces and assigns points to each piece, based upon concepts to be demonstrated.

Personally, I often draft my own solution, divide the solution into sections, and allocate points based upon the relative difficulty and importance of each section. With even a simple rubric, I grade student solutions to a problem in a single pass, so the work for grading a problem is roughly $O(n)$, where n is the number of students. In contrast, if I must revisit earlier work as I grade new solutions, grading is likely $O(n^2)$ or worse.

In allocating points within a rubric, single sections with many points can make grading consistency difficult. For example, for a 40-point question, assigning 15, 20, or 25 points to partially correct answers may take considerable time, but 10-point variations impact consistency. To address such issues, large questions might be subdivided into several parts. If the 40-point question has 8 identifiable components, then a grader only must allocate 5 points at a time—a task that is likely more consistent and faster than keeping all 40 points in mind at once.

Difficulties with point allocations also arise at a high level; how many points should an assignment be worth? Allocation of relatively few points to individual parts often works well. For example, suppose a course contains 10 labs that count 25% of the overall grade. In determining the number of points for each lab, one suggestion is to allocate points in a way that makes grading easy. For example, the allocation of 5 points to each step can provide reasonable student feedback while allowing fast grading. If labs have 6 steps and the 10 labs count equally, the resulting 300 points for a semester provide adequate data to compute statistics for a course grade.

As a final consideration (particularly for tests), some faculty have observed motivational advantages in scaling an overall exercise to somewhat over 100 points (e.g., 105 or 108 points/test); similarly, motivation can suffer with tests totaling just 80 or 9 points. Many students think in percentages, but do not carefully convert points to percentages. Thus, students may feel discouraged with scores of 83 of 90 points—they only earned an 83 on the test! In contrast, students may feel fresh motivation after receiving 95 of 108 points—after all, they earned a 95.

Reinforcing Goals and Themes

Students typically view points as identifying what matters. If students lose credit, then an error is significant; if students receive a written comment but do not lose credit, the matter has secondary importance. Several examples illustrate likely messages sent by potential point allocations:

- If all points on a test come from writing code, then computer science seems to equal coding.
- If many points on a test come from basic syntax and programming constructs, then students understand that mechanics (not problem solving) are most important.
- If grading a program depends solely upon its processing of instructor test data (black-box testing), then design, structure, and maintainability are unimportant.
- If no points are lost for variable names b1, b2, b3, b4, b5,and if the context is not vitamins, then descriptive variable names are unnecessary.

- If internal program documentation can receive full credit in spite of grammatical or spelling errors, then writing details are not important.

Overall, point allocation provides a measure of importance, and lack of points indicate qualities with no value. The allocation of points is complicated, however, in that students should lose credit for bad practices, but not earn credit for following standard practice.

One approach for handling a range of basic elements is sometimes called a "usage sheet", which I first heard about in 1987 from Stuart Reges, then Chief Reader for the Advanced Placement Computer Science (AP CS) Exam. The idea, which has evolved over the years, identifies a basic standard for the work at hand. Writing to that standard does not earn points, but failure to meet the standard yields a deficiency and may lose points.

A common analogy involves grading history papers. Historians expect students to write in a logical structure, organized paragraphs, grammatical sentences, etc. Students do not earn points just for writing a subject and predicate in a grammatical construction. However, students may lose points for using sentence fragments, lack of organization, etc.

Within AP CS grading, usage sheets have evolved considerably since 1987. For example, at this writing, the usage sheet for 2006 is publicly available on page 6 of

http://apcentral.collegeboard.com/apc/public/repository/_ap06_compsci_a_sg.pdf[1]

This AP CS usage sheet considers likely typographical errors unimportant in the context of a stressful, tightly-timed test. Compilers would detect the omission of some semi-colons in a Java program, so such difficulties yield no point deductions. In contrast, points are deducted for code with undesired side effects (e.g., print statements in functions that are just supposed to compute). However, code with no side effects (e.g., code that does absolutely nothing) does not automatically earn credit.

In summary, the division of points into small sections aids fair and consistent grading, and the allocation of points suggests the practical course goals and themes. Thus, examining grading scores and practices can be helpful in determining de facto course priorities.

57.2 SUBSEQUENT REFLECTIONS

The original column highlighted three key steps when grading papers, either by an individual instructor or by a team including the instructor and assistants.

- Prepare one or more solutions ahead of time to make a first approximation of the range of solutions that might arise,
- Develop a rubric that allocates points to the various elements anticipated in solutions.
- Once actual grading begins, supplement the grading rubric by making notes as you decide what points to take off or add for various mistakes or special insights.

Particularly when reviewing many papers, the aim must be to grade by making a single pass through each problem. A grader may refer to the rubric and supplemental notes, but the grader should not refer back to previously graded problems. In the terms of algorithmic analysis, making a single pass through student work yields an $O(n)$ process, whereas frequent referring to previous work likely has $O(n^2)$ or worse algorithm (as noted earlier in this chapter).

[1]Since this column appeared in December 2009, the URL of the 2006 usage sheet has changed. The URL given here reflects the new location of this Web page (viewed May 12, 2017).

In high-enrollment classes, the single-pass approach is particularly important, even if the initial start-up time for grading requires moderate effort. The following comments may further amplify points made in the original column

In some situations, particularly when stress levels are high, rubric development may be dominated by a natural tendency to focus on details of grading, and sometimes high-level problem-solving approaches and algorithms may seem secondary. With this in mind,

- After developing a draft rubric, try applying it to several sample solutions (e.g., a few student papers—photocopied to allow note taking).
- Try to avoid separate rubrics for different approaches to the same problem.
  - ◇ In principle, solutions might follow separate paths (e.g., iterative or recursive), but in practice a student's solution might contain elements of multiple paths. In such cases, a grader may have substantial trouble deciding which rubric to apply.
  - ◇ When different approaches might be expected, a rubric might identify key elements of each at a general level. For example, either iteration or recursion involves a starting point (e.g., assignment of initial values or an initial function call), a stopping condition (e.g., an exit condition or base case), and processing through cases (e.g., a loop body or processing with a recursive call).
- In reviewing preliminary scores,
  - ◇ do the scores generally reflect the full range of topics addressed by the problem?
  - ◇ can a student earn most or all of the credit by getting some pieces properly formed, but where those pieces do not fit together effectively to make substantial progress in a program?
  - ◇ to the extent possible, do the draft points allocated address high-level processing? For example, does processing within a 2-dimensional array actually handle the two subscripts independently?

Such preliminary work for rubric development may be particularly useful when a grading team involves numerous grading assistants within a large-enrollment class. As noted in the original column, each grader may have a different perspective, but grading can be consistent and fair if the grader applies her or his detailed interpretations to all students' work equally.

- An instructor might review a rubric with a grader for a specific problem, looking at a few sample solutions to clarify how the rubric might be interpreted.
- If at all possible, one person should do the grading for the same problem(s) for the entire class. Each person may have some inherent tendencies (e.g., applying a guideline loosely or firmly), but the grading likely will be consistent if the same person applies the same perspective to work for all students.
- If feedback is desired as well as the assignment of points, one approach is to develop during grading a numbered list of comments. When grading student work, the grader simply references the comment number rather than transcribing a full comment. As grading progresses, new entries are added to the comment list, and when papers are returned, the entire numbered list can be distributed to all students. Rarely will all comments be relevant to any individual student, but a student can use the comment list as a key to interpret notations made specifically on the student's work.

CHAPTER 58

Selected/annotated references for assessment and grading

ONCE ASSIGNMENTS, PROJECTS, AND TESTS have been developed, they must be graded, and the chapters in this Part XI focus upon this grading process. Several broad areas of interest include:

- Principles and practices for grading by an individual teacher or a small team,
- Software tools that automate parts of the grading process, and
- Academic dishonesty: automated detection impact.

The following notes suggest some starting places for further reading in these categories.

Principles and Approaches for Grading by an Individual Teacher or a Small Team

A search of the Web for "instructor grading principles" returns millions of references for effective grading. The first of the following references discusses the general grading process, and the next two come from centers for teaching at large universities.

- Victoria Smith PhD and Stephanie Maher Palenque, "Ten Tips for More Efficient and Effective Grading", *Faculty Focus from Magna Publications*, https://www.facultyfocus.com/articles/educational-assessment/ten-tips-efficient-effective-grading/, accessed May 12, 2017 [176] .
  A listing of simple, but effective, techniques that can help instructors and teaching assistants in their grading. The Web site also contains several references for additional reading.

- Office of Graduate Studies, University of Nebraska–Lincoln, "Fairness in the Classroom", http://www.unl.edu/gtahandbook/fairness-classroom, accessed May 12, 2017 [139] .
  A review of three components of "fairness:" "interactional," "procedural," "outcome" [139].

- Center for Teaching, Vanderbilt University, *Grading Student Work*, https://cft.vanderbilt.edu/guides-sub-pages/grading-student-work/, accessed May 12, 2017 [33] .
  A wide ranging discussion of several key elements of grading, including purposes, developing criteria, efficiency, feedback, consistency, and handling complaints. Additional links reference Bloom's Taxonomy, student learning and learning styles, flipped classrooms, teaching large classes, and many more topics.

Software Tools that Automate Parts of the Grading Process

Many tools have emerged in recent years to automate elements of the grading process. These references suggest the range of of packages now available.

- Yoav Yair, "Did You Let a Robot Check My Homework?", *ACM Inroads*, Volume 5, Number 2, June, 2012, pages 33-35 [293].
  A short piece providing an interesting perspective on manual versus automated grading, particularly with the use of techniques from artificial intelligence.

Question/Evaluation Libraries

- Valerie Barr and Deborah Trytten, "Using Turing's Craft CodeLab to support CS1 students as they learn to program", *ACM Inroads*, Vol. 7, Number 2, June 2016, pages 67-75.
  A comparison of Turing's Craft CodeLab environment with such tools as CodingBat, Problets, zyBooks, and Python Tutor. In this context, "the student write[s] an expression, a statement, a code fragment, a method, or a program," and the software provides immediate feedback regarding whether the student answer solves the problem given [22, p. 67].

Support for Grading Teams

- Arjun Singh, Sergey Karayev, Kevin Gutowski, Pieter Abbeel , "Gradescope: A Fast, Flexible, and Fair System for Scalable Assessment of Handwritten Work", *L@S '17: Proceedings of the Fourth (2017) ACM Conference on Learning @ Scale*, April 2017, pages 81-86 [171].
  Daniel D. Garcia, "TECH launch with GradeScope", *ACM Inroads*, Vol. 6, Number 2, June 2015, pages 82-83 [69].
  Two discussions of Gradescope, a team-grading tool which allows instructors to develop and modify rubrics, display different parts of scanned tests to various graders, and allows graders to record scores by checking boxes within a rubric. With such systems, graders still must review each question of each test, but mechanics are streamlined: "No paper shuffling, collaborative not-necessarily-in-person grading, dynamic rubric creation with change-anytime points and categorization, no data entry, immediate grade returns, and no post-grading cheating!" [69, p. 83]

Multi-faceted Grading Packages

- Richard Lobb and Jenny Harlow, "CodeRunner: A tool for assessing computer programming skills", *ACM Inroads*, Vol. 7, Number 1, March 2016, pages 82-83.

A discussion of the comprehensive grading package, Coderunner, that allows automated grading of assignments, labs, programs, and a variety of types of test questions (e.g., multiple choice, short-answer, matching, programming problems). Templates are available for common types of questions, such as writing a function, program, or class; and additional templates may be defined. In addition to templates that specify programs to evaluate student answers, Coderunner also utilizes a pre-processor that can check a submission against style sheets or other constraints. Although this type of automated tool can handle grading tasks for many types of problems, an instructor may need to spend substantial time refining a template for analyzing answers to new types of questions. [111],

Academic Dishonesty: Issues, Detection, and Correlations

- Simon, Judy Sheard, Michael Morgan, Andrew Petersen, Amber Settle, Jane Sinclair, Gerry Cross, and Charles Riedesel, "Negotiating the Maze of Academic Integrity in Computing Education", *ITiCSE '16 Proceedings of the 2016 ITiCSE Working Group Reports*, Arequipa, Peru, July, 2016, pages 57-80. [170].
  A careful and extensive review of academic integrity, with comparisons between practices in academia and in industry.

- Z. A. Al-Khanjari, J. A. Fiaidhi, R. A. Al-Hinai, and N. S. Kutti, "PlagDetect: a Java programming plagiarism detection tool", *ACM Inroads*, volume 1, issue 4, December 2010, pages 66-71. [14]
  A good overview of common student techniques used in plagiarism, approaches for detecting plagiarism during the grading process, and results in applying these approaches within a course context.

- K. W. Boyer and L. O. Hall, "Experience using "MOSS" to detect cheating on programming assignments", 29th annual *Frontiers in Education Conference*, Volume 3, 1999, pages 13B3/18-13B3/22 [28]
  An interesting account of the incidence of plagiarism and cheating patterns over six semesters at one institution, based on the use of the MOSS software package to detect plagiarism on programming assignments.

- Jonathan Pierce and Craig Ziles, "Investigating Student Plagiarism Patterns and Correlations to Grades", SIGCSE '17 Proceedings of the 2017 ACM SIGCSE Technical Symposium on Computer Science Education, pages 471-476.[148].
  An interesting review of factors that do and do not seem to impact occurrences of plagiarism.

XII

Outreach and Public Relations

COMPUTING IS WIDELY MISUNDERSTOOD. Many popular images of computing include typists working in isolation at a keyboard—often playing games or working on activities that have little benefit to society. With such negative stereotypes, computing educators must work steadily to replace misconceptions with positive, constructive, and realistic viewpoints regarding the computing discipline.

Of course, effective communication requires several elements, including

- What themes or content do we want to communication?
- How would we describe our target audience(s)?
- What media should we utilize in our communications?
- How should we phrase our communications for maximum impact (or what wording or topics should we avoid as potentially confusing or counter productive)?

The chapters in this section address several components of these questions.

- Chapter 59 takes a broad view of advertising, particularly in the context of recruiting students and communicating with the general public.
- Within a curriculum, course descriptions provide a high-level view of main elements of a program (e.g., courses within a program), and Chapter 60 considers messages presented by these course descriptions.
- Chapter 61 observes that different audiences ascribe contrasting meanings to the term "programming", and as a result, narratives using this word may be misunderstood by one group or another. The chapter suggests the term be avoided in most discussions to avoid confusion and possible unintended interpretations of narratives.
- Chapters 62 and 63 consider messages sent by specific educational settings (CS1/CS2 courses and computing labs). In these cases, students gain perspectives regarding the nature of computing by experiencing coursework throughout an entire course, and reactions to direct contact with the discipline may well be stronger than messages received through public-relations literature.
- The topic of computer games arises frequently in the context of introductory computing. Many [most?] K-12 students and students in beginning computing classes have had contact with games, and many of these students may be experienced gamers. Chapter 64 examines messages that may be sent and received when computer games are incorporated as central components of introductory courses.

Altogether, this part of book examines messages sent, not just in written narratives or online postings, but also through on-going contact students might have with a curriculum itself and its courses.

CHAPTER 59

Advertising and recruiting

This column first appeared in the *SIGCSE Bulletin*, Volume 40, Number 2, June 2008, pages 16–17[224] Subsequent reflections on this subject appear later in this chapter.

59.1 ORIGINAL COLUMN

MUCH IS SAID THESE DAYS regarding low enrollments in computer science, reports indicate significant difficulties in recruiting women and other underrepresented groups to the field, and many studies discuss difficulties with the public image of computing. Clearly many deep societal and cultural issues underlie much of the problem, and many organizations (including ACM, ACM-W, CSTA, and SIGCSE) are considering ways to address these systemic issues.

With such wide-scale difficulties, some computing faculty seem resigned to the inevitable and may have given up in attracting students. However, global problems need not imply that nothing at all can be done locally.

Instead, this column considers several practical, local, small-scale actions for reaching out to prospective students and to students already on campus, with the goal of attracting new students to computer science. Much of what follows is oriented specifically for college programs, but many of these ideas also would seem adaptable for other venues.

Space constraints require a limited focus for this discussion:

Vital Assumption: This column assumes that courses and programs are exciting and appealing. Numerous presentations at SIGCSE conferences discuss approaches for the introductory sequence (e.g., multimedia, Alice) that reach out to diverse audiences; and the following discussion expects that computer science courses are stimulating, attractive, and well taught. I assume that the first course(s) will hook a moderate number of students—if students are willing to enroll. Thus, for this column, the primary issue is how to get students to try a first computer science course.

Two Target Audiences: It seems fairly obvious that there are at least two main sources of computing students and majors: (1) prospective students who are considering computer science as a possible major, and (2) students already enrolled at our school who are undeclared or considering other majors. We consider each of these in turn.

Prospective Students: When prospective students are considering computer science as a possible major, there are at least two challenges: (A) we want these students to come to our school and (B) these students should have reasonable expectations when they arrive.

Thus, the main challenges for recruiting interested, prospective students involve visibility and image: We need to get our programs on their radar screens, so they will consider our programs, AND we need to present an attractive and honest picture of what the students will find when they arrive.

Why Study CS? When I talk to prospective students interested in computing, a very large fraction seem to believe that computer science is some combination of surfing the Web, playing games, and hacking. There also is a perception that the field of computing offers few jobs. Of course, none of these images is valid. For example, as far as I can tell from reports of gross revenue from games and from the IT industry, video games account for somewhere around 1% of the IT industry receipts. Thus, computer games make up only a very small component of this industry. At the same time, students should be told that the industry is very broad—not a niche. For example, all projections from governmental agencies (e.g., the U.S. Bureau of Labor Statistics) indicate that the demand for computing jobs is very high, and demand will increase further in the coming years.

In considering the message to be promoted, I suggest three possible themes:

- intellectual challenge/stimulation,
- problem solving to help people through applications, and
- excellent job prospects.

In my experience, only a subset of these themes connect with each student, but together these themes resonate with many potential prospective students. As an additional exercise, one might hold a department meeting when faculty discuss why they decided to enter the field of computer science. I expect most faculty enjoy the field (or they would be teaching or doing something else). Faculty experiences can help shape a department's message for recruiting.

Advertising: With this spadework done, the main challenge is to get this recruiting message to the prospective students. Of course, the details of what might work at one school may be rather different from what might work at another.

Here are some possibilities that might be considered:

1. Develop a departmental flyer, highlighting why study CS, the strengths of the department, the opportunities of the CS program(s), and perhaps what graduates do. This flyer might be in paper and pdf format for both hard copy and electronic distribution.
2. Be sure copies of the flyer are available in the Admission Office.
3. Send letters (and the flyer) to prospective students.
4. E-mail prospective students, with names obtained from the Admission Office.
5. Get names and telephone numbers of prospective students from the Admission Office and call them. (For example, I call a few dozen prospective students each year.)
6. Talk to prospective students during their campus visits.
7. Host a reception in the CS department during prospective student days, so potential students can talk to CS majors and faculty.

8. Send departmental flyers or newsletters to alumni and request distribution to friends, neighbors, and any other folks who might be interested in computing.

9. Visit nearby schools to talk to students and guidance counselors about opportunities in computing (the ACM has some nice materials to help).

10. Consider developing a departmental question/answer blog to allow prospective students to interact electronically with faculty.

Of course, some of these likely will not fit a specific school, and doing all of these might well be beyond what any individual department can accomplish. In any case, I hope these ideas will suggest some possible actions—even among those who might feel great frustration and have given up.

Recruiting Students Already On Campus: For this column, we are assuming that a department has exciting courses and opportunities. With such a program in place, the challenge is visibility. Students throughout the school should think of computer science as something to try. Then, once they try a first course, we expect many will become hooked and want to take more. Thus, we need to consider how to keep computer science part of the general discussion of subjects on campus; whenever students are shopping for courses, computing should be in the forefront.

Here are some possibilities.

1. As a department, decide what makes the introductory courses and the CS major exciting (e.g., intellectual stimulation, problem solving, applications that address needs of people). If you cannot describe what is exciting, then why would you think students would find the courses exciting? Capture this spirit in widely-distributed venues:

   (a) Prepare/update the departmental Web site to show exciting projects, activities, student-faculty collaborations, and highlights of beginning courses;

   (b) Develop posters to advertise the first couple courses, and have a departmental student assistant put the poster up in dorms and student spaces around campus. (At Grinnell, we tried this approach and immediately doubled enrollments in the non-majors course while slightly increasing enrollments in CS1.)

2. Send letters to each incoming [accepted] student, indicating appropriate placement, describing the suggested first course, and encouraging follow up discussions.

3. Consider developing a T-shirt for CS1 students with image(s) of project(s) from the course.

4. Participate in campus-wide poster exhibits, showing neat student-faculty projects.

5. Be sure that Admission Office tour guides and Admission Officers are well briefed on CS; give tours of CS areas to Admission folks, and be sure they have copies of the departmental flyer.

6. Talk to all advisors of incoming students, so they will be encouraged to sign up their students for introductory CS courses.

7. Be sure a departmental bulletin board has attractive posters of computing, opportunities on campus and for careers, faculty, courses, CS Major internships, careers, REUs, etc.

8. Organize open houses/receptions to bring students to the CS area.
9. Talk to the Art Department about having a rotating exhibit of student art work in the CS area. (This not only can bring students to the department for a look, but also shows that computing folks are multi-dimensional – breaking some stereotypes.)
10. Post flyers, posters, seminar announcements through the Science Building and/or central campus buildings.
11. At any opportunity within faculty or committee meetings (e.g., approval of curricular proposals, change of major requirements, restructuring of departments), use required presentations as a forum to talk about what CS is and how it fits.
12. If your school has distribution requirements or other general graduation requirements, be sure some CS courses fulfill some requirements (typically either a lab requirement or a quantitative reasoning requirement). This gives particular incentive for students to try CS at an early stage.

With this column, I have tried to encourage readers to think about getting the word out regarding the excitement and opportunities of computer science—targeting prospective students and current students. Pragmatically, we have tried many of these approaches at Grinnell, and we have received much positive feedback from our students. (For a Web-based introduction, go to our home page: http://www.cs.grinnell.edu, and click on the link "For visitors and prospective students".)

Beyond these suggestions, I am confident that readers will have many other wonderful ideas as well. If you are willing to share, let me know. I would love to include your suggestions in another column.

59.2 SUBSEQUENT REFLECTIONS

Although computing enrollments today are booming—in sharp contrast to enrollments in June 2008, when the original column first appeared, recruitment and advertising remain vital to support at least three on-going goals:

- Meet demands from industry for students with strong computing backgrounds. Although computing enrollments have increased greatly, projections suggest that current enrollments in computing still will not meet the demand from industry, business, government, and academia.
- Promote diversity among computing students. Computing continues to suffer from low numbers of women and members from under-represented groups; programs should not assume that strong numbers among some populations imply a strong representation among all groups.
- Enhance possibilities for faculty recruitment. In today's hiring market, departments often have difficulty in filling positions—even tenure-track positions. Naturally, faculty candidates review on-line materials, and they may want to talk about departmental practices and activities.

Altogether, the demand for computing graduates remains very high, and current increased enrollments are unlikely to meet that demand. Students within under-represented groups may represent a vital largely-untapped resource, and increasing diversity is strategically important for many reasons.

Promoting Diversity

Although enrollments in computing are booming at many schools, statistics regarding the [lack of] diversity of computing students remain grim.

For example, the Taulbee Survey from the Computing Research Association (CRA) provides comprehensive statistics for undergraduate and graduate programs at Ph.D.granting universities in North America. For the most recent year data available (2014-2015), selected statistics highlight current challenges [294]: only 15.7% of BS/BA degrees in computing are earned by women, 0.4% to American Indians or Alaska Natives, 3.5% to Blacks or African-Americans, 7.0% to Hispanics, and 2.1% to multiracial (non-Hispanics), whereas 55.0% are earned by whites. Anecdotal evidence suggests that undergraduate, liberal arts colleges may have modestly better, but often not distinguished, records.

With a national emphasis on increasing diversity within the field of computing, ongoing efforts involve numerous activities to understand and address various components of the problem. Two important elements of this work involve recruiting new students and working to retain existing students—areas highlighted in the original column.

In recent years, recruiting efforts have expanded greatly, with outreach to students at many levels.

- Contacts with students in elementary school and middle school can generate initial interest in computing.
- Involvement with students in secondary school can help students understand opportunities within computing, experience the joys of creative problem solving, appreciate potential career paths, and provide a foundation for further work.
- Outreach to students starting college can encourage students to try computing at an early stage of their post-secondary education—in time for them to following up initial experiences with majors or minors in the computing field.

Much of this work falls squarely in the realm of advertising and recruiting. With the need for diversity within the computing field, the focus may have shifted over the years, but the need for advertising and recruiting continues. Four examples illustrate current activities for reaching out to diverse populations.

- Many efforts, reported at various SIGCSE-sponsored conference and at regional conferences, highlight outreach programs to middle school and secondary school students who are just starting to consider their long-term career possibilities.
- The newly-designed course, *AP Computer Science Principles*, developed by a collaboration between The College Board and the National Science Foundation, "introduces you to the essential ideas of computer science and shows how computing and technology can influence the world around you" [26]. By connecting with students at the secondary level, the course explicitly seeks to reach out to diverse students who might not have been thinking about computing previously.
- Some computing departments send personalized letters invitations to all incoming students from under-represented groups (e.g., women, ethnic minorities). The goal is to begin contact with students who might be interested in computing—given some initial encouragement.

- Some departments send notes of encouragement to all students [from under-represented groups] with passing grades in CS1—encouraging them to sign up for CS2. In some cases, all students [from under-represented groups] who have passing records in CS2 receive letters of encouragement for CS3.

Altogether, the foci of recruitment and advertising activities may have evolved over the years, but the need for those activities continues.

The Role of the Web in Recruiting Students and Faculty

Many potential students and faculty examine online resources in considerable detail when researching undergraduate schools, programs, and faculty positions. For example, most prospective students I talk with have already read departmental and college materials on the Web carefully, and almost all candidates for faculty positions have studied materials in depth regarding academic programs and faculty activities. When students and faculty have choices, they often want to go to places that are energetic, engaged, and forward looking.

In such a context, efforts to recruit students also may demonstrate attitudes and activities that may appeal to prospective faculty—possibly providing some positive factors during recruiting. Although some online elements were mentioned in the original column, the role of online materials seems to have become much more prominent in recent years.

In considering a program's Web site, an important consideration is what message or story one wants to highlight.

- What programmatic elements are interesting or different or noteworthy?
- How is the curriculum organized, and what are its strengths?
- What activities (perhaps outside the classroom) are significant?
- What scholarship or student-faculty collaborations are active?
- What strengths might bring students and/or faculty to the school?

Of course, each school and program has its own message and story. In the past, such narratives often were in paper form, and in some contexts those materials might still be relevant today. However, with the widespread use of the Web, advertising and public relations likely should include the Web as well.

CHAPTER 60

Course descriptions and public relations for computer science

This column first appeared in the *SIGCSE Bulletin*, Volume 41, Number 2, June 2009, pages 74–75[228] Subsequent reflections on this subject appear later in this chapter.

60.1 ORIGINAL COLUMN

CONSIDER THE FOLLOWING TWO [hypothetical] course descriptions for a Java-based CS1-course:

CS 1A: Java Programming (4 credits)
An introduction to programming in Java, including syntax and semantics of classes, objects, methods, declaration, initialization, computation, control structures (conditionals, loops). Lab allows students to apply constructs in Java to small-scale problems and video games.

CS 1B: Object-oriented Problem Solving with Java and Lab (4 credits)
An introduction to the solving of problems with computers, using object-oriented concepts of classes, objects, methods, and encapsulation. Emphasis on the design, implementation, and analysis of solutions, with lab exercises that integrate concepts and practice.

Both descriptions have exactly the same length: 35 words; both might accurately describe many introductory courses; and both emphasize important course elements. However, they communicate rather different messages regarding the nature of the course. In this column, I explore the relative merits of these descriptions and offer some conclusions about how our language can have a significant impact on how outsiders and prospective students view our discipline and courses.

Course Title

Computer scientists often use the word "programming" to refer to the entire process of problem solving, including specification formulation, design, algorithm development, coding, testing, and analysis. From this perspective, "programming" covers a broad range of topics – much of the entire field of computer science. Likely, course description 1A above arises from this view of "programming".

However, "programming" also may refer to the limited task of translating an algorithmic outline or pseudo-code into a computer-based language. In this context, general problem solving comes first, and "programming" may be one small step in an overall process. In my personal experience, many people outside computing may consider "programming" in this limited sense.

Unfortunately, course descriptions circulate widely, and we cannot clarify our meaning of "programming" to each reader. The title of CS 1A invites some confusion and may reinforce a common prejudice that computer science is narrow and reasonably trivial. In contrast, the title of CS 1B emphasizes conceptual problem solving and high-level thinking skills; the word "programming" does not actually appear anywhere in the description.

Purpose of Labs

The title "Java Programming" also leaves open the question as to whether students will receive guided, hands-on instruction within a laboratory environment. The term "programming" suggests some coding, but the title leaves open how this experience might be gained. In contrast, the CS 1B title clarifies that the course contains a structured laboratory experience.

Altogether, title CS 1A has the potential to reinforce stereotypes of the narrowness of coding, syntax, and semantics; while title CS 1B emphasizes higher-order intellectual skills and clarifies the role of a supportive lab environment.

Syntax and Semantics

Course description CS 1A focuses on the many low-level details that must be mastered as part of the programming process. Fluency in problem solving certainly requires that practitioners master syntax and semantics, control structures, data structures, etc. However, a natural question is whether such programming mechanics are an end or a means; what really is the purpose of a programming language in an introductory course?

As an analogy, consider the content and emphasis of an introductory course in French (or Spanish or Chinese or ...). At least in my experience, arguments for taking a foreign language often focus on understanding culture, exploring new literature, and allowing personal interactions (e.g., through travel or commerce). In order to develop comfort in reading, writing, and speaking a foreign language, students must master vocabulary, grammar, and the like; but language mechanics are rarely touted as the ultimate goal. Rather, students seek to converse freely with those in a new region and to gain insights about a society.

Similarly, for an introductory computer science course, CS faculty often comment that the emphasis should be on high-level concepts, such as problem solving, algorithms, design, abstract data types. As part of this work, students must develop a precision of thinking and writing; for example, algorithms must be precise, complete, logically composed, and correct. Writing instructions in a programming language supports this endeavor by providing a framework for careful expression and allowing proposed solutions to be examined carefully and tested. As with courses in introductory French, beginning computer courses cover syntax, semantics, and other details; but these mechanics are employed to support higher-level thinking skills.

Course description CS 1B emphasizes high-level skills and concepts. The course mentions implementation, and programming naturally provides the means to realize algorithms.

Overall, however, description CS 1B highlights course goals, while description CS 1A highlights the means to get there.

Mention of Problems and Games

Realistically, programs in an introductory course will likely be rather short. Starting at the very beginning, students simply cannot produce large and complex packages in a single semester. Thus, the reference to "small-scale problems" in description CS 1A may be promoted as truth in advertising. Further, some computing faculty argue that [many?] computing students have an interest in games, and the mention of video games may help recruit committed gamers.

Another view suggests that an important goal of an introductory course is to expand horizons and to anticipate exciting developments in the field. To achieve these objectives, a course should provide pointers to interesting future topics. Thus, rather than focus on "small-scale problems", the emphasis might be on suggesting possible important applications. For example, an introductory course may not be able to develop a full-scale simulation, robot, e-commerce application, etc.; but a course could choose elements of such applications to get started.

From another perspective, students are often drawn to the field of computing, because they want to contribute to society—students want to help society solve important problems. Unfortunately, this audience often does not consider games as providing such a contribution; and furthermore [almost] all students in today's society will know that computers can be used for games. In recruiting students, therefore, the reference to "games" in description CS 1A does not provide insight, but it might reinforce a misconception that computing focuses on entertainment that has little positive societal impact. Course description CS 1B avoids the pitfalls of "small-scale problems" and "video games".

Overall, description CS 1B may help create a positive image of the discipline, although this description may not publicize how computer science can contribute to society. If the introductory course draws primarily upon a few general areas to motivate problem solving, methodology, and techniques, however, perhaps description CS 1B provides a good overview of appropriate high-level goals and objectives.

However, if examples and assignments provide hints at a wide range of applications, then a further revised course description (also exactly 35 words) might provide additional visibility beyond both versions given at the start of this article:

> *CS 1C: Object-oriented Problem Solving with Java and Lab (4 credits)*
> An introduction to object-oriented problem solving, including concepts of classes, objects, and methods. Lab exercises emphasize design, implementation, and analysis of solutions. Problems may draw from domains such as simulation, robotics, multimedia, e-commerce, and bioinformatics.

Acknowledgment

Thanks to my colleague, Marge Coahran, for her feedback on a draft of this article and for suggesting a refinement of wording for description CS 1C.

60.2 SUBSEQUENT REFLECTIONS

The opening of this part of the book (Part VIII: *Outreach and public relations*) highlights that effective communication requires an identification of the themes or content to be communication, an understanding of the target audience, a determination of the media for the communication, and a sensitivity to specific wording. In this context, course descriptions can be challenging.

- A course description normally provides a brief, high-level overview of a course. The narrative can outline key elements of themes and content, but space constraints prevent discussion of most details.
- Audiences for course descriptions normally are diverse:
  - ◇ Prospective students likely do not understand many technical terms (these may be introduced in the course), so course descriptions must be written for a general, non-technical audience.
  - ◇ Graduate schools and prospective employers likely want to know what material is covered at a reasonably technical level.
- At many schools, each course description emerges from an involved approval process, so there may be an incentive for broad statements that are sufficiently flexible to allow the course to evolve without the need for another approval proceeding.

Altogether, in many contexts, a course description should avoid technical language (to be understandable to general audiences), provide technical content (to provide information needed for graduate schools and employers), and allow flexibility for course evolution. Further, course descriptions should emphasize themes regarding the nature of computing as a discipline and present constructive images of each course and the overall curriculum.

With so many challenging demands, developing course descriptions often will require an iterative process. One reasonable approach is to create a draft and then send it to representatives of different target audiences for their feedback. Can prospective students understand the narrative? Does the description contain adequate information for professional audiences? Will descriptions allow adjustments that faculty can foresee?

Overall, narratives, even narratives with extensive editing, may have blemishes for one group or another. However, a conscious review of draft descriptions in the light of multiple audiences can identify potential difficulties and help communicate needed information to diverse constituencies.

CHAPTER 61

Resolved: ban 'programming' from introductory computing courses

This column first appeared in ACM *Inroads*, Volume 2, Number 4, December 2012, pages 16–17[238] Subsequent reflections on this subject appear later in this chapter.

61.1 ORIGINAL COLUMN

THIS COLUMN ARGUES the word "programming" should be stricken from course descriptions, syllabi, and any other materials associated with introductory courses in computing. This conclusion follows from three main arguments:

- The word "programming" can mean different things to different people, so its use can be misunderstood.
- Use of the word "programming" can suggest priorities that do not reflect the real emphases of most introductory courses.
- Other words better communicate intended foci of courses, objectives, and goals.

The rest of this column expands these arguments, targeting introductory courses. The column concludes with a note about the use of "programming" in advanced courses.

"Programming" has Different Meanings for Different Audiences

Computer scientists often use the word "programming" to refer to the entire problem-solving process. In software development, this includes identification of specifications, refinement of human-computer interfaces, design of structures and classes, development of algorithms, the translation of those algorithms into a machine-readable form (coding), testing of programs, and perhaps even maintenance. From this perspective, "programming" is a broad term that covers many activities and skills.

However, the general public often considers "programming" as a narrow (and uninteresting) task of producing code by typing at a workstation. Also, "programming" might have negative connotations, such as disruptive hacking and mindless coding.

"Programming" Can Suggest Incorrect and Misplaced Priorities

Many contemporary discussions highlight the need for computational thinking and the coverage of high-level computer science principles in introductory computing courses. Altogether, the real emphasis of many introductory courses is problem solving.

Often these courses devote substantial time and energy to the solving of problems carefully and precisely. As part of this work, it likely will be important to analyze potential solutions for correctness and compare alternative solutions. To promote communication, reduce ambiguity, and allow analysis, solutions may be written in a specific computer-readable language, and this activity may involve coding and running computer programs. However, at the introductory level, the focus often is on high-level thinking skills, computational thinking, and constructive uses of computers. Coding may be involved, but its role is secondary to problem solving.

Over the years, I have heard arguments that students need to cover all of programming language at the introductory level, or that students should develop considerable fluency writing in a specific language during a first course. Although local needs may differ, I continue to be skeptical about these arguments.

In solving problems, beginning students need to learn common techniques (e.g., conditionals, iteration and/or recursion, basic data types and structures); and precise communication requires some basic syntax and semantics. However, such coding requires only a small part of most modern programming languages.

Beyond these basics, students need to learn how to find out more as needs arise. Thus, students likely should learn to use manuals, documentation, and online aids. Further, a focus on problem solving will help students learn how to think about problems and their solutions. The high-level challenge is not writing code — it is easy to churn out line after line of syntactically correct nonsense. Rather, the high-level challenge is developing solutions (algorithms, code) that address problems in a helpful way. Helping beginning students to learn to think seems much more valuable than extensive rules of syntax.

In addition, extensive emphasis on coding may have negative side effects. Students may believe that computing really is the same as coding (programming in the narrow sense), and students may not understand how computing can be an intellectually challenging enterprise that helps people. Further, covering all of a language likely will emphasize syntax and semantics, which beginning students will not have occasion to use. In addition, languages evolve, and numerous details memorized during the first year of college may become obsolete by graduation. And, if students cram many syntactical constructions quickly for one course, those details are likely to be forgotten just as quickly. The phrase, "easy come, easy go" can provide warning in this regard.

Learning to code can help students get short-term jobs and aid in later courses, but learning basics plus the use of manuals and documentation also would be likely to meet these short-term goals while providing long-term benefits.

Altogether, in beginning courses, students focus on sharpening their problem-solving abilities, practice communicating their solutions, and learn to analyze potential solutions; and coding is largely a side effect of higher-level goals and objectives.

Other Words Can Communicate Ideas more Clearly

This column as already observed that the word "programming" means different things to different people. Depending on the intended meaning, two alternative wordings can largely take the place of "programming" wherever it appears in introductory courses.

- When "programming" is used to mean the entire process of problem solving or software development, "problem solving" can usually be used in place of "programming" to clarify the broad scope of this activity.
- When "programming" is used to indicate attention to low-level syntax, semantics, and code development, "coding" can be more descriptive than "programming".

Upper-level Courses:

This column focuses on introductory courses. Some upper-level courses focus on the design and implementation of programming languages. Others may explore subtleties of specific languages and environments. For such cases, the term "programming" may indeed be appropriate. However, such courses normally target specialists and relatively advanced students, and these people likely will understand the use of "programming" intended at this advanced level.

61.2 SUBSEQUENT REFLECTIONS

In the five years since the original column appeared, it seems little has changed:

- The term "programming" still is widely use both by the general public and by computing professionals.
- Typical citizens often consider "programming" to mean "coding".
- Many computing professionals consider "programming" to include a full range of problem solving activities.

As argued in the original column, this difference of perspective confuses discussions in the public press regarding the computing discipline, and inadvertently reinforce stereotypes of solitary "programmers" as staring at computer monitors for hours at a time.

Upon reflection, this confusion with "programming" raises a more general issue regarding the understanding of terms by different audiences. For example, the term "induction" is interpreted by many mathematicians as meaning "induction over the non-negative or positive integers", whereas "induction" for many computer scientists means "induction over structures and substructures." As a result, conversations about discrete mathematics may involve use of common terms, but computer scientists and mathematicians may understand different content. On some campuses, such confusion yields frustration and misunderstandings regarding topics covered in specific courses.

As a second example, Chapter 64 discusses issues related to games in the computing classroom. For some, "computer games" may be considered a type of "fun and games:" students may be envisioned as coming to class to play minecraft or tetris or other favorite video games. For others, "games" may suggest underlying design and development principles and techniques, such as game design, game development, techniques from artificial intelligence, principles of GUIs that impact usability, and other techniques that may be used in many applications. In considering the earning of academic credit, the former "fun and games" may seem frivolous, but the latter may highlight intellectual challenges and explore algorithms and analysis.

Such interpretations of the word "game" in this second example may suggest a need for clarify of intent and meaning. One approach might be to avoid the word, "game" or "games", when used as a noun, but allow the word as an adjective, such as "game design", "game development", etc. In considering discussions of "computer games", such a distinction might be overly prescriptive, but care in word choice seems prudent.

More generally, any communication should convey an intended message from sender to audience. The terms, "programming", "induction", and perhaps "game" (as a noun), may be subject to multiple interpretations. As with any communication, the sender should be careful that the message received will be consistent with the message intended. Care in word choice seems an important principle in all communications.

CHAPTER 62

What image do CS1/CS2 present to our students?

This column first appeared in the *SIGCSE Bulletin*, Volume 39, Number 4, December 2007, pages 18–19[223] Subsequent reflections on this subject appear later in this chapter.

62.1 ORIGINAL COLUMN

IN MY DISCUSSIONS with prospective students and their parents, many or most high school students seem to believe that computer science focuses on some combination of three elements: surfing the Web, playing games, and hacking. Further, these potential students often believe that computing is solitary, mindless button-pushing, and anti-social. In addition, many of these folks have the impression that there are few jobs in the computing industry (most are thought to be outsourced), and the few that remain lack the potential for creativity or societal impact.

In contrast, keynote speakers at the ITiCSE 2007 conference highlighted perspectives what can make computer science exciting and attractive to potential students and professionals alike.

- Paul Curzon talked about the intellectual excitement of problem solving and solving puzzles. [52, 53]
- Chris van der Kuyl discussed his experiences in creating games and the need for diversity in the computer-games industry. Further, he indicated that he and his company are being driven to consider outsourcing, because there simply are not enough computer scientists being produced locally. [195]
- Vicki Hanson talked about applications that can assist people with disabilities and can transform society. [81]

Of course, this contrast of perspectives has been widely observed, and many suggestions highlight possible approaches to content (e.g., breadth-first, multi-paradigm, multi-media, etc.). Further, exciting approaches within CS1/CS2, such as Alice and Georgia Tech's multimedia project, utilize interesting contexts for many of the fundamentals (conditionals, iteration, recursion, data structures) that beginning students must learn.

Steve Cunningham, former SIGGRAPH Chair and NSF Program Director, uses the term "looking outward" for interesting courses that connect with external themes and applications. In contrast, courses that focus on internal details are "looking inward". He further suggests that adjusting the "looking inward/looking outward" balance was "one of the major themes of the 1980s calculus reform."

Since discussions of content typically involve extensive review of the scope, shape, and details of the curriculum, this short column considers a different, but related, matter: how is content presented, and what image do students see of the discipline of computer science in CS1/CS2?

Although few faculty admit shortcomings in courses at their local institution, reports suggest many introductory courses may be deadly dull. Some textbooks and syllabi suggest that numerous introductory courses focus largely on language syntax and routine programming exercises. Here are several examples I have learned about in the last year:

- Depth-first CS1 courses that drill endlessly on basic control structures; e.g., how many ways can one translate a "for" loop to a "while" loop?
- Breadth-first CS1 materials that emphasize linear search as the main (only?) algorithm for countless applications.
- Objects-first CS1 syllabi that overwhelm simple problems with extensive syntax and machinery; e.g., convert a temperature in Fahrenheit to Celsius in 300 lines of code, using I/O classes, multiple classes with inheritance, and mathematics libraries.
- Discussions of implementations of CC2001 that consider how to fit every topic into a constrained set of courses, but may omit whether courses are coherent, interesting, or at an appropriate level.

In contrast, conversations with faculty suggest that almost everyone's courses are doing well: interesting, effective, stimulating, etc. As in Garrison Keillor's Lake Wobegone, it seems all CS1/CS2 courses are above average.

Similarly, in the past year or so, I have heard criticisms of the Advanced Placement Computer Science (AP CS) syllabus as being unexciting, language driven, and narrow in focus. Critics complain that AP CS discourages women and other under-represented groups. And yet, the AP CS case study provides fine opportunities for moving beyond syntax and considering simulations, design alternatives, and other applications. Further, the goal of AP CS is to identify what colleges and universities actually do in their first year, so that high school students can get college credit. While we might debate some details, it seems plausible that AP CS is quite successful in capturing what colleges expect in CS1/CS2, and the syllabus is more focused than in many university courses.

If our courses are all above average, if AP CS successfully outlines what most CS1/CS2 courses actually do, and if we hear about introductory (or AP CS) courses not connecting with students, then we may need to rethink how our courses are really going. If we think "it's not my problem", then we should ask how we know that. Are we engaging in wishful thinking, or do we have solid evidence that our courses actually are going well?

In probing evidence and possible success further, the keynote talks at ITiCSE 2007, together with recent discussions, have raised issues for me about what images are being presented in CS1/CS2. For example, here are some questions for consideration:

- To what extent do introductory courses reinforce or challenge the cultural image that computing is surfing the Web, playing video games, and hacking?

item Do CS1/CS2 courses emphasize programming details to the exclusion of other topics, even though professionals commonly say that computer science is not the programming?

- ◇ Do assignments only involve writing programs?
- ◇ Do tests ask for code (not explanations, same as comparisons, analysis, ideas for experimentation, possible simulations, design, etc.)?
- ◇ Do evaluation questions focus on idiosyncrasies of language syntax or ask students to merely mimic the same constructs they have done numerous times previously?

- Do CS1/CS2 courses and laboratory exercises highlight solitary, individual work (with collaboration not allowed), or do some activities promote the image that teamwork is a fundamental environment for software development?
- Do examples and assignments in CS1/CS2 appeal to a wide range of students or only a narrow group? For example, an emphasis on games is well known to turn off many women and folks from under-represented groups—a population many departments are trying to attract.

As one considers image, it is worthwhile to observe that, in some cases, just a recasting of a problem can have a significant impact on the image being presented. To illustrate, consider the image presented in these two equivalent problems:

- Conduct 1000 simulations to count the number of times a coin should be tossed before both a head and a tail occur.
- A couple plans to have children until they have at least one boy and one girl. Conduct a simulation of 1000 couples to determine the number of children the couple might have.

In summary, I invite readers to consider what images of the discipline CS1/CS2 really present at their institutions. For example,

- What is actually required on assignments and tests?
- What experiences do students really have?
- How much of the course is syntax and straightforward practice?
- If simulations are important, are these in CS1/CS2?
- Do CS1/CS2 discussions and assignments hint at real applications? (Actual applications may be too complex for beginners, but connections and directions are possible.)
- Do CS1/CS2 examples suggest possibilities for computing to meet social needs, help people, improve the quality of life?

Of course, it is not feasible to pack every possible topic and theme into just a couple courses. However, every CS1/CS2 course inevitably portrays an image of the discipline, and considering that image may have an important impact on the recruitment and retention of students.

Acknowledgment

Many thanks to Steve Cunningham for his feedback on an early draft of this column and for his insights and contributions.

62.2 SUBSEQUENT REFLECTIONS

Since the original column appeared about a decade ago in 2007, much has changed, but perhaps much also has stayed the same.

- Multiple significant efforts over the past several years have sought to change the image of computer science among middle-school and high-school students. Outreach activities often include workshops, after-school activities, summer camps, speaker series in K-12 schools, etc.

- "Code.org is a non-profit dedicated to expanding access to computer science, and increasing participation by women and underrepresented minorities. Our vision is that every student in every school should have the opportunity to learn computer science, just like biology, chemistry or algebra. Code.org organizes the annual Hour of Code campaign which has engaged 10% of all students in the world, and provides the leading curriculum for K-12 computer science in the largest school districts in the United States." [39]

- The course, *AP Computer Science Principles*, originally envisioned when the original column appeared, is now fully developed through a collaboration between The College Board and the National Science Foundation. This course, targeting high school students, "introduces you to the essential ideas of computer science and shows how computing and technology can influence the world around you" [26]. By connecting with students at the secondary level, the course explicitly seeks to reach out to diverse students who might not have been thinking about computing previously.

Such efforts are reaching out effectively to K-12 students, and the *AP Computer Science Principles* course [26] is impacting courses for non-computing majors at some colleges and universities. For example, both code.org and the *AP Computer Science Principles* developers (e.g., NSF and The College Board) have developed many resources to support courses in high schools and courses for non-computing majors in college. Further, various reports indicate that these course successfully convey the range of topics spanned by computer science and the impact that computer science can have upon societal needs.

However, the success of beginning courses for non-computing majors does not address issues identified in the original column regarding CS1 and CS2. In this regard, the track record seems spotty.

- At one liberal-arts college, enrollments in the course for non-computing majors (generically called CS0) are respectable, but the enrollments in CS1 are a full order of magnitude higher, and both CS2 and CS3 have enrollments many times those in CS0—even though no college requirements compel students to take any of these courses. Also, drop out rates for CS0, CS1, CS2, and CS3 at this school are all at or below 2%. Regarding diversity, at this school, roughly 40–45% of the students in CS1 are women, and that record generally continues through later computing courses.

- The Best Presentation Award at ITiCSE 2016 went to a paper, "Learning to Program is Easy," by Andrew Luxton Reilly and reported drop out rates as high as 50% to

> 90% [157]. As the paper's abstract indicates "By revisiting the expected norms for introductory programming we may be able to substantially improve outcomes for novice programmers, address negative impressions of disciplinary practices and create a more equitable environment."

From my perspective, many approaches have produced exciting, engaging, and technically strong CS1 and CS2 courses that connect well with diverse audiences and that highlight something of the range of possible applications. Models of successful introductory courses with components that describe the breadth of computing are presented at conferences regularly. However, conference reports also describe other courses that are narrow and tedious and that focus on unmotivated and artificial problems that highlight syntax.

As the original column concludes, "every CS1/CS2 course inevitably portrays an image of the discipline, and considering that image may have an important impact on the recruitment and retention of students." Models for CS1/CS2 indicate that CS1/CS2 courses can reach out to diverse populations, just as the new *AP CS Principles* course is designed for outreach. An ongoing challenge for the computing-education community is to incorporate these successful models broadly—not discouraging beginning students, but capturing their imaginations and intellect.

CHAPTER 63

Computing teaching labs can communicate negative messages

This column first appeared in ACM *Inroads*, Volume 1, Number 3, September 2010, pages 13–14[230]. Subsequent reflections on this subject appear later in this chapter.

63.1 ORIGINAL COLUMN

OVER THE YEARS, I have visited numerous campuses to observe computing programs and their facilities. When viewing teaching labs, I continually consider what messages these labs are sending to prospective students, what priorities these labs demonstrate, what the labs communicate about the nature of computing, and how lab environments might be made more comfortable and appropriate. In several cases, the teaching labs for computing sent **a strong negative message** to prospective students and current students. Further, uninviting lab environments sometimes seem to undermine efforts to recruit students!

Four Disappointing Observations

1. A few labs were dark and dreary, with poor lighting, low ceilings, dark woodwork or dark paint, dirt marks on the walls, layers of yellowed wax on the well-used floor, etc. Sometimes students called these facilities "the cave" or "the catacombs". Prospective students ask themselves if they would want to spend several years studying in such an environment, and the answer almost certainly would be "No".

2. Walls in several labs were bare, with few bulletin boards, pictures, announcements, or other materials. At best, these labs seemed dull and boring; more commonly, the few posters present were old and worn, suggesting computing is a staid discipline with little excitement or energy.

3. In several labs, spaces for students were small and confining, and the environment effectively prevented group activities. Further, high partitions sometimes separated the workstations, reinforcing the misconception that computing is a solitary enterprise with little communication with others. Such facilities cannot support modern

pedagogy that involves collaboration, pair programming, or team projects, On several visits, I commented that I could not use these labs for my own courses, since I regularly utilize lab-based exercises that emphasize collaboration.

4. At one facility, workstations were named for the Knights of the Round Table; at another, workstations had names of characters in computer games. Although some might find such names cute, the names have limited appeal. For example, few women or people from under-represented groups are included in any listing of Knights. Further, students who do not play specific games may feel ignorant or out of touch with the field of computing—the students may not know who the characters are. And students who do know the characters may not relate to the qualities represented by these figures. Altogether, many computer games connect primarily with a restricted subset of white males.

Positive Messages are Possible

In contrast to these examples, well-designed teaching labs can send positive signals and can help recruiting. Some achievable goals for teaching labs might include:

- a bright, open environment with good lighting, clean surfaces, fresh paint;
- good sight lines to see walls, boards, and other people;
- room for books and notes around each computer;
- sufficient space for collaborative teams to work together;
- interesting pictures, posters, and materials on the walls to capture some of the excitement of computing;
- wide, well-placed aisles that allow instructors to circulate easily from student team to team; and
- visual clues (including posted materials, images, naming conventions, practices) that computing is an inclusive discipline.

The achievement of these goals requires motivation, creativity, thought, and effort. The challenge is to best utilize local spaces and possibilities. To help the process, a companion article in this *ACM Inroads* issue [rewritten as Chapter 27 in this book] discusses possible lab configurations. Some configuration changes may require substantial creativity and analysis. For example, fundamental change may require an architectural review of the room, including identification of bearing walls, electrical outlets, heating and cooling ducts.

However, massive renovation is not always needed, and substantial improvements often depend primarily upon identifying the issues and taking simple steps. For example, clean surfaces, fresh paint, interesting bulletin boards, and visual elements require very little money, but can have a significant impact. Although these lab environments demonstrated difficulties, many problems can be remedied at low cost. Walls can be painted, bulletin boards installed, high partitions removed, pictures refreshed, and posters updated on a very modest budget.

Outside Perspectives

After we live in an environment for several years, we tend to accept the facilities as being natural and comfortable—this is where we live professionally. To obtain different perspectives, **I strongly encourage readers to invite outsiders to visit their teaching labs, asking for reactions to these facilities.**

For example, whenever I participate in an external review, one of my early activities involves touring the facilities on my own. As an outsider, what do I see? In some cases, I have noted that all pictures are of old white men, a lack of clear signage makes the computing department largely invisible, posters show technology that is at least a decade old, lists of graduates suggest no alumni have entered the computing field for 15 years, etc. When I raise these observations, computing faculty commonly indicate they had not really looked at the spaces and signage in recent years. Sometimes faculty indicate that there had been a project for bulletin boards awhile back, but they had not realized how dated the materials had become.

Overall, teaching labs suggest to both prospective students and current students something about the nature of the computing curriculum and discipline Messages about computing may be positive or negative, but it can be difficult for computing faculty to realize the impact of the facilities that the faculty have seen daily for years. An outsider can look at teaching labs with a fresh eye to help identify both good and bad points of these facilities.

Acknowledgments

Many thanks to John Impagliazzo and Marge Coahran for their helpful suggestions for this column.

63.2 SUBSEQUENT REFLECTIONS

The original column highlighted environments and circumstances that can significantly impact students' perception of computing and computing courses—either positively or negatively. In all of those comments, the nature of the user experience is an over-arching theme—both from the perspective of prospective students and for current students (particularly at beginning levels). At the risk of over-generalizing, an environment should allow students to focus on learning and the developmental tasks that support learning. In contrast, an environment should not impose obstacles that distract students or add unnecessary elements to students' cognitive load.

The following comments reinforce this over-arching theme, based on my own observations on several campuses over the past decade. Some of these comments may seem obvious, but not all lab environments yield positive user experiences.

- Chairs, tables, and other furniture should be in good condition.
  - ◇ wheels on chairs should move smoothly.
  - ◇ adjustable seat backs and seats should be comfortable and secure (Chairs should be sufficiently comfortable that students can work for a few hours without cramps, aches, and pains, but chairs likely should not be so comfortable that students fall asleep regularly).
  - ◇ levers for adjusting chair height and angles should be in good condition (not broken).

- Software and programming environments should be stable from session to session.
  - ◇ compilers, editor, integrated environments, etc. should work consistently and reliably.
  - ◇ needed files (e.g., Makefiles, initialization scripts) should be readily available .
  - ◇ software updates should not be installed during a semester (unless in an emergency), as new versions often yield altered procedures for even the most common tasks.

- Programming environments should be user/beginner friendly.
  - ◇ problem solving and program development should be reasonably straightforward (e.g., low cognitive load), so that beginners (really, all students) can focus on new ideas and techniques, rather than being distracted by complex menus, esoteric options, and long command sequences.
  - ◇ environments should work reliably, with hardware, software, and network connectivity available with few malfunctions.
  - ◇ processing (e.g., compiling, running programs) should proceed at reasonable speed (fast speeds versus very fast speeds may not be an issue, but slow or very slow speeds can undermine other efforts for support and encouragement)

- Students should experience consistent interfaces and environments between the classroom and lab.
  - ◇ when students observe a demonstration or example in class, their experience in the lab should be similar.
  - ◇ font sizes in the classroom may need to be larger than in the lab, but the functionality and displayed work should work the same (to avoid adding to cognitive load).

Over the years, I have seen labs on some campuses where chairs were in disrepair, workstations regularly malfunctioned (e.g., a different 5-7 workstations out of 20 would not run needed software each class session), Bluetooth connections regularly crashed, etc. Although each of these situations could be addressed with day-to-day workarounds, the clear results involved high student frustration and steadily decreasing student interest and motivation.

Altogether the user experience in teaching labs contributes to student interest and performance. Smoothly functioning hardware, software, and environments can reduce stress and allow students to focus upon active learning, while complex procedures, malfunctioning equipment, and broken furniture can distract students from their work and add to cognitive load. The success of teaching-lab environments likely have an impact on both prospective and currently-enrolled students at all levels.

CHAPTER 64

Do computer games have a role in the computing classroom?

This column first appeared in ACM *Inroads*, Volume 35, Number 4, December 2003, pages 18–19[214] Subsequent reflections on this subject appear later in this chapter.

64.1 ORIGINAL COLUMN

FOR SEVERAL YEARS, I have been struck by two conflicting themes that arise regularly at conferences. Some speakers promote games as excellent motivators that provide opportunities to introduce modern technology and encourage creativity. Other speakers describe games as discouraging underrepresented groups and distracting students from fundamental principles, structures, algorithms, and methodologies. This column explores several common arguments on each side.

Due to space limitations, this column focuses on the use of games within introductory courses, although similar issues may apply at other levels as well. As with previous columns, I invite feedback on any side of the argument.

Types of Games

Of course, games encompass many styles and subjects. For example, games may be competitive or cooperative, be played by individuals or groups, and touch on numerous themes, such as adventure, education, social interactions, science fiction, violence, sexual circumstances, etc. Simulations sometimes are considered games as well. With this diversity, discussions of advantages or disadvantages of games may apply only to some some types or application areas; and one must be careful not to over generalize in making comments.

Positives

Several arguments are commonly cited for the use of games in the computing classroom. While the following list is NOT ordered according to importance, I have numbered the main points for later reference.

1. Motivation:

   a. Some (many?) students find games very motivating.
   b. Many students have prior experience with a variety of computing games, so using games in courses may connect with students' background.

2. Fancy graphics can capture students' interest and imagination.
3. Games are often easy to understand, so developing programs that play games can highlight problem solving, data structures, classes/methods, and other high-level skills.
4. Games provide options for creativity in assignments, possibilities for extensions, and opportunities to develop projects through a sequence of assignments.
5. Games allow assignments to be described in layers, where a moderate level of functionality is required for a "C", additional features constitute a "B", and extensive refinements yield an "A".
6. Games provide opportunities for the early introduction of elements of modern technology, such as client/server computing, concurrency, and object-oriented programming.

Negatives

While games have various constructive elements within the classroom, various reports suggest that many positive elements also have counterbalancing negatives. Here are some commonly cited problems for the use of games. Numbering is keyed to the points in the list of positives.

1. Motivation:
   a. Some groups, particularly women and other underrepresented groups, are often turned off by competitive games. Students in these groups often want an emphasis on socially-constructive applications. (See note 1.)
   b. An emphasis on games may reinforce the popular misconception that video games represent a major component of computer science. (See [41] and note 2.)
   c. Games and game playing can be quite addictive, so emphasizing games in the classroom can reinforce anti-social behavior.
   d. Assignments utilizing games can encourage distractions during class sessions, as students show off their programs.
2. Graphics:
   a. Much class and/or student time can be devoted to graphics and I/.O. If there is not a corresponding emphasis on HCI, game interfaces could focus on personal idiosyncrasies rather than principles and analysis.
   b. Extensive time devoted to I/O can limit time available for such fundamentals as algorithms, data structures, and software engineering.
4. Extensions and Creativity:
   a. Encouragement to add features may undermine a sense of writing to specifications and considering actual customer needs.
   b. Options for extended functionality may encourage program bloat and unnecessary complexity.

5. When grades depend on multiple levels of functionality, true beginners (who often are students from underrepresented groups) can be at a significant disadvantage for the best grades and building self confidence.

Overall, this discussion suggests strong counter arguments may apply to at least four of the six arguments cited previously in favor of the use of games in computing courses.

Additional Notes

The remaining pro-games arguments involve (3) high-level problem solving and (6) opportunities for the early discussion of modern technology. In both cases, other applications may be equally effective as games in addressing desired objectives. Sometimes a simple rewriting of a problem can broaden its appeal. For example, consider the following problem for a CS1/2 course:

- *Original Problem:* Use a random number generator and a conditional loop to count the minimum number of tosses of a coin needed to get at least one head and one tail. Then embed this code in an outer loop that prints this minimum number of rolls for 20 simulated experiments.

While this problem provides helpful experience with random number generators, conditionals, loops, and simple simulations, my students found the problem somewhat contrived. The following restatement generated much more student interest.

- *Restated Problem:* A couple decides to have children until they have at least one boy and one girl. Then they will stop having children. Assume the likelihood of having a girl is 50% and the probability of the next child being a girl is independent of the genders of previous children. Use a random number generator to simulate the gender of a child, and a conditional loop to count the number of children for a specific couple. Then embed this code in an outer loop that prints the number of children for each of 20 couples.

While the two resulting programs are functionally identical, students often found the context of family size considerably more interesting than flipping a coin.

Example A

As a second example, the new Java-based Marine Biology Case Study of the Advanced Placement Computer Science course utilizes simulation on several levels—not games—to provide a wonderful context for problem solving, data structures, and object orientation.

Example B

As a third example, the SIGCSE Project grant for a client/server framework for CS1/2 supports both games and multi-user applications—promoting the use of modern technologies in many contexts, not just games.

Concluding Comments

While space prevents this column from touching all arguments either for or against the use of games in computing courses, it suggests that some arguments for games have corresponding counter arguments, while other arguments apply to various simulations and other contexts as well as games. With the potential negatives—particularly related to the discouragement of women and underrepresented groups, this preliminary discussion might suggest applications other than games be used in courses; at least, instructors should be careful before incorporating games into their courses. Of course, faculty on either side of this debate might bring other points forward as well to swing the debate to one side or the other.

Footnotes and Bibliography

1. Negative perspectives by women of computer games are discussed in several articles in the "Women and Computing" special issue of the SIGCSE *Bulletin inroads*, June 2002.[32].

2. While gathering comparable statistics is challenging, here is one estimate of the relative size of the computer gaming industry. The Interactive Digital Software Association reported that "2001 U.S. sales of computer and video games grew 7.9 percent year-on-year to $6.35 billion, ..." (See http://www.idsa.com/2001SalesData.html). Also, the Information Technology Association of America combines information and communications technology (ICT) products and services within its definition of the information technology industry and reports that "U.S. spending in ICT has increased almost 70 percent since 1992, to almost $813 billion in 2001." (See http://www.itaa.org/news/gendoc.cfm?DocID=120). Putting these numbers together, computer and video games made up 0.78% of total IT sales for the year 2001.

64.2 SUBSEQUENT REFLECTIONS

Although the role and applications of computing in contemporary society have evolved over the years, many observations in the original article remain valid.

Scope of Video Games within the Information and Communications Technology (ICT) Commercial Sector

As when the original column appeared, gathering comparable statistics continues to be challenging. For example, statistics in the original article regarding revenue from video games and from overall information and communications technology (ICT) came from the Interactive Digital Software Association and the Information Technology Association of America, respectively, but neither association exists today—at least in its previous form with these names.

Today, the United States Department of Commerce may serve as a consistent source:

- "More than a quarter of the $3.8 trillion global IT market is in the United States" [193], and this suggests that the total IT market in the United States is about $1 trillion.

- "Combined revenues in entertainment software from computer and video games $15.4 billion in 2014" [193].

Altogether, these statistics suggest that in the United States video games revenue comprise about 1.5% of software and information technology revenue. As in 2001, sources vary, but most reports suggest that video games revenue represents 1%–2% of ICT revenue for the United States. Although this percentage may be about double the 2001 figure, video games continue to be a very small component of the information technology sector.

Games as Student Motivators

Today, games clearly continue to motivate some enrolled and prospective students. Although some groups gravitate to competitive or violent games, non-competitive games often connect with a wider audience. Also, many young people today collaborate using social-media sites. In response, various courses, programs, and workshops have emerged to serve this audience. Anecdotal reports at conferences suggest interest in game-based programs may be constant or waning somewhat, but enrollments often are healthy.

Overall, courses related to video games seem to continue to attract certain groups of students. However, past studies have suggested that many women and students from under-represented groups have motivations related to helping others and making a difference in addressing societal issues. For these students, courses based on video games may be far from motivating—just as discussed in the original column.

Public Relations and Promotion

Beyond the risk of alienating prospective students with topics that may seem frivolous, two issues may arise with courses highlighting video games. The first was discussed in the original column, but the other may be somewhat new.

- From a public relations perspective, most students know about video games as a computing application, and yet the statistics reported above indicate that this use of computing captures less than 2% of the overall ICT field. Thus, discussions of video games seems unlikely to increase awareness of the range of applications within the field of computing. The vast majority of opportunities within the field of information and telecommunications technology lie outside the realm of video games, but how will student learn about such possibilities?
- In recent years, numerous news reports have raised concerns about expenditures of funds on research projects and on educational courses that seem frivolous or wasteful. As a made-up example, studies of inappropriate behaviors among intoxicated individuals may provide insights in psychology or social work, but a press report might ask, "why are we funding people to get drunk?" Although I have not heard such news reports regarding courses related to video games, it does not seem a stretch that future headlines might ask, "why are we funding students to play video games?"

Both of these issues may suggest that promotional pieces, course descriptions, grant proposals, research reports, etc. related to video games be worded carefully. For example, narratives may need to distinguish between playing games and themes of game design, game development, techniques from artificial intelligence, principles of GUIs that impact usability, and other techniques that may be used in many applications. When discussions connect elements of game construction and algorithmic development to other computing applications, students may discover components of the wide world of computing, and potential critics may learn how game-oriented courses and projects may have broad application in many problem domains.

XIII

Additional Topics

YEARS AGO, my conversations with faculty in one computing program included consideration of a course that included some discrete mathematics and some formal program verification. After some probing, a general criterion for content emerged: this was the course that would cover those important topics that did not fit very well elsewhere. Although such a categorization of the course likely is not completely fair, the listing of topics did not seem obviously coherent or integrated. On the positive side, by placing topics in one catch-all course, other courses in the curriculum could be coherent with well-identified themes and topics that allowed a natural progression of study. On the negative side, the one course was a challenge for both the instructor and the students; some parts seemed largely disjointed from other parts, and pieces did not seem to fit together.

Turning to this book, most parts have an overarching theme. The book progresses from high-level curricular principles, to courses and course planning, to the role of theory, to course formats and pedagogy, to various low-level course components, and to assessment and grading. Within this structure, each part identifies an overall subject, and chapters explore the subject. Hopefully the reader has found this progression thoughtful and worthwhile. Each part has coherence, and chapters connect with each other in a natural framework.

And then this part appears. As with other parts, each chapter here has a clear focus, and each provides ideas central to teaching computer science. However, the original columns presented here do not fit naturally with other themes, and they do not fit easily with each other either.

Altogether, this part presents several topics which generally fall under the heading "none of the above." Together, these chapters fill in gaps from earlier discussions and provide further background for ongoing considerations of teaching and learning.

CHAPTER 65

Sorting algorithms: when the Internet gives out lemons, organize a course festival

This column first appeared in ACM *Inroads*, Volume 6, Number 1, March 2015, pages 28–29 [260]

65.1 ORIGINAL COLUMN

YEARS AGO, when I was writing my first books, my series editor, Gerald Weinberg, put forward the principle that any example for class or a textbook should be valid on its own merits. Of course material introduced early may be refined, but students should not learn a technique initially, then discover the approach was wrong. With so much for students to learn, students should not be devoting time to examples that don't work or are not valid in any context. Students should not have to unlearn an early example and then learn something to replace it. Rather, every example should be valid as presented—at least in some context within current understandings of the discipline. To illustrate, an insertion sort is a fine choice for the first example of a simple sorting algorithm: it is easily understood and is extremely efficient when the data set is almost ordered. Of course, much better (but more complicated) algorithms are available for random data or data in descending order—but the insertion sort is an algorithm of choice in at least some situations. (In contrast, the bubble sort is never the algorithm of choice in any context, so should never be used or taught.)

Traditionally, sorting algorithms have been part of introductory courses for decades. Often, the pedagogy has involved some of the following elements:

- Trace the algorithm when applied to one or more data sets
- Code the algorithm in whatever language is used in the course
- Time the algorithm on data sets of different types (e.g., ascending, random, descending) and different sizes (e.g., 40,000, 80,000, or 160,000 elements)

For example, at the end of a unit on sorting, a common assignment might be to write code for several algorithms, run them on various data sets to determine execution time, and graph the run times. In principle, this approach gives practice working through the algorithms and provides experimental data to support algorithmic analysis of efficiency. For those interested, simple Java code to time an algorithm might have the form

```
// time and check insertion sort without swapping
start_time = System.currentTimeMillis();
call_algorithm_method (data_set);
end_time = System.currentTimeMillis();
System.out.print (end_time - start_time);
```

Enter the Internet: Lemons

In recent years, however, numerous versions of standard sorting algorithms can easily be found on the Web. Unfortunately, many of these examples violate Gerald Weinberg's principle—the code is simply awful. I would be embarrassed to have my name associated with many examples found on the Internet, but apparently the writers of these solutions have other perspectives. Some common difficulties include

- Unnecessary memory allocation
- Excessive data movements
- Calls to procedures that perform very little work
- Use of overly-generalized mechanisms to perform specific, simple tasks

Three examples may illustrate the widespread difficulties, The first two of these comes from class notes available on the Web, and the second comes from Web-based material to support a published textbook. In an insertion sort, instead of taking an element out of an array, sliding elements up, and putting the element in its place, code on the Web swaps the element down one position at a time. As shown in Table 1, column 2 shows time required for extraction, sliding up, and reinsertion of successive items in an insertion sort. Column 3 swaps elements down one position at a time, and column 4 implements the swap by separate procedure calls. Altogether, the swap-function version ran 7-8 times slower than a simply-coded insertion sort.

In a merge sort, each recursive step may involve creating subarrays, copying data from an original array to the subarrays, and then merging the subarrays back into the original array. In some cases, the merge may put data into a third array, and then the results may be copied back to the original array. In many cases, the online code may contain separate functions to compare values in sorting—but for integers in Java this requires additional autoboxing and adds calls for a simple ?less than? comparison. (Generality may be wonderful in some circumstances, but it may come at a price.) From timings comparing two versions, the multiple-copy-with-comparator approach ran 3-5 times slower than a simply-coded merge sort.

In a radix sort of integers, using Java and an int array based on decimal digits, one approach on the Web creates an array of 10 linked lists for each int digit. With autoboxing and autounboxing, new objects are created for every decimal digit of every data element. As a conservative estimate, for an initial int array of n elements, where the integers contain 8 decimal digits, memory allocation is required at least 16n times. Further, the Math.pow function is used to compute the power of 10 needed to extract each decimal digit from an integer. (Math.pow is wonderful for fractional exponents, but often inefficient for small positive integer powers that do not need to be recomputed at each step.) Timings indicated that the online code ran 10 times longer than a simply-coded radix sort.

Organize a Course Gala

With such awful code easily available, many traditional coding assignments have limited usefulness. When asked to write a specific sorting algorithm, students can draw upon hundreds (thousands?) of sources. When asked to time and compare algorithms, bad implementations may indicate that some algorithms work relatively well, when a better implementation might highlight shortcomings.

An alternative approach is to utilize online examples as starting points to highlight algorithms, implementation inefficiencies, and timing issues. The basic idea is that one can begin with awful implementations and organize an assignment gala (or festive celebration), in which students consider how to turn misguided code into efficient and effective implementations. Here are several examples I have used in the recent semesters. In each case, initial code is given, possibly limiting students trying to find solutions on the Web—student work must be based on the code given.

- After highlighting common inefficiencies, give students one or two bad implementations and ask them to make improvements. Then students can time versions of the code on various data sets to determine what, if any, speed up has been achieved.

- Give students several implementations of the same sorting algorithm and ask them to compare and contrast. Based on these different versions, students might write a new version that builds on the strengths of the given versions, but avoids the weaknesses.

- Give students a list of potential inefficiencies in code, as well as several implementations of a sorting algorithm. Then ask the students to find which, if any, of those difficulties are present in several implementations

- Start with a bad implementation, and ask the students to time it on various data sets. Then the exercise could identify 3-6 improvements, and students could time the resulting code when each adjustment was made. In addition to the code, students might produce a table similar to Table 1, which shows timings (in milliseconds) of several sorting algorithms using comparably-coded Java methods.

- Give two or three versions of a merge sort (perhaps changing what happens in a merge when two values are equal), and ask students to analyze which version(s) are stable.

- Give students a specific sorting implementation and ask them to adjust it so that one version uses a comparator for determining order and another version compares elements directly (no comparator function parameter). Then ask students to time the two versions to determine the extent to which a comparator adds noticeable overhead.

- After reviewing several sorting algorithms, ask students how testing might be automated, so that a user will know that an implementation is working correctly.

- In the spirit of a gala or festival, ask students to examine implementations of a specified sorting algorithm from the Web. Awards might be given to students who find the most efficient or the least efficient implementations.

All of these activities openly acknowledge that a simple Web search will generate numerous implementations of various algorithms. Rather than ignore these sources, these approaches ask students to focus on specific algorithms, analyze available code, make improvements, and time results. Not only may these activities help students learn about specific algorithms, but students also may gain insights on the quality or lack of quality found on the Internet.

Altogether, each of these activities allow students to examine examples at an early stage, by contrasting inefficient code with well designed code. Further, students gain direct experience with qualities that separate well constructed code from misguided code. In this context, even awful examples can provide insights—although not necessarily about the problems the examples might have been trying to address.

Acknowledgments

Some of the ideas for the exercises mentioned in this column build upon approaches suggested by my colleague, Samuel Rebelsky. Thanks also to Marge Coahran for her suggestions on this column.

Table 1: Timings of Two Implementations of an Insertion Sort (time in milliseconds for comparable code written in Java)

	Insertion Sort		
Array Size	NoSwaps	WithSwaps	SwapFunc
Ascending Data			
10000:	0	1	1
20000:	1	1	3
40000:	0	0	0
80000:	0	0	1
160000:	1	0	0
Random Data			
10000:	30	30	197
20000:	116	155	779
40000:	466	702	3125
80000:	1863	2961	12490
160000:	7418	12048	49801
Descending Data			
10000:	59	72	389
20000:	232	368	1557
40000:	928	1552	6228
80000:	3749	6271	24908
160000:	14886	24749	99670

CHAPTER 66

1000 (binary) thoughts for developing and using examples

This column first appeared in ACM *Inroads*, Volume 4, Number 3, September 2013, pages 40–41[246] Subsequent reflections on this subject appear later in this chapter.[1]

66.1 ORIGINAL COLUMN

I OFTEN CONSIDER developing and using examples within courses as something of an art form. Examples have great potential to add interest, excitement, and sense of fun, as they promote learning. Thus, I expect every educator uses examples to motivate topics, illustrate the use of algorithms and data structures, highlight important points, encourage readers to integrate concepts, etc.

However, when I consider the use of examples, my enthusiasm for challenging or intriguing details sometimes can interfere with long-term goals and learning objectives. This column identifies $1000_{(binary)}$ thoughts and suggestions regarding the development and use of examples.

000. Any example must be valid on its own terms.

When writing my first books, series editor, Gerald Weinberg, observed that students should never have to unlearn an example—even after they learn more about a topic later on. At least in some context, the example should illustrate an appropriate way to proceed. (If an algorithm is never the approach of choice—in any context, why should the students learn it and then discover they need to disregard what they learned? There is too much material in computing to waste time on ill-conceived or misguided approaches.)

To illustrate, I never teach Bubble Sort, because that algorithm is never the algorithm of choice in any context. Instead, I might introduce Insertion Sort, since it has about the

[1] *Regarding the Specification of Number Bases:* The title of this column begins, "1000 (binary)", and the question may arise as to why not write "1000(2)". As I have thought about numeric format over the years, it strikes me as odd and inconsistent that a base ten number is commonly used to identify the base for another number. As an analogy, if I were asked what language I used in the writing of this column, I would not reply in French (anglais), German (Englisch), or Russian(по-английски). Rather, I would state that I wrote in English for this column. So why is the number eight in base two often written as 1000(2)?

same level of coding difficulty, but Insertion Sort also may be a good choice if data are already almost ordered.

001. The scope of example should be consistent with educational objectives at hand.
Sometimes after reading student solutions to a homework assignment, I decide to write up my own solution to illustrate a clean and efficient approach. Sometimes, I get carried away, making up four or more solutions, using different data structures, different problem-solving perspectives, different algorithms, and different language features. Sometimes, a brief discussion may provide students insights about different ways to approach problems. However, a 3-hour, in-class example is likely not appropriate to support a few minor points.

010. The level of detail for an example should not obscure the main points to be learned.
Students likely should work through some details to solidify how an algorithm or data structure works. However, I have heard a number of talks through the years, in which (esoteric) details obscured the main points of a presentation, and the audience lost any sense of how the details fit into a broader framework.

011. An example should be connected with context—not just "here is a nifty topic to talk about".
When presenting an example (or most anything else), I consider how this example fits—what content(s) do I want the audience to learn, and how does this example contribute to my goals and objectives? Often, I find it useful to state this connection explicitly at the beginning of an example, or at the end, or (most likely) both.

100 . Although one example may be good, this does not imply that ten examples are better.
As an extreme example, some mathematics textbooks seem to be structured with one formula dominating a section, followed by 80 or more exercises. The idea seems to be that a student can plug numerous numbers into the given formula to get an answer. Although this structure might provide practice in arithmetic, I may wonder how such massive drill and practice actually contribute to understanding and insight.

101. Examples should not be used to define terms, present specifications, or state proofs.
In defining a binary search tree (BST), for example, a presentation might begin with four examples, but the presentation also should formally identify the defining properties of BSTs. Similarly, specifications need to explicitly state assumptions rather than rely upon examples. Examples of a binary search within an array may show an array to be sorted, but students should be told explicitly that an ordered array is a pre-condition rather than inferring this from examples. Also in stating a homework or test question, there may be a temptation to use an example to clarify specifications or suggest a general principle.

Overall, over-reliance on examples can have at least two pitfalls:

- When reading an example within a specification or an assignment, the student may have no way to know what assumptions are being made. Does the example illustrate a special case, or are assumptions made within the specifications?
- Sloppiness in stating specifications or proofs may undermine student perceptions of what constitutes a formal specification or a proof. Examples should not reinforce the common student belief that they can prove a general result by giving an example,

Certainly it is fine for an example to illustrate a general approach or technique, but the example should be careful to distinguish between a comment about a specific case and a broad assertion of a general result.

110. Instead of one extensive example, a sequence of focused examples can highlight alternative problem solving approaches, algorithms, assumptions, data structures, etc.

For example, when I introduce the concept of stacks in CS2, a student exercise may involve a stack of strings. Questions can arise:

- Should the stack maintain a pointer to the original string?
- Should the stack be a two-dimensional array (an array of strings), with a string-copy for both push and pop?
- Should the stack be a two-dimensional array, with a string copy for push and a reference to the string on the stack on return?
- Should a push operation create a new copy of the original string that can then be returned with pop?

Each example in a coordinated series can focus attention on a single approach, and the full sequence can highlight advantages and disadvantages of alternative design options.

111. Although instructors often develop examples, students also can be challenged to create their own examples.

- An assignment might ask students to develop their own examples—for themselves, as a challenge to a partner or small group, or for the class. For example, working in pairs, each student might bring a data set or example to class for the other student in the pair to tackle.
- An instructor might challenge students to find examples that have certain properties (e.g., find a sequence of data, which when inserted into a tree in the specified order, produce a given tree).
- An instructor may challenge students: either develop an example of a given circumstance, or show such a circumstance cannot exist. (Of course, sometimes this type of problem can be overwhelming—give an example of a problem in Class NP that is not in Class P, or prove that no such problem exists).

Overall, examples provide wonderful opportunities to promote learning and insight, but they also can be overused or mis-used. Care in selecting and developing examples—particularly considering how examples relate to goals and objectives—can help avoid pitfalls.

Acknowledgment

Many thanks to Marge Coahran for her insights and feedback during the development of this column!

66.2 SUBSEQUENT REFLECTIONS

The original column reviews the development, role, and use of examples within a course setting. But what pedagogy might be employed?

When using examples, I try to connect the example with student activities (either upcoming or just completed), so students will be engaged in the example and work through many of its details. Following the format of the original column, here are $100_{(binary)}$ techniques I use periodically, selected from the many approaches possible.

00: Class presentation/discussion

An audience may find long or complicated examples hard to watch, and listeners' minds may tune out when presented many lines of details. Instead of lecturing or discussing extensive examples, a full account might be circulated to a class (either on paper or electronically), and class time can be devoted to illustrative examples at a high or moderate level of detail. For example, in balancing a red-black tree or maintaining a heap for a priority queue, an example can follow insertions and deletions to clarify the underlying principles. Follow-up homework can then ask students to work through the details for a specific data set or for developing an actual program. *01: Real-time interactive development during class*

After discussing a variety of concepts and techniques in class, I may state a new problem that incorporates many of those ideas. Once stated, my plan is to draw students into a discussion of how to approach the problem, ask questions about this approach versus that one, and interactively develop a working program in real time—with constant involvement of the students. During the class, students observe and contribute as I type and the program evolves in front of them during class. Pragmatically, this approach can take substantial time, but when used sparingly, students often comment that they appreciate observing how I think about a problem and what techniques might be used in program development.

10: Integration with lab exercises

When an example appears within a reading, I worry that students may skim over the narrative and move on quickly. However, when a subsequent lab exercise explicitly references the example, asks students to modify the example, and considers alternative approaches, students must engage the example directly. Even if a lab requires a modest change (e.g., sort in descending order rather than ascending order or search an array of strings rather than an array of numbers), then students must think about what part of an algorithm performs what task and why.

11: Wrap-up after a class assignment or exercise

Students sometimes believe there is only way to solve a problem. Thus, after they have submitted programs to solve a reasonably open problem, I may present 4-8 different types of solutions. (I tend to go overboard, but I have been know to solve the same problem with 1-dimensional arrays, 2-dimensional arrays, dynamic memory, stream input, line-by-line input, various functional decompositions, etc.) Since the students have already worked on the underlying problem, they can compare their approach to the others and consider both advantages and disadvantages of each sample solution.

Although these in-class techniques suggest only a few of the many ways examples might be incorporated into a class, each highlights some ways to foster student engagement. As with so many elements of pedagogy, if an instructor drones on through a detailed example, students may take the opportunity to catch up with sleep. However, if an example highlights concepts and techniques that have clear relevance to student work, then students can appreciate connections and payoffs between the example and their own work.

CHAPTER 67

How to prepare students for lifelong learning

This column first appeared in ACM *Inroads*, Volume 3, Number 2, June 2012, pages 10–11[244]

67.1 ORIGINAL COLUMN

WITH THE RAPID EVOLUTION of science and technology, I have heard much talk about the need for students to be prepared to learn about new directions, innovations, techniques, and skills throughout their lifetimes. For example, at a high-level, the European Commission defines lifelong learning as "all learning activity undertaken throughout life, with the aim of improving knowledge, skills and competence, within a personal, civic, social and/or employment-related perspective." [65]

The need for lifelong learning within computing is particularly important, because technology evolves at a rapid pace. For example, in 2002, William Wulf, former President of the National Academy of Engineering, reported upon a 2000 workshop that determined that "half-life of engineering knowledge" as between 2.5 years and 7.5 years. Wulf concluded, "half of what we are teaching our students in some fields (computer science, by the way, was the field of 2.5 years) is obsolete by the time they [students] graduate." [290, p. 6] Altogether, in order to stay current and remain productive professionals, people within the computing field need to undertake a lifetime of learning.

However, even with high-level comments about the need for lifelong learning, I have heard few discussions about what learning skills students might need to be successful in this fast-paced field; and I have heard even fewer discussions about how those skills might fit within an undergraduate curriculum. I hope this column stimulates readers to consider how the need for lifelong learning might translate into skills and curricula.

What Skills do Students Need for Lifelong Learning?

The ACM/IEEE Computing Curricula [7] and similar curricular recommendations present an extensive collection of knowledge areas that are important for professional activities within the field of computing. To supplement such specifics, computing faculty should consider what skills students need for lifelong learning. To begin this discussion, here is my initial list of essential lifelong-learning skills.

- ability to read technical materials: articles, manuals, documentation
- knowledge of where to search literature, together with the ability to determine the validity, relevance, and biases of material found
- strong analytical skills, so students can examine claims and determine advantages and disadvantages of new approaches, techniques, hardware
- knowledge of core subjects, so students can relate new ideas to underlying concepts and frameworks
- experience in learning new approaches (e.g., languages, paradigms)
- perspectives on social and ethical dimensions of computing, and experience in applying those perspectives, so students can assess the impact of what they develop—and even determine whether a product should be built
- perspectives on the importance of developing and maintaining contacts within the professional computing community, so students appreciate the importance of the constant upgrading of technical skills

By creating this list, I am not suggesting that other skills (e.g., knowledge of algorithms, ability to debug, problem-solving and coding techniques) are unimportant. However, I believe students must develop the above list of skills to be successful lifelong learners.

Where in a Curriculum do Students get the Needed Skills?

If computing faculty agree that the above skills are essential, it is natural to ask where students will gain proficiency with these skills. Of course, all skills need not be covered in every course, but students should practice each skill somewhere. Further, many of these skills require on-going practice—it likely is not adequate for subjects to be introduced in one course without follow-up in later courses. For example, if students should be able to read articles, manuals, and documentation, then students should be reading articles and relying upon manuals in one or more courses. Careful spoon-feeding may be appropriate at the introductory level, when students are just getting started (we do not want to scare beginners off by making them sink or swim). However, if all courses broke all content into bite-sized pieces, students would never have to dig into a new topic on their own, discover new advances in technology from primary sources, or explore new elements of a programming language or environment.

More generally, faculty might consider how much emphasis to place on specific details versus life-long learning skills. Technical details may be quite important for a specific short-term job, but the need for life-long learning skills is essential for a career. If students lack the ability to grow and develop, computing education would seem to produce a strong credential for unemployment in no more than five to ten years.

Turning to assessment, it would seem that some assessment should include opportunities for students to demonstrate these needed lifelong-learning skills. For example, short-answer and multiple choice likely are not very helpful in assessing student abilities in many of these areas (e.g., ability to read, analyze, synthesize, and apply new material; skill in performing literature searches and critically analyzing the results; etc.). Often, lifelong learning arises in the context of open-ended problem solving, so it seems likely that some assessment should address situations in which narrow answers may not be adequate.

Four Examples

In reviewing curricula, faculty will need to identify what types of lifelong learning skills might be incorporated where. A few examples can illustrate some possible approaches.

1. Rather than provide separate handouts or utilize lectures for all language constructs, labs and assignments can refer to standard documentation, such as the Java API, the Linux help facility (man pages), C header files, the current Scheme Revised Report, etc. Introductory students may need some initial help using these materials, but students at all levels can work with primary manuals without undue difficulty.

2. Students in an upper-level course (e.g., operating systems, languages, algorithms, theory) might read a current article on a topic that complements course themes. Working individually or in groups, the students could prepare a report on their reading or present their findings to the class. For example, when I teach the Theory of Computation, I ask groups of 2 or 3 students to investigate an NP complete problem from outside sources. The groups then present to the class the nature of the problem, why the problem is in Class NP, and what argument demonstrates that all other NP problems can be reduced to it.

3. If a course cannot fully cover a textbook (e.g., an algorithms course using Cormen et al), students could pick chapters that would not otherwise be covered, and then outline the material in a lecture to the rest of the class. Although these reports take class time and might reduce slightly the coverage of other topics, this exercise forces students to engage and synthesize new material at a deeper level than is typical in simply working problems or writing programs.

4. Typical of many departments, my department organizes a weekly series of talks, with presenters including outside speakers, faculty, and students. In some cases, students discuss results from their student-faculty research projects. In other instances, students explore a topic not in the regular curriculum, and then present their findings and conclusions in the weekly departmental series. (Although a presentation in the weekly series is not a requirement in the computing major, such a talk does help a student fulfill a requirement for graduation with honors.)

Overall, courses and programs can provide a foundation for students to understand current, state-of-the-art technology; but we cannot anticipate what technology and skills students will need five, ten, or twenty years from now. To work successfully within a technological environment, today's students must prepare for a lifetime of learning, and today's curricula should provide students with the needed lifelong-learning skills. With this in mind, I encourage computing faculty to review how these skills fit within curricula. Since many current details will be obsolete in just a few years, it is possible that some courses might best prepare students for the future by replacing some specifics for today with practice in learning new topics for the future.

Acknowledgment

The author gratefully acknowledges several suggestions from Marge Coahran regarding this column. This article has benefitted greatly from her feedback.

CHAPTER 68

How to challenge students

This column first appeared in ACM *Inroads*, Volume 2, Number 3, September 2011, pages 10–11[236] Subsequent reflections on this subject appear later in this chapter.

68.1 ORIGINAL COLUMN

SEVERAL MONTHS AGO at a multi-disciplinary conference, I had an extensive conversation with several computing graduate students about challenging undergraduates in courses. Apparently this was a new idea to this group of teaching assistants, and we spent some time discussing benefits of challenging students, successful strategies, and misguided approaches. This column expands that discussion.

Why Bother to Challenge Students?

Over the years, my students consistently have given me positive end-of-course evaluations if the course moves along reasonably, but evaluations consistently are higher if my students feel they have been challenged.

Of course, at an extreme, students can become discouraged if they rarely make progress on many assignments, or if they perceive that a high work load does little to promote meaningful learning. Rather, I want assignments to stretch students' abilities and understanding with problems that generally are within the realm of possibility.

The rest of this column identifies several successful strategies for challenging students to stimulate brainstorming and also discusses a few cautions.

Some Successful Strategies

Challenging students can provide motivation and promote learning in several ways.

Facilitate Learning at all Levels of Bloom's Taxonomy

Bloom's Taxonomy identifies six levels of learning from basic "knowledge of specifics" through use of a concept to synthesis and evaluation [25, p. 201-207]. Recently, I read two introductory books—both involving programming. One emphasized extensive drill and practice, covering only the lowest levels of Bloom's Taxonomy. The other outlined syntax briefly and then emphasized the use of ideas in various applications. Solutions to problems in the second book required students to learn basic concepts and skills but also challenged students to increase their knowledge and understanding.

Also, collaborative exercises and team projects often require students to explain their thinking and review ideas with a partner. A jointly-developed solution not only uses ideas but encourages students to state approaches in their own words—challenging students to work through middle- or upper-levels of Bloom's Taxonomy.

Encourage Needed Practice

Pragmatically, different students often need varying amounts of practice in order to master specific skills, concepts, and techniques. One way to address this diversity is to require a few solutions be turned in, but suggest additional problems for further practice. In my courses, I do not collect these "suggested problems", but I advertise that 1/3 to 1/2 of each test will be drawn from this collection. Thus, although students need not work out all details, they have incentive to think about solutions. Also, since several dozen "suggested problems" are identified for several weeks leading to a test, students covering all the problems almost certainly will have sharpened their problem-solving skills. Furthermore, conscientious students are rewarded by knowing much of a test—increasing confidence and reducing stress.

Stretch Student's Abilities, Understandings, and Skills

A challenging problem does not always have a long, involved solution. Problem solving is a multi-faceted endeavor that proceeds on many levels. Often, thoughtfulness and insight yield better solutions than hasty initial guesses. Thus, I tell my students, "if your are developing a program that looks long and complex, stop and think more deeply. Much [most?] of the time, you have missed something that yields a better solution." For example, my first column in 1997 presented a problem to simulate racquetball or volleyball. Although the solution only requires conditional expressions, assignments, and iteration or recursion (recursion is simpler), the solution requires thought and insight, and a complete solution may be only a page or two in length.[204]

However, even though the best solutions often come from insight and analysis, sometimes one must dig into the pieces. When I was in graduate school, a mathematics faculty member stated that a problem was "trivial" if there was an obvious way to proceed and that approach worked—possibly after writing out 20-30 pages of details. Similarly software developers must be able to work through all of the ramifications of an idea to obtain a full solution. It certainly is not necessary or even advised for every problem in a course to have a long and involved solution, but students should occasionally have the experience of working through all of the details—even if the basic problem is indeed "trivial".

Highlight Social and Ethical Issues

Students regularly wonder how computing might benefit people and how course work relates to real applications. Such questions provide opportunities to discuss social and ethical issues. What ideas underlie automated telephone-based systems for airline or train reservations? What data might be collected in a database, who has access to that data, what procedures ensure personal data remains private, etc.? What computing-related issues are highlighted in newspapers and magazines in the past week? Assignments based on such questions might utilize paper assignments, short news reports presented by students, group discussions in class, etc. Altogether coverage of social and ethical issues can provide a natural, on-going theme in courses, challenge student perspectives, and connect assignments with the broader community [57].

Help Develop Lifelong Research Skills

Although students often think they are proficient in finding information by looking online, students typically are poor at analyzing correctness and bias in sources. Lecturing on the evaluation of sources typically has little impact. Rather students can be asked to find answers to apparently simple questions that have subtle or surprising answers. Students often find such research exercises motivating,while learning about weaknesses in their traditional research strategies [279].

Some Cautions

Although challenging students can yield benefits, care is needed to be certain student efforts will motivate students and encourage learning rather than have unintended negative consequences.

For example, over the years, I have heard many talks promoting open-ended assignments to provide motivation and encourage creativity. Students may feel ownership of a project, if they can shape various details and directions within an assignment. However, open-ended assignments also are subject to at least two drawbacks. First, open-ended assignments have the potential to consume vast amounts of student time. Students may not know when they are done; and they may continue to tinker on a project (e.g., make countless adjustments to an output display), when that work may contribute little to promote learning of the topic at hand. Second, much practical software development must center on clients and their specifications and needs. If an open-ended project depends only upon a student-developer's feelings without reference to others, the student may not encounter the service-orientation that often comes with working with an actual client.

As another example, time is limited for in-class activities, so extensive analysis likely must be assigned as homework. In addition to projects, take-home tests provide a natural environment for multi-step problems, for questions that integrate ideas, and for challenging assignments that require in-depth problem solving. However, take-home tests also carry possible disadvantages. Students may feel compelled to continually refine an answer, and work may expand to fill all available time—undermining work in other courses and other activities of life. Setting page limits can address such problems. For example, on a take-home test with 5 problems, I may limit an answer to 1.5 pages, and the total work must not exceed 6 pages (with specified type fonts, margins, etc.). In this context, students may refine their answers, but they cannot write dozens of pages when a relatively short answer will suffice. In contrast, time limits can be problematic, since actual time spent can be difficult to track, and students may feel pressure to misrepresent their times.

Altogether, challenging students can yield many advantages that promote learning, although an instructor must think through details to alleviate any potential negative consequences.

Acknowledgments

This column has benefitted greatly from suggestions from Marge Coahran. Her comments were particularly helpful in improving the structure of this presentation, and the author extends many thanks to Marge for this feedback.

68.2 SUBSEQUENT REFLECTIONS

The original column focused upon an instructor's role in challenging students. How can instructors help students expand their horizons, hone their skills, and move to new levels of understanding and ability? How can faculty help prepare students for future endeavors within an evolving world where new subjects and perspectives constantly arise?

The strategies presented in the original column have worked well in some environments, but each school or classroom is different. A difficulty for an instructor, therefore, is to identify what motivates specific students in a particular class, in an effort to get students to accept the challenges presented. The possibilities are vast, so the following list is intended to help identify just some of the options. In each case, I have encountered at least one course and school where the approach has worked well, but no approach can be expected to work everywhere.

- At some schools, students have substantial trust in the faculty to highlight those skills and techniques that are needed to obtain jobs. At these places, students understand that the challenges will be good for them, and trust the instructor's judgment. For one program, I found both students and faculty wanted to make each potential subject and challenge a graduation requirement, so no one graduate would have missed any important insight.
- At other schools, students may be motivated by intellectual curiosity and the excitement of problem solving. In these settings, students may have relatively little interest in routine exercises, but rather want activities that will stretch their backgrounds and interests.
- Some students have an over-riding interest in helping others. If an exercise or activity can be framed within a context where solutions will help a group on the campus or in the community, then students may eagerly embrace the challenges at hand.
- Still other students may be motivated by grades.
    - ◇ If tackling a challenge may yield points toward a course grade (either as required work or as extra credit), then students are interested.
    - ◇ If the instructor advertises that 35% or 50% or some other fraction of the problems on a test will come from labs or assignments that have not been turned in, this possibility may provide motivation for some.
    - ◇ On the other hand, activities which will not be graded may generate little student interest. In these settings, an instructor may need to adjust grading scales to include a remarkably wide range of activities.
- At one school I know, students were motivated only by grades on required work; for many, extra credit seemed irrelevant, and they simply would not attempt any lab or assignment above the minimum required. In this setting, an instructor might need to require every activity. Handling the logistics for such work can be daunting for instructors (perhaps selected problems, chosen at random, could be graded carefully), but the students may ignore any material not officially required.

Altogether, challenging students seems an important component of courses, but means to motivate students to take up these challenges may depend greatly upon local circumstances and cultures. Further, as the students in a course change from one semester to the next, motivations also may change; and finding mechanisms to motivate students to tackle challenging activities may be an ongoing puzzle for instructors.

CHAPTER 69

Wellness and the classroom

This column first appeared in ACM *Inroads*, Volume 1, Number 1, December 2010, pages 27–30[235] Subsequent reflections on this subject appear later in this chapter.

69.1 ORIGINAL COLUMN

WHEN CONSIDERING TEACHING AND LEARNING, we naturally focus on academics and the classroom. We ask questions, such as "what content will we teach?", "what pedagogy will we use?", "will we utilize tests and/or other approaches to assessment?", "what textbooks are available?", etc. However, in my experience, we rarely consider how the classroom experience connects with the rest of life. But academic success often depends directly on non-academic factors as well. Here are two motivating examples:

- Years ago, I was teaching calculus I. About an hour before the final exam, a student came to see me with a few questions. The student had a "B" average going into the final and had stayed up all night to nail down every last detail in the hopes of ending the semester with an "A". The first question the student asked was "what is the antiderivative of x?" (In discrete mathematics, the equivalent would be to name all subsets of $\{a, b\}$.) As this question suggests, the student was so mentally and physically exhausted as to be unable to think. As a result, the student's exam was a disaster.
- More recently, I talked with a CS faculty member at another institution. In trying to provide the best courses and projects for students, the faculty member consistently taught overloads (often 2-3 extra courses per semester), each course involving substantial preparation. Eventually, after not feeling well, the faculty member consulted a physician, who determined either the faculty member must go home and to bed for 2 weeks, or the physician should check the faculty member into a hospital for extensive treatment.

Such examples remind us that academic success for both students and faculty depends upon mental and physical well being. Our bodies simply rebel after too much abuse. This column begins a discussion of how issues of wellness might connect to the classroom.

Introduction: Opportunities and limitations

As teachers, we want our students to learn, develop insights, master skills, etc.; and we want to create an environment that promotes intellectual development.

However, from the start, we also must recognize our limitations. Students typically take several courses; they have jobs, extra curricular activities, and family responsibilities; they have varying interests and priorities. Ultimately, the students must learn themselves; we can try to facilitate, encourage, and guide; but we can shape only a small part of each student's environment.

When considering wellness and the classroom, therefore, teachers cannot expect to shape all parts of the learning environment; many elements are beyond our control. Our courses represent only a small part of a student's activities, but we can examine that our part promotes healthy approaches that encourage constructive behaviors.

Components of Wellness

The American Heritage Dictionary defines "wellness" as "The condition of good physical and mental health, especially when maintained by proper diet, exercise, and habits." [147] Expanding upon this definition, wellness includes several key components, including:

- maintenance of a healthy diet,
- regular exercise,
- adequate sleep,
- management of stress, with built-in times for stress reduction (e.g., physical exertion, extra-curricular activities, fun and recreation),
- good time management,
- development of constructive personal and social relationships,
- control of stimulants (e.g., caffeine, sugar) and responsible use of drugs (e.g., medications, alcohol).

A full list of healthy traits could easily extend for pages, but these items highlight some of the range of qualities that both students and faculty need to build into their environments.

Course Policies and Wellness

Although many components of wellness fall outside the influence of the classroom, some course policies and practices can make a substantial difference. For students, three vital and inter-related areas involve time management, stress, and sleep. Students often need help in learning to manage their time, but time management requires planning—often over a few weeks or even the entire semester. Pragmatically, students need to know what to expect for each course. Here are some suggestions for course planning:

- At the start of a course, write and distribute a tentative class-by-class schedule. This allows you as instructor to gauge how topics might be organized to maintain a reasonably constant pace through a semester, and it tells students when to expect readings, tests, projects, etc.
- To the extent possible, develop a regular pattern for assignments to be due. Students can plan, if they know when assignments will be distributed and when they will be due. For example, if CS1 meets 4 times per week, I have a lab or assignment due about every third class day. When teaching the Theory of Computation, assignments might be due every Monday or every Monday and Friday.

- Announce dates of tests and projects at the start of the semester. For example, include the dates for tests, exams, and major projects in the course syllabus. Students often want substantial time to prepare, and early notification allows students to juggle this work with other responsibilities.

Overall, even well-organized students experience high stress when they encounter surprises, such as unexpected assignments with short due dates. Maintaining a regular schedule of course work can encourage the development of good time-management skills.

Similarly, course policies can have a direct impact on both stress and sleep. The following suggestions may suggest possible approaches.

- If assignments are due in the first part of the morning (e.g., 8:00 am or 9:00 am), then students may be encouraged to stay up very late to complete the work. All nighters might be the norm. However, work due at 11:00 pm or midnight might strongly encourage students to get to bed, Logistically, a midnight deadline might be difficult to enforce for paper-based assignments (someone must be willing to collect the materials at midnight). For electronic submissions, however, logistics may be straight forward, either via time stamps or by setting software to block submissions after a specified deadline.

- When taking take-home tests, students may feel pressure to work constantly to cover all aspects of every problem. Adding page limits for each problem and for the entire test, however, can help enforce manageable bounds on student work. For example, for a 6-problem take-home exam, I might specify no answer can exceed 1 page in length (typed with minimum margins and type font indicated), and an entire student answer can take no more than 5 pages. Such limits inhibit students from spending exorbitant amounts of time on the test (and also limit the time required for grading).

- Alternatively some people suggest placing time limits on take-home assignments as a mechanism to create bounds for students on out-of-class work. Although this system might work, this approach also can reward students who decide not to follow the rules. Dishonest students may have more time than honest ones, and time limits may tempt stressed students. In addition, time limits seem unenforceable (how can I really know how long students work on a problem?), so I use page limits rather than time limits in take-home exercises.

- When using virtual community software (e.g., a Wiki or Blackboard) for a course, advertise that the instructor and any assistants will not respond to questions during specified night-time hours. If students can expect answers at 2:00 am, they will have an incentive to skip the sleep that they desperately need. (More about this later, from the faculty perspective.)

In general, class policies regarding homework, deadlines, scope of work, timing of feedback, and the like can encourage time management, control of stress, and sleep schedules; or class policies can compound problems that already may be difficult for students.

Of course, in practice, individual instructor policies can only impact the work of students in a specific course. Students might choose to spend more time on other courses, employment, recreation, or other activities. In any case, carefully-crafted course policies allow an instructor to shape activities in one realm and possibly set a tone for other arenas of student life.

Student Advising

Outside the classroom, faculty often advise students. Typically, the faculty role focuses on academic issues, but issues of wellness also arise. For example,

- in helping a student plan a semester schedule, discussions might include work load and the balance among types of courses. Of course, details will depend upon student circumstances, institutional requirements, and times courses are offered. However, stress likely will be relatively high if all courses have a similar style (e.g., all lab-based courses, all courses with high levels of reading or essays, etc.). Also, in planning a schedule, an advisor can talk to students about building in time for extra-curricular activities, exercise, work, etc.

- if students become overloaded, an advisor can help students look realistically at options. Without constraints, students may be able to do many activities. However, life always involves constraints, and students must make choices. Advisors can help students review a daily or weekly schedule to determine how to organize study, work, and other responsibilities. Faculty can take this opportunity to talk to students about issues of wellness as well.

- Either as advisors or when talking to students about our own courses, an instructor can help students assess realistic possibilities. If a student has been ill or has gotten far behind, considerations of overall physical or mental health might dictate advising a student to drop a course. Under different circumstances, a student might be able to do fine work in a course, but within a current context the student may need to drop something.

- When talking to students about performance within a course, a common cycle is for students to stay up late to do required work. By staying up late, students then get relatively little sleep, and their efficiency can become quite low. With less ability to work effectively, students may stay up even later, get even less sleep, and become even less effective. An advisor can help students recognize this problematic cycle that effects both academic performance and wellness.

- When students are having extensive troubles with their academic work or other parts of their lives, faculty can play a role in referring students to resources throughout the campus. Faculty are rarely trained as counselors, but faculty often know what assistance might be available on campus, and can help connect students with counselors or other specialists. In addition, students may be more open to counseling if this is mentioned by a trusted faculty member rather than by other members of the community.

Students typically have direct, regular contact with faculty and faculty advisors, and this may allow faculty to observe problems related to academics or wellness at a relatively early phase. Also, many students tend to listen to their advisors. Thus, an alert advisor can be particularly helpful in referring students for specialized counseling or medical attention to an Office of Academic Advising, a Health Services Department, or Counseling Center.

Faculty as Role Models

Students, colleagues, staff, and administrators observe faculty members all the time. What activities are we involved in, what are our priorities, how do we conduct our lives and professional activities, how do we interact with others? Further, students and others see us as being successful professionals. In effect, faculty members are modeling the activities of an academic life. This function as role models has several consequences:

- If faculty pay attention to their mental and health needs, then we are encouraging students to do likewise. However, if faculty skip exercise, skip meals, eat convenience food regularly because it's quick, stay in their offices at all hours of the day and night, etc., then faculty are sending a clear message that various elements of wellness are not important.
- The lifestyles faculty live can send either positive or negative messages, and students notice the message. At one institution I know, a junior CS faculty member was highly regarded as extremely effective and productive—and the number of CS majors dropped dramatically. When students were asked why they declared majors in other disciplines, the main theme was, "I don't want to live like that."
- Faculty actions show priorities and set boundaries. If faculty respond to email at 2:00 am, faculty may be providing quick feedback, but they also are demonstrating that it is acceptable to skip sleep in favor of course work. Similarly, if faculty eat lunch and dinner at their office desks, they may be sending the message that junk food is fine or a healthy diet is not important.

Over the years, I have been guilty of not setting the best example regarding work hours and overcommitment to activities. However, I am learning. Here are a few simple ideas to consider.

- Leave the office at a reasonable hour each day. Even if you do some work at home, your students will not observe you working at all hours of the day and night, and you might even be able to find some balance.
- Do not answer the telephone or respond to email outside of reasonable hours. Early in my career, I received a telephone call from a student at 1:00 am, asking about a homework assignment. My response was that office hours were posted for the next day, and both the student and I needed to get our sleep.
- If you do need to be in the office or lab outside normal hours, be clear that you are working on your own projects. A pet peeve of mine is for a student to appear at my office door at 10:00 pm and ask, "are you busy?" To be an effective teacher, you want to be helpful and available. However, as a person and role model, you also need to establish parameters for health and well being.

Being a role model is both an opportunity and a responsibility. Faculty can demonstrate what is important, what being a professional means, and how to effectively collaborate. However, there also is the potential to encourage counter-productive behaviors—if faculty follow unhealthy patterns (e.g., skip meals, regularly work through the night, wait until the last minute to start projects), students may be encouraged to emulate the same problematic behaviors.

Faculty Wellness

During my undergraduate days, I remember hearing comments about "the quiet, contemplative life of academia." Since that time, I have talked to many faculty; none has described their lives as either quiet or contemplative. Rather, I hear a litany of problems related to time, stress, institutional pressures, and the like. Periodically I see evidence of burnout, exhaustion, ill health, mental strain, etc.

Some of this difficulty may be institutional. A school identifies expectations for tenure, promotion, or merit salary increases. If faculty want to succeed at an institution, they must perform in various ways.

Further, most faculty I know want to do the very best they can. They want their students to succeed, they want to be helpful, they want to be productive scholars, and they want to be fine citizens to their school and the broader community. However, in my experience, it is rather rare for faculty to consider the impact of too many activities on themselves. Here are a few examples.

- A faculty member may include regular exercise on a weekly schedule. However, whenever a student needs to meet, a colleague has a question, or a committee must meet, the faculty member may cancel the exercise. It's easy to take an hour out of exercise in order to accommodate an immediate need.
- In discussing local needs, it is common for a department or school to form a new committee. The discussions are expected to have limited scope, and there will be some benefits.
- More generally, a faculty member may see some advantage in a new assignment, lab, exercise, program, etc., and the incremental time for the work seems limited.

Taken individually, each conference, meeting, or handout may have some value. However, faculty can become caught up in the same work-late/reduce-sleep/reduce-efficiency cycle already described for students. At a more basic level, faculty must identify priorities—not only for professional work, but for other components of life as well.

With issues of tenure and promotion, administrators and senior faculty may need to take a lead, but schools and departments also should consider the long-term impact of stresses and work loads on wellness. In the short term, individuals can push to meet demands, but in the long term such overcommitment can lead to both mental and physical health problems.

Conclusions

Each faculty member can set a tone within each class to encourage wellness. Since both students and faculty members work within a larger environment, class policies and practices cannot guarantee mental and physical health. However, some course practices can encourage wellness, while others can send a different message. In setting policies, therefore, faculty members might consider what behaviors they are encouraging.

Of course, the comments here only begin to address the topic of wellness and the classroom. As a next step, I encourage readers to discuss connections between academics and wellness with colleagues and with administrators and staff. As a next step, Grinnell College Wellness Coordinator, Jen Jacobsen recommends *The Smart Student's Guide for Healthy Living* by M. J. Smith and Fred Smith [175].

Acknowledgment

This column has evolved after a discussion on this topic by the Science Teaching and Learning Group at Grinnell College, led by Jerod Weinman and Ben DeRidder. Special thanks to Jerod and to Jen Jacobsen, Grinnell College Wellness Coordinator, for their insights and suggestions.

69.2 SUBSEQUENT REFLECTIONS

A course instructor's influence is inherently limited, but opportunities still can arise to support wellness for the whole person. The original column included numerous suggestions, and the following thoughts complement those suggestions within three categories:

- Wellness for the whole person
- The instructor as a role model
- Class scheduling

Promoting Wellness for the Whole Person

In my experience, classroom discussions and activities commonly focus only on work for the course. With the many learning objectives and subject-based activities that are part of each course, both the instructor(s) and the students may ignore all other aspects of life. Although such focused attention may highlight course goals, it also may send unintended messages regarding health and wellness.

For mental and physical health, both faculty and students need to have multiple interests, activities, and outlets. Within the classroom, an instructor might regularly highlight the value of whole-life activities by mentioning events, opportunities, and healthy practices.

- The first 1-3 minutes of a class might mention an upcoming presentation, workshop, or health tip. Although a few announcements take little time, they can set a tone and raise visibility regarding appropriate activities outside the course.
- During announcements, students in class are invited to announce their upcoming performances, etc.
- In many classes, I offer extra credit (but only a point or two), if students attend a talk, presentation, or other on-campus activity. To receive credit, a student must reflect on the event and email me a 4-6 sentence summary or response within a week.
    - ◇ Talks within a departmental lecturer series certainly are included.
    - ◇ Events on campus (e.g., theater, dance, music, etc.) may be eligible for extra credit, if a student in the class is an active participant/leader.
- Beyond forthcoming events, tips might be given from time to time regarding the need to eat properly, get enough sleep, etc.
- Suggestions regarding time management can be particularly useful when assignments or projects will be coming due in a few days.

Overall, I seek to encourage students in their outside activities, make visible aspects of life beyond the classroom, and remind students of tips and practices for healthy living. Such comments take little time (e.g., 1-3 minutes per class), but they can raise awareness.

Additional Thoughts on the Instructor as a Role Model

The original article noted that faculty often serve as role models to students and others. In practice, this position suggests that faculty not only pay attention to their own needs, but they also should be [somewhat] visible about their responses to their responsibilities.

For example, the original column suggests the need for faculty to take care of themselves, their families, and their other commitments.

Take walks or play racquetball or swim or ..., but do something!

Each faculty member should consider how to maintain some type of exercise regime.

- Early in my career, I planned to swim each day at noon. However, since I was responsible only to myself, it was easy to skip this swim when a student wanted to talk, a colleague requested a conference or a committee needed to meet. As a result, I skipped swimming 1-2 days a week, and then more. Eventually, I was in sub-optimal physical shape, swimming was no longer fun, and I stopped.
- For the past decade or more, I have been more successful in scheduling racquetball, when I must commit to a partner. Now, when a student or colleague wanted to meet, I cannot simply give up the racquetball session—at the very least, I must to contact my partner. As a result, exercise has became much more common.
- I know one faculty member who walks regularly, as part of office hours. Any interested person can sign up, and they talk during the hour scheduled for the walk.

Family and Community Responsibilities

Naturally, faculty also must plan time for family and other responsibilities. With the stresses of academia, immediate demands of the job may seem pressing, but other responsibilities are pressing as well. Sometimes working with a partner or friend can be helpful, but such collaborations require on-going communication and action.

Visiblity

Taking time for health needs, wellness activities, and family responsibilities is vital. As a role model it is equally important to be visible about these activities. For example, a faculty member might post on a weekly schedule such commitments as "racquetball" or "family outing". Such notations highlight the need for everyone to pay attention to health and wellness. As role models, faculty openness for these activities is important for everyone!

Additional Thoughts regarding Scheduling Help within a Course

The original column described the need for courses to clearly identify a tentative day-by-day schedule that presents readings, in-class activities, assignments, projects, and tests through the semester. Such a schedule provides students with clear tasks and milestones, so they manage their time to include both personal and outside responsibilities and commitments. Some sample schedules, showing forthcoming dates, deadlines, and day-by-day activities, are shown in Chapter 33.

In addition, some students may appreciate a narrowly-focused list of specific assignments and tests for a semester. While some students find the day-by-day schedule provides a helpful context, others prefer a more detailed list of specific dates and deadlines.

Chapters 32 and 33 describe the value of including time within a daily class schedule to "pause for breath" and "practice time" for students needing to give demonstrations or class presentations. In principle, such class sessions allow students to catch up, review their work, obtain help when difficulties have arise, and reflect somewhat on where they have been and what might come next. For many, such sessions may help relieve stress and lessen tension.

However, I am aware of at least one course in which such pauses were viewed by students as a type of recess. Student attendance was low, and student apathy was widely observed. In such a context, an instructor might consider how to keep students engaged, even if the time is not covering new material.

CHAPTER 70

Selected/annotated references for additional topics

THIS CHAPTER highlight several basic references for further reading. Some references complement topics and materials in this section, while others highlight subjects that do not fit well elsewhere in this book. A brief annotation provides context or commentary for each reference.

Broadening Participation in Computing

- Allison Scott, Alexis Martin, Freda McAlear, Sonia Koshy, "Broadening Participation in Computing" Examining Experiences of Girls of Color", *Proceedings of the ITiCSE 2017 Conference*, July, 2017[162]. Reprinted in *ACM Inroads*, Volume 8, Number 4, December 2017, pages 48-52.
  A careful study and report of "the outcomes of a rigorous out-of- school culturally relevant computer science intervention designed to engage underrepresented students in computing." The narrative provides insights regarding the recruitment and participation of girls of color within the field of computing.

- ACM Retention Committee and Colleen M. Lewis, "Twelve tips for creating a culture that supports all students in computing", *ACM Inroads*, Volume 8, Number 4, December 2017, pages 17-20[108].
  ACM Retention Committee and Henry M. Walker, "Retention of Students in Introductory Computing Courses: Curricular Issues and Approaches", *ACM Inroads*, Volume 8, Number 4, December 2017, pages 14-16 [274].
  In November 2016, the ACM Retention Committee was created by the ACM Education Committee to "address the current issue of retention in 4-year, post-secondary CS education programs, specially of the retention of women and URM students following CS1 and CS2 (where the pipeline is most leaky)"[275, p. 12]. The above references are the first of several planned by this committee to examine issues of recruitment and retention of students in computing (particularly women and those from underrepresented groups). The first focuses upon cultural issues, while the second examines curricular matters. Additional articles are under consideration for future publication.

Accessibility

- DO-IT, " Promoting inclusion and success for people with disabilities,", *DO-IT (Disabilities, Opportunities, Internetworking, and Technology) Center*, University of Washington, https://www.washington.edu/doit/, 2018 (accessed February 20, 2018) [59].
  The DO-IT Center at the University of Washington provides extensive programs, resources, videos, and other materials for support for teaching, learning, and career development for individuals with disabilities. This home page serves as a fine starting place for exploration.

- World Wide Web Consortium (W3C), *Web Content Accessibility Guidelines (WCAG) 2.1*, https://www.w3.org/TR/WCAG21/, January 30, 2018 (accessed February 20, 2018) [196].
  The W3C develops standards for Web-based materials throughout the Internet. WCAG 2.1 contains recommendations to help make Web content accessible to people with a wide range of circumstances and disabilities. As a formal technical document, this material presents best practices and recommended approaches. A reader may wish to search the Web for WCAG 2.1 for expanded discussion of the principles given here.

- Sheryl Burgstahler, "Universal Design: Implications for Computing Education,", *ACM Transactions on Computing Education (TOCE)*, Volume 11, Issue 3, October 2011, Article 19 [30].
  This article provides a fine overview of Universal Design (UD), as a "paradigm for designing instructional methods, curriculum, and assessments that are welcoming and accessible to students with a wide range of characteristics, including those related to race, ethnicity, native language, gender, age, and disability." [30, Abstract]. Rather than focusing on accommodations for students outside some norm, with Universal Design "; disadvantages associated with disabilities are considered, for the most part, to be imposed by the inaccessible design of products and environments" [30, Section 3]. The narrative includes relevant history, practices, and an extensive bibliography.

- Kristen Shinohara, Saba Kawas, Andrew J. Ko, and Richard E. Ladner, "Who Teaches Accessibility? A Survey of U.S. Computing Faculty," *SIGCSE 2018, Proceedings of the 2018 SIGCSE Technical Symposium on Computer Science Education*, February 2018 [166].
  Although faculty may be aware of issues related to accessibility, "there is little knowledge about the prevalence of higher education teaching about accessibility or faculty?s perceived barriers to teaching accessibility" [166]. This paper examines what courses are taught related to accessibility, who teaches them, what challenges/barriers may be encountered, and what resources might help.

Student Enrollments and Quality

- National Academies of Sciences, Engineering, and Medicine. 2017. *Assessing and Responding to the Growth of Computer Science Undergraduate Enrollments,* Washington, DC: The National Academies Press. https://doi.org/10.17226/24926[136].
  This report provides perspectives on the recent increase in undergraduate course enrollments and degree production that often strains faculty and resources. The report's description notes, "There is also significant interest about what this growth will mean for the future of CS programs, the role of computer science in academic institutions, the field as a whole, and U.S. society more broadly" [136, Description].

- Mehran Sahami, "Statistical Modeling to Better Understand CS Students", *ITiCSE '16 Proceedings of the 2016 ACM Conference on Innovation and Technology in Computer Science Education*, Arequipa, Peru, July, 2016 [159].
  Mehran Sahami and Chris Piech. "As CS Enrollments Grow, Are We Attracting Weaker Students? A Statistical Analysis of Student Performance in Introductory Programming Courses Over Time", SIGCSE '16 Proceedings of the 47th ACM Technical Symposium on Computer Science Education, pages 54–59, New York, NY, 2016 [160].
  As enrollments in computing courses increase, questions naturally arise regarding the quality of the students. This paper presents data from eight years at Stanford University, develops a statistical model consistent with the data, and reports results from a careful statistical analysis. According to the study, the authors "find that the distribution of student performance during this period, as reflected in their programming assignment scores, remains remarkably stable despite the large growth in enrollment" [160, Abstract].

Code Boot-Camps

- Graham Wilson, "Building a new Mythology: The Coding Boot-camp Phenomenon", *ACM Inroads*, December, 2017 [288].
  With the shortage of qualified candidates to fill jobs in the computing workplace, coding boot-camps have developed as a relatively quick and inexpensive alternative for training potential employees. As noted at the end of the abstract, this article discusses, "Do they [coding boot-camps] deliver on their promises? Are they transparent enough? And, are they truly addressing the skills shortage in the sector" [288, 66].

- Bryan Clark, "Ex-Googler warns coding bootcamps are lacking in two key areas," *TNW, The Next Web*, https://thenextweb.com/dd/2018/02/09/ex-googler-warns-coding-bootcamps-are-lacking-in-two-key-areas/, posted February 9, 2018 (Accessed 20 February 2018).
  In reviewing reports of graduates from bootcamps, the author identifies a lack of depth in background related to algorithmic analysis (Big-O), and he notes, "What is important, however, is that Big-O is often used to quantify ability, particularly in problem solving and executing on tougher design challenges" [36]. He recommends two additional courses, beyond usual coding boot camps, that can be of particular value: data structures and algorithms, and probability and statistics.

Program Verification, Validation, and Usefulness

- Henry M. Walker, "Software Correctness and Usefulness in the Classroom", *ACM Inroads*, March, 2018 [277].
  The correctness of software often has been judged by how well the code meets its specifications, and two approaches for determining correctness has involved program verification (using formal methods to prove code always meets its specifications) and program validation (developing test cases to check code produces desired output in a range of situations). Both verification and validation, however, base the concept of correctness on the ability to write careful specifications. In some settings, precise specifications may not be available or even possible, in which case another measure of software usefulness may be appropriate. This article describes "software usefulness" as another way to measure the value of software—at least in some contexts.

Faculty Issues

- American Association of University Professors (AAUP), *Issues in Higher Education*, https://www.aaup.org/issues-higher-education, accessed January 10, 2018 [16]. AAUP provides resources and support for faculty in high education. Headings on its "Issues in Higher Education" include "academic freedom"," governance of college & university", "resisting the harassment of faculty", "contingent faculty positions", "civility", "collective bargaining", the "sanctuary campus movement", "tenure" ", "women in the academic profession", and "intellectual property and copyright".

Social and Ethical Responsibility

- The Pledge of the Computing Professional, *The Pledge of the Computing Professional*, http://pledge-of-the-computing-professional.org, accessed January 11, 2018[141]. As noted on the organization's Web site, "*The Pledge of the Computing Professional* is solely intended to promote and recognize the ethical and moral behavior of graduates of computing-related degree programs as they transition to careers of service to society." Although simple in concept and implementation, the organization and the corresponding pledge that may be taken by students, faculty, and others, can provide a framework for thinking about one's professional activities and its consequences.

Bibliography

[1] Hal Abelson, Mike Fischer, Danny Weitzner, Keertan Kini, Jessie M. Strickgold-Sarah, and Michael Trice. *M.I.T. course 6.805/STS085/STS487: Foundations of Information Policy*, 2016 (accessed May 20, 2017.

[2] Accreditation Board for Engineering and Technology (ABET) and its Computing Accreditation Commission (CAC). *Criteria for Accrediting Computing Programs, 2017-2018.*

[3] ACM: Association for Computing Machinery. *ACM Code of Ethics and Professional Conduct*, October 1992 (accessed June 20, 2017). https://www.acm.org/about-acm/acm-code-of-ethics-and-professional-conduct.

[4] ACM Committee on Computers and Public Policy, Peter G. Neumann, moderator. *The Risks Digest: Forum on Risks to the Public in Computers and Related Systems.* http://catless.ncl.ac.uk/Risks/.

[5] ACM Curriculum Committee on Computer Science. Curriculum '68: Recommendations for academic programs in computer science. *Communications of the ACM*, 11(3):151–197, March 1968.

[6] ACM SRC. *ACM Student Research Competition*, 2017. http://src.acm.org.

[7] ACM/IEEE-CS Interim Review Task Force. *Curriculum 2008: An Interim Revision of CS 2001.* ACM and the IEEE Press, New York, December 2008.

[8] ACM/IEEE-CS Joint Curriculum Task Force. *Computing Curricula '91.* Association for Computing Machinery, New York, 1991.

[9] ACM/IEEE-CS Joint Task Force on Computing Curricula. *Computer Science Curriculum 2013: Strawman Draft.* ACM and the IEEE Press, New York, February 2012.

[10] ACM/IEEE-CS Joint Task Force on Computing Curricula. Computer science curricula 2013. Technical report, ACM Press and IEEE Computer Society Press, December 2013.

[11] ACM/IEEE-CS Joint Task Force on Computing Curricula. *CS2013 Curricular (Learning Outcome) Spreadsheet*, 2013.

[12] ACM/IEEE-CS Task Force on the Curriculum. *Computing Curricula 2001.* ACM and the IEEE Press, New York, December 2001.

[13] Alex Aiken. *Moss: A System for Detecting Software Similarity*, 2017 (accessed: April 15, 2017). https://theory.stanford.edu/~aiken/moss/.

[14] Z. A. Al-Khanjari, J. A. Fiaidhi, R. A. Al-Hinai, and N. S. Kutti. Plagdetect: a java programming plagiarism detection tool. *ACM Inroads*, 1(4):66–71, dec 2010.

[15] Alice.org. *Tell Stories, Build Games, Learn to Program*, 2017 (accessed July 7, 2017. http://www.alice.org.

[16] American Association of University Professors (AAUP). *Issues in Higher Education*, 2018 (accessed January 10, 2018). https://www.aaup.org/issues-higher-education.

[17] Richard Austing, Bruce Barnes, Della Bonnette, Gerald Engel, and Gordon Stokes. Curriculum '78: Recommendations for the undergraduate program in computer science. *Communications of the ACM*, 22(3):147–166, March 1979.

[18] Sara Baase and Timorhy M. Henry. *A Gift of Fire: Social, Legal, and Ethical Issues for Computing Technology, Fifth Edition.* Pearson Education, Inc, Hoboken, NJ, 2018.

[19] Christan Balch, Michelle Chung, and Karen Brennan. *An introductory computing curriculum using Scratch*, 2017 (accessed July 7, 2017). http://scratched.gse.harvard.edu/guide/.

[20] Douglas Baldwin, Henry M. Walker, and Peter B. Henderson. The roles of mathematics in computer science. *ACM Inroads*, 4(4):74–80, December 2013.

[21] William Barker. *The Curriculum Foundation Project*, 1999.

[22] Valerie Barr and Deborah Trytten. Using turing's craft codelab to support cs1 students as they learn to program. *ACM Inroads*, 7(2):67–75, June 2016.

[23] Vasilisa Bashlovkina, Anita DeWitt, Anqing Liu, Nicolas Knoebber, and Henry M. Walker. A refined c-based infrastructure and curriculum to support robots in introductory cs. *Journal of Computing Sciences in Colleges*, 30(5):136–143, May 2015.

[24] Maŕia Cecilia Bastarrica, Daniel Perovich, and Maíra Marques Samary. What can students get from a software engineering capstone course? *IEEE/ACM 39th International Conference on Software Engineering: Software Engineering and Education Track*, pages 137–145, 2017.

[25] Benjamin S. Bloom and et al. *Taxonomy of Educational Objectives: Handbook 1 Cognitive Domain.* Longmans, Green and Company, 1956.

[26] College Board. *AP Computer Science Principles*, 2017 (accessed: February 28, 2017). https://apstudent.collegeboard.org/apcourse/ap-computer-science-principles.

[27] Ernest L Boyer. *Scholarship reconsidered: Priorities of the professoriate.* Carnegie Foundation for the Advancement of Teaching, 1990.

[28] K. W. Boyer and L. O. Hall. Experience using "moss" to detect cheating on programming assignments. *29th annual Frontiers in Education Conference*, 3:13B3/18–13B3/22, 1999.

[29] Bo Brinkman, Don Gotterbarn, Keith Miller, and Marty J. Wolf. Making a positive impact: Updating the acm code of ethics. *Communications of the ACM*, 59(12):7–13, December 2016.

[30] Sheryl Burgstahler. Universal design: Implications for computing education. *ACM Transactions on Computing Education (TOCE)*, 11(3), October 2011. Article 19.

[31] Sheryl Burgstahler, Richard E. Ladner, and Scott Bellman. Increasing the participation in computing of students with disabilities. *ACM Inroads*, 3(4):42–48, December 2012. Special Issue on Broadening Participation.

[32] Tracy, Editor Camp. Special issue: Women and computing. *SIGCSE Bulletin*, June 2002.

[33] Center for Teaching, Vanderbilt University. *Grading Student Work*, 2017 (accessed May 12, 2017).

[34] Michelene T. H. Chi and Ruth Wylie. The icap framework: Linking cognitive engagement to active learning outcomes. *Educational Psychologist*, 49(4):219–243, 2014.

[35] Pauline Rose Clance and Suzanne Imes. The imposter phenomenon in high achieving women: Dynamics and therapeutic intervention. *Psychotherapy Theory, Research and Practice*, 15(3):241–247, Fall 1978.

[36] Bryan Clark. *Ex-Googler warns coding bootcamps are lacking in two key areas*, February 2018 (accessed February 20, 2018). https://thenextweb.com/newsletter/2018/02/20/tnws-big-spam-show-google-eyeballs/.

[37] J. M. Clement. A call for action (research): Applying science education research to computer science instruction. *Computer Science Education*, 14(4):343?364, 2004.

[38] Code.org. *Computer Science Fundamentals*, 2017 (accessed July 7, 2017). https://code.org/educate/curriculum/elementary-school.

[39] Code.org. *Code.org: About Us*, 2017 (accessed: March 1, 2017). https://code.org/about.

[40] G. L. Cohen, J. Garcia, N. Apfel, and A. Master. Reducing the racial achievement gap: A social-psychological intervention. *Science*, 313(5791):1307–1310, September 2006.

[41] Commission on Technology, Gender and Teacher Education. *Tech-Savvy: Educating Girls in the New Computer Age*. American Association of University Women Educational Foundation, 2000.

[42] MAA Committee on Departmental Review. *Program Review*, 2017 (accessed June 12, 2017). http://www.maa.org/programs/faculty-and-departments/curriculum-department-guidelines-recommendations/program-review.

[43] Committee on the Undergraduate Program in Mathematics (CUPM). Undergraduate Programs and Courses in the Mathematical Sciences: A CUPM Curriculum Guide, Draft 3.2, January 2003.

[44] Committee on the Undergraduate Program in Mathematics (CUPM), Carol S. Schumacher and Martha J. Siegel, Co-Chairs, and Paul Zorn, Editor. 2015 CUPM Curriculum Guide to Majors in the Mathematical Sciences, 2015. http://www2.kenyon.edu/Depts/Math/schumacherc/public_html/Professional/CUPM/2015Guide/Program%20Reports/CompSciSummaryReport.pdf.

[45] Computer Science Accreditation Commission (CSAC) of the Computer Science Accreditation Board (CAB). *Criteria For Accrediting Programs In Computer Science In The United States*, June 1996. http:\\www.csab.orglcriteria96.html.

[46] Computer Systems Research Group. *Sorting out Sorting.* Dynamic Graphics Project, University of Toronto, 1981. 30:56 minute film.

[47] Stephen Cooper and Wanda Dann. Programming. *ACM Inroads*, 6(1):50–54, March 2015. Special issue on "The Role of Programming in a Non-Major, CS Course".

[48] Cooperative Education and Career Development, Northeastern Development. *Grow. Adapt. Thrive.*, 2017 (accessed: July 5, 2017). https://www.northeastern.edu/coop/.

[49] Thomas H. Cormen, Charles E. Leiserson, Ronald L. Rivest, , and Clifford Stein. *Introduction to Algorithms, Third Edition.* M.I.T. Press, 2009.

[50] David Cowden, April O'Neill, Erik Opavsky, Dilan Ustek, and Henry M. Walker. A c-based introductory course using robots. *SIGCSE '12: Proceedings of the 43rd ACM technical symposium on Computer Science Education*, pages 27–32, February 2012.

[51] Jeff Cramer and Bill Toll. Beyond competency: a context-driven cso course. *SIGCSE '12 Proceedings of the 43rd ACM technical symposium on Computer Science Education*, pages 469–474, 2012.

[52] Paul Curzon. Serious fun in computer science. *ITiCSE '07 Proceedings of the 12th annual SIGCSE conference on innovation and technology in computer science education*, page 1, jun 2007.

[53] Paul Curzon and Peter W McOwan. *The Power of Computational Thinking: Games, Magic and Puzzles to Help You Become a Computational Thinker.* World Scientific, 2016.

[54] Andrea Danyluk and Stephen Freund. *CSCI 134: Introduction to Computer Science (Events Version)*, 2017 (accessed July 8, 2017). http://dept.cs.williams.edu/~freund/cs134-171/.

[55] Data.gov. *The home of the U.S. Government's open data*, 2017 (accessed April 15, 2017). https://www.data.gov.

[56] Janet Davis and Samuel A. Rebelsky. Food-first computer science: starting the first course right with pb&j. *SIGCSE '07 Proceedings of the 38th SIGCSE technical symposium on computer science education*, pages 372–376, March 2007.

[57] Janet Davis and Henry M. Walker. Incorporating social issues of computing in a small, liberal arts college: a case study. *SIGCSE '11 Proceedings of the 42nd SIGCSE technical symposium on computer science education*, pages 69–74, March 2011.

[58] Dabin Ding, Mahmoud Yousef, and Xiaodong Yue. A case study for teaching students agile and scrum in capstone courses. *Journal of Computing Sciences in Colleges*, 32(5):95–101, May 2017.

[59] DO-IT. *Promoting inclusion and success for people with disabilities.* DO-IT (Disabilities, Opportunities, Internetworking, and Technology) Center, 2018 (accessed February 20, 2018). https://www.washington.edu/doit/.

[60] Herbert Dreyfus, Stuart Dreyfus, and Tom Athanasiou. *Mind Over Machine.* The Free Press, New York, 1986.

[61] Ed Dubinsky, Barbara Reynolds, and David Mathews. *Readings in Cooperative Learning for Undergraduate Mathematics*, 1997. MAA Notes Series #44.

[62] David Dunning. We are all confident idiots. *Pacific Standard*, 2014. https://psmag.com/we-are-all-confident-idiots-56a60eb7febc.

[63] Joyce Ehrlinger, Kerri Johnson, Matthew Banner, David Dunning, and Justin Kruger. Why the unskilled are unaware: Further explorations of (absent) self-insight among the incompetent. *Organizational Behavior and Human Decision Processes*, 105:98–121, 2008.

[64] Heidi J. C. Ellis, Gregory W. Hislop, StoneyJackson, and Lori Postner. Team project experiences in humanitarian free and open source software (hfoss). *ACM Transactions on Computing Education (TOCE) - Special Issue on Team Projects in Computing Education*, 15(4), December 2015. Article Number 18.

[65] European Commission. What is lifelong learning? *ESAE Headquarters Magazine, the European Society of Association Executives*, August 2007. http://ec.europa.eu/education/policies/III/III_en.html.

[66] Lance Fortnow. *The Golden Ticket: P, NP, and the Search for the Impossible*. Princeton University Press, Princeton and Oxford, 2013.

[67] Clinton P. Fuelling, Anne-Marie Lancaster, Mark C. Kertstetter, R. Waldo Roth, William A. Brown, Richard K. Reidenbach, and Ekawan Wongsawatgul. Computer science undergraduate capstone course. *SIGCSE '88 Proceedings of the nineteenth SIGCSE technical symposium on Computer science education*, page 135, 1988.

[68] Ursula Fuller, Arnold Pears NS June Amillo, Chris Avram, and Linda Mannila. A computing perspective on the bologna process. *ITiCSE-WGR '06 Working group reports on ITiCSE on Innovation and technology in computer science education*, pages 115–131, 2006.

[69] Daniel D. Garcia. Tech launch with gradescope: Exam grading will never be the same again! *ACM Inroads*, 6(2):82–83, June 2015.

[70] Norman Gibbs and Allen B. Tucker. A model curriculum for a liberal arts degree in computer science. *Communications of the ACM*, 29(3):202–210, March 1986.

[71] Michael Goldweber. Programming should not be part of a cs course for non-majors. *ACM Inroads*, 6(1):55–57, March 2015. Special issue on "The Role of Programming in a Non-Major, CS Course".

[72] Michael Goldweber, John Barr, Steve Cooper, and Henry M. Walker. Panel: What everyone needs to know about computation. *Proceedings of the 41st ACM/SIGCSE Technical Symposium on Computer Science Education*, pages pp. 127–128, 2010.

[73] *Google BigQuery Public Datasets*, 2017 (accessed April 15, 2017). https://cloud.google.com/bigquery/public-data/.

[74] Donald Gotterbarn. *Software Engineering Ethics Research Institute*. Computer and Information Science Department, East Tennessee State University, 2009 (accessed June 20, 2017). home page: http://csciwww.etsu.edu/gotterbarn/.

[75] L. Graham and P. T. Metaxas. "Of course it's true; I saw it on the Internet!": critical thinking in the Internet era. *Communications of the ACM*, 46(5):70–75, May 2003.

[76] David Gries. *The Science of Programming.* Springer-Verlag, New York, 1981.

[77] Grinnell College. *Mission and Values.* Mission Statement, as described in https://www.grinnell.edu/about/mission, retrieved from the Web 3 February 2017.

[78] Mark J. Guzdial and Barbara Ericson. *Introduction to Computing and Programming in Python: A Multimedia Approach, Fourth Edition.* Pearson, 2015.

[79] Nancy L. Hagelgans, Barbara E. Reynolds, G. Joseph Wimbish, Mazin Shahin, and Ed Dubinsky. *Practical Guide to Cooperative Learning in Collegiate Mathematics*, 1995.

[80] R. R. Hake. A six-thousand-student survey of mechanics test data for introductory physics courses. *American Journal of Physics*, 66(1):133–137, January 1998.

[81] Vicki Hanson. Inclusive thinking in computer science education. *ITiCSE '07 Proceedings of the 12th annual SIGCSE conference on Innovation and technology in computer science education*, page 1, jun 2007.

[82] Harvey Mudd Computer Science. *Computer Science Clinic*, 2017 (accessed July 2, 2017). https://www.cs.hmc.edu/clinic/.

[83] Benjamin Haytock, Zaven Karian, and Stanley Selter. Teaching computer science within mathematics departments. *Computer Science Education*, 1(3):181–203, 1990.

[84] Jesse M. Heines, Gena R. Greher, S. Alex Ruthmann, and Brendan L. Reilly. Two approaches to interdisciplinary computing +music courses. *IEEE Computer Special Issue on Computers and the Arts*, 44(12):25–32, December 2001.

[85] Peter B. Henderson. Mathematics in the curricula. *SIGCSE Bulletin*, 37(2):20–22, June 2005.

[86] Peter B. Henderson. Mathematical reasoning in computing education. *ACM Inroads*, 1(3):22–23, September 2010.

[87] Eugene A. Herman and Michael D. Pepe. *Visual Linear Algebra.* Wiley, 2005.

[88] Diane Horton, Michelle Craig, Jennifer Campbell, Paul Gries, and Daniel Zingaro. Comparing outcomes in inverted and traditional cs1. *ITiCSE '14 Proceedings of the 2014 conference on Innovation and technology in computer science education*, pages 261–266, 2014.

[89] Peter Hubwieser, Michal Armoni, TTorsten Brinda, Valentina Dagiene, Ira Diethelm, Michael N. Giannakos, Maria Knobelsdorf, Johannes Magenheim, Roland Mittermeir, and Sigrid Schubert. Cs/informatics in secondary education. *Proceedings of the 16th Annual Conference Reports on Innovation and Technology in CS Education-Working Group Reports*, page 19?38, 2011.

[90] Peter Hubwieser and Andreas Zendler. How teachers in different educational systems value central concepts of computer science. *WiPSCE '12 Proceedings of the 7th Workshop in Primary and Secondary Computing Education*, pages 62–69, 2012.

[91] Chuck Huff. *Home page*, 2012 (accessed June 20, 2017). https://www.stolaf.edu/people/huff/.

[92] Marija Ivica, Sara Marku, Thu Nguyen, Ruth Wu, and Henry M. Walker. Student-faculty collaboration in developing and testing infrastructure for a c-based course using robots. *Journal of Computing Sciences in Colleges*, 31(1):57–64, October 2016.

[93] Scott Jaschik. Majoring in a professor. *Inside Higher Ed*, August 2013. https://www.insidehighered.com/news/2013/08/12/study-finds-choice-major-most-influenced-quality-intro-professor (accessed June 10, 2013).

[94] Michael Jipping. *Private communication*, circa 1997. Computer Science Department, Hope College.

[95] Michael Jonas. Capstone experience — achieving success with an undergraduate research group in speech. *SIGITE '14 Proceedings of the 15th Annual Conference on Information technology education*, pages 55–60, October 2014.

[96] Lisa C. Kaczmarczyk. Welcome. *ACM Inroads*, 6(1), March 2015. Special issue on "The Role of Programming in a Non-Major, CS Course".

[97] Shalini Kesar. Including teaching ethics into pedagogy: preparing information systems students to meet global challenges of real business settings. *ACM SIGCAS Computers and Society - Special Issue on Ethicomp*, 45(3):432–437, September 2015.

[98] Richard Kick and Frances P. Trees. Ap cs principles. *ACM Inroads*, 6(1):42–45, March 2015. Special issue on "The Role of Programming in a Non-Major, CS Course".

[99] Maria Klawe. Impostoritis: A lifelong, but treatable, condition. *slate.com*, March 2014. http://www.slate.com/articles/technology/future_tense/2014/03/imposter_syndrome_how_the_president_of_harvey_mudd_college_copes.html.

[100] Jeffrey Koperski. *Philosophy 210B Online?Engineering and Computer Ethics*. Department of Philosophy, Saginaw Valley State University, 2009 (accessed June 20, 2017). syllabus: http://www.svsu.edu/~koperski/Syllabi.htm.

[101] Justin Kruger and David Dunning. Unskilled and unaware of it: How difficulties in recognizing one's own incompetence lead to inflated self-assessments. *Journal of Personality and Social Psychology*, 77(6):1121–1134, December 1999.

[102] Deepak Kumar. *Learning Computing with Robots*. Institute for Personal Robots in Education, 2011. http://wiki.roboteducation.org/Introduction_to_Computer_Science_via_Robots.

[103] Suthikshn Kumar. A skit-based approach to teaching web and networking protocol. *ACM Inroads*, 2(4):30–32, December 2011.

[104] Clif Kussmaul. Guiding students to discover concepts and develop process skills with pogil. *SIGITE '14 Proceedings of the 15th Annual Conference on Information technology education*, pages 159–160, 2014.

[105] Amy N. Langville and Carl D. Meyer. *Google's PageRank and Beyond: The Science of Search Engine Rankings*. Princeton University Press, Princeton and Oxford, 2006.

[106] Patricia Lasserre. Adaptation of team-based learning on a first term programming class. *ITiCSE '09 Proceedings of the 14th annual ACM SIGCSE conference on Innovation and technology in computer science education*, pages 186–190, 2009.

[107] Priscilla W. Laws. Calculus-based physics without lectures. *Physics today*, pages 24–31, December 1991.

[108] Colleen M. Lewis. Twelve tips for creating a culture that supports all students in computing. *ACM Inroads*, 8(4):17–20, December 2017.

[109] Liberal Arts Computer Science Consortium. A 2007 model curriculum for a liberal arts degree in computer science. *Journal on Educational Resources in Computing (JERIC)*, 7(2), June 2007. Article 2.

[110] David Lindquist, Tamara Denning, Michael Kelly, Roshni Malani, William G. Griswold, and Beth Simon. Exploring the potential of mobile phones for active learning in the classroom. *SIGCSE '07 Proceedings of the 38th SIGCSE technical symposium on Computer science education*, pages 384–388, 2007.

[111] Richard Lobb and Jenny Harlow. Coderunner: A tool for assessing computer programming skills. *ACM Inorads*, 7(2):47–51, March 2016.

[112] W. Ted Mahavier, E. Lee May, and G. Edgar Parker. *A Quick-Start Guide to the Moore Method*, 2009 (accessed February 2, 2018). http://legacyrlmoore.org/reference/quick_start-3.pdf.

[113] Craig Marais and Karen Bradshaw. Towards a technical skills curriculum to supplement traditional computer science teaching. *ITiCSE '16 Proceedings of the 2016 ACM Conference on Innovation and Technology in Computer Science Education*, pages 338–343, 2016.

[114] Jane Margolis and Allan Fisher. *Unlocking the Clubhouse: Women in Computing.* MIT Press, November 2001.

[115] Jane Margolis and Allan Fisher. *Stuck in the Shallow End, Updated Edition: Education, Race, and Computing.* MIT Press, March 2017.

[116] Cindy Marling and David Juedes. Cs0 for computer science majors at ohio university. *SIGCSE '16 Proceedings of the 47th ACM Technical Symposium on Computing Science Education*, pages 138–143, March 2016.

[117] F. Martin, G. R. Greher, J. M. Heines, J. Jeffers, H.-J. Kim, S. Kuhn, K. Roehr, N. Selleck, L. Silka, and H. Yanco. Joining computing and the arts at a mid-size university. *Journal of the Consortium for Computing Sciences in Colleges*, 24(6):87–94, 2009. https://jesseheines.com/~heines/academic/papers/2011ieee/ieee2011paper-v33-forWebsite.pdf.

[118] Kent McClelland. *Assessment Criteria for Class Presentations.* http://web.grinnell.edu/Dean/Tutorial/Skills/Oral/aco.pdf.

[119] Jeffrey J. McConnell. Active and cooperative learning: more tips and tricks (part ii). *SIGCSE Bulletin*, 37(4):34–38, December 2005.

[120] Jeffrey J. McConnell. Active and cooperative learning: tips and tricks (part i). *SIGCSE Bulletin*, 37(2):27–30, June 2005.

[121] Jeffrey J. McConnell. Active and cooperative learning: further tips and tricks (part 3). *SIGCSE Bulletin*, 38(2):24–28, June 2006.

[122] Jeffrey J. McConnell. Active and cooperative learning: tips and tricks (part iv). *SIGCSE Bulletin*, 38(4):25–28, December 2006.

[123] L. C. McDermott and E. F. Redish. Rl-per-1: Resource letter on physics education research. *American Journal of Physics*, 67:755–767, September 1999.

[124] Wilbert McKeachie and Marilla Svinkicki. *McKeachie's Teaching Tips: Strategies, Research And Theory for College And University Teachers, 14th edition.* Wadsworth, Cengage Learning, 2011, 2014.

[125] *Merriam-Webster Dictionary*, 2017 (accessed June 2, 2017). https://www.merriam-webster.com/dictionary/curriculum.

[126] L. K. Michaelsen, A. B. Knight, and D. L. Fink. *Team-Based Learning: A Transformative Use of Small Groups in College Teaching.* Stylus Publishing, Sterling VA, 2004.

[127] Radu P. Mihail, Beth Rubin, and Judy Goldsmith. Online discussions: improving education in cs? *SIGCSE '14: Proceedings of the 45th ACM technical symposium on Computer science education*, pages 409–414, March 2104.

[128] Akira Miyake et al. Reducing the gender achievement gap in college science: A classroom study of values affirmation. *Science*, 330(6008):1234–1237, November 2010.

[129] Akira Miyake et al. *Supporting Online Material: Reducing the Gender Achievement Gap in College Science: A Classroom Study of Values Affirmation*, 2010. http://science.sciencemag.org/content/suppl/2010/11/22/330.6008.1234.DC1?_ga=1.96127065.4589012.1491152262.

[130] Phatludi Modiba, Vreda Pieterse, and Bertram Haskins. Evaluating plagiarism detection software for introductory programming assignments. *CSERC '16 Proceedings of the Computer Science Education Research Conference*, pages 37–46, July 2016.

[131] Sriram Mohan, Stephen Chenoweth, and Shawn Bohner. Towards a better capstone experience. *SIGCSE '12: Proceedings of the 43rd ACM technical symposium on Computer science education*, pages 111–116, 2012.

[132] R. S. Moog and Editors J. N. Spencer. *Process-oriented Guided Inquiry Learning (POGIL).* Oxford University Press, 2008.

[133] Murdough Center of the Texas Tech University. *Ethics Cases*, 2017 (accessed June 20, 2017). http://www.depts.ttu.edu/murdoughcenter/products/cases.php.

[134] Thomas Murtaugh. *CSCI 134 - Digital Communication and Computation: An Introduction to Computer Science (Networking Version)*, 2017 (accessed: March 13, 2017). http://www.cs.williams.edu/~cs134/.

[135] Arvind Narayanan, Joseph Bonneau, Edward Felten, Andrew Miller, and Steven Goldfeder. *Bitcoin and Cryptocurrency Technologies: A Comprehensive Introduction.* Princeton University Press, Princeton and Oxford, 2016.

[136] National Academies of Sciences, Engineering, and Medicine. *Assessing and Responding to the Growth of Computer Science Undergraduate Enrollments.* The National Academies Press, Washington, DC, 2017 (accessed January 11, 2018).

[137] The National Center for Women NCWIT and Information Technology. *Computer Science-in-a-Box: Unplug Your Curriculum*, 2017 (accessed July 7, 2017). https://www.ncwit.org/resources/computer-science-box-unplug-your-curriculum.

[138] Peter Norvig. *The Gettysburg Powerpoint Presentation*, 2000. http://www.norvig.com/Gettysburg/.

[139] Office of Graduate Studies, University of Nebraska-Lincoln. *Fairness in the Classroom*, 2017 (accessed May 12, 2017).

[140] Pedagogy Focus Group 2 of the Computing Curricula 2001 Task Force. *Report of Group 2: Supporting Courses*, July 2000. Draft 5.2.

[141] The Pledge of the Computing Professional. *The Pledge of the Computing Professional: home page*, 2018 (accessed January 11, 2018).

[142] Office of Science and Technology Policy (OSTP). *CS for All.* https://www.nsf.gov/news/special_reports/csed/csforall.jsp.

[143] Mike O'Leary. Innovative pedagogical approaches to a capstone laboratory course in cyber operations. *SIGCSE '17: Proceedings of the 2017 ACM SIGCSE Technical Symposium on Computer Science*, pages 429–434, 2017.

[144] Cathy O'Neil. *Weapons of Math Destruction: How Big Data Increases Inequality and Threatens Democracy.* Penguin Random House, New York, 2016.

[145] Youwen Ouyang, Ursula Wolz, and Susan H. Rodger. Effective delivery of computer curriculum in middle school – challenges and solutions (panel). *SIGCSE '10 Proceedings of the 41st ACM technical symposium on Computer science education*, pages 327–328, 2010.

[146] J. Parker, R. Cupper, C. Kelemen, D. Molnar, and G. Scragg. Laboratories in the computer science curriculum. *Journal of Computer Science Eduction*, pages 205–221, 1990.

[147] Joseph P. Pickett, Executive Editor, editor. *American Heritage Dictionary of the English Language, Fourth Edition.* Houghton Mifflin Company, 2009.

[148] Jonathan Pierce and Craig Ziles. Investigating student plagiarism patterns and correlations to grades. *SIGCSE '17 Proceedings of the 2017 ACM SIGCSE Technical Symposium on Computer Science Education*, pages 471–476, 2017.

[149] L. Porter and Beth Simon. Retaining nearly one-third more majors with a trio of instructional best practices in cs1. *SIGCSE '13 Proceedings of the 44th ACM technical symposium on computer science education*, pages 165–170, March 2013.

[150] Leo Porter, Dennis Bouvier, Quintin Cutts, Scott Grissom, Cynthia Lee, Robert McCartney, Daniel Zingaro, and Beth Simon. A multi-institutional study of peer instruction in introductory computing. *SIGCSE '16 Proceedings of the 47th ACM Technical Symposium on Computing Science Education*, pages 358–363, 2016.

[151] The POGIL Project. *POGIL: Process Oriented Guided Inquiry Learning*, 2017 (accessed June 24, 2017).

[152] Michael J. Quinn. *Ethics for the Information Age, Seventh Edition*. Pearson/Addison-Wesley, 2016.

[153] Anthony Ralston. Do we need any mathematics in computer science curricula? *SIGCSE Bulletin*, 37(2):6–9, June 2005.

[154] Anthony Ralston and Mary Shaw. Curriculum'78 iscomputer science really that unmathematical? *Communications of the ACM*, 23(2):67–70, February 1980.

[155] Raghu Raman, Smrithi Venkatasubramanian, Krishnashree Achuthan, and Prema Nedungadi. Computer science (cs) education in indian schools: Situation analysis using darmstadt model. *ACM Transactions on Computing Education (TOCE) - Special Issue II on Computer Science Education in K-12 Schools*, 15(2), May 2015.

[156] Samuel A Rebelsky, Janet Davis, and Jerod Weinman. Building knowledge and confidence with mediascripting: a successful interdisciplinary approach to cs1. *SIGCSE '13 Proceedings of the 44th ACM technical symposium on computer science education*, pages 483–488, March 2013.

[157] Andrew Luxton Reilly. Learning to program is easy. *Proceedings of the 2016 ACM Conference on Innovation and Technology in Computer Science Education*, pages 284–289, July 2016.

[158] Rebecca H. Rutherfoord and James K. Rutherfoord. Flipping the classroom: is it for you? *SIGITE '13 Proceedings of the 14th annual ACM SIGITE conference on Information technology education*, pages 19–22, 2013.

[159] Mehran Sahami. Statistical modeling to better understand cs students. *ITiCSE '16 Proceedings of the 2016 ACM Conference on Innovation and Technology in Computer Science Education*, July 2016.

[160] Mehran Sahami and Chris Piech. As cs enrollments grow, are we attracting weaker students? a statistical analysis of student performance in introductory programming courses over time. *SIGCSE '16 Proceedings of the 47th ACM Technical Symposium on Computer Science Education*, pages 54–59, 2016.

[161] Dorothy Sayers. *The Mind of the Maker*. Harcourt, Brace and Company, New York, 1941.

[162] Allison Scott, Alexis Martin, Frieda McAlear, and Sonia Koshy. Broadening participation in computing" examining experiences of girls of color. *ITiCSE '17 Proceedings of the 2017 ACM Conference on Innovation and Technology in Computer Science Education*, pages 252–256, July 2017.

[163] Scratch. *Create stories, games, and animations*, 2017 (accessed July 7, 2017). https://scratch.mit.edu/.

[164] Amazon Web Services. *AWS Public Data Sets*, 2017 (accessed April 15, 2017). https://aws.amazon.com/public-datasets/.

[165] Amber Settle, John Lalor, and Theresa Steinbach. Computer science linked-courses learning community. *Proceedings of the 2015 ACM Conference on Innovation and Technology in Computer Science Education*, pages 123–128, July 2015.

[166] Kristen Shinohara, Saba Kawas, Andrew J. Ko, and Richard E. Ladner. Who teaches accessibility? a survey of u.s. computing faculty. *SIGCSE 2018, Proceedings of the 2018 SIGCSE Technical Symposium on Computer Science Education*, February 2018.

[167] SIGCAS: Special Interest Group on Computers and Society within the Association for Computing Machinery (ACM). *Welcome to ACM SIGCAS*, 2017 (accessed June 25, 2017).

[168] SIGCSE Committee on the Implementation of a Discrete Mathematics Course, Bill Marion and Doug Baldwin Committee Co-Chairs. *On the Implementation of a Discrete Mathematics Course*, April 2007. http://sigcse.org/sigcse/files/documents/pdfs/DiscreteMathReport.pdf.

[169] SIGCSE: Special Interest Group on Computer Science Education within the Association for Computing Machinery (ACM). *Special Interest Group on Computer Science Education*, 2017 (accessed June 25, 2017).

[170] Simon, Judy Sheard, Michael Morgan, Andrew Petersen, Amber Settle, Jane Sinclair, Gerry Cross, and Charles Riedesel. Negotiating the maze of academic integrity in computing education. *ITiCSE '16 Proceedings of the 2016 ITiCSE Working Group Reports*, pages 57–80, July 2016.

[171] Arjun Singh, Sergey Karayev, Kevin Gutowski, and Pieter Abbeel. Gradescope: A fast, flexible, and fair system for scalable assessment of handwritten work. *L@S '17: Proceedings of the Fourth (2017) ACM Conference on Learning @ Scale*, pages 81–86, apr 2017.

[172] Abraham Sinkov. *Elementary Cryptanalysis: A Mathematical Approach.* The New Mathematical Library, Random House and the Mathematical Association of America, 1968.

[173] Abraham Sinkov and Todd Feil. *Elementary Cryptanalysis, Second Edition.* The New Mathematical Library, the Mathematical Association of America, 2009.

[174] Michael Sipser. *Introduction to the Theory of Computation, Third Edition.* Cengage Learning, 2014.

[175] M. J. Smith and Fred Smith. *The Smart Student's Guide to Healthy Living: How to Survive Stress, Late Nights, and the College Cafeteria.* New Harbinger Publications, 2006.

[176] Victoria Smith and Stephanie Maher Palenque. *Ten Tips for More Efficient and Effective Grading*, feb 2015 (accessed February 6, 2018).

[177] British Computing Society. *Academic Accreditation*, 2017 (Accessed July 7, 2017. http://www.bcs.org/category/5844.

[178] TIOBE Software. *TIOBE Programming Community Index for April 2010*, 2010 (accessed April 2010).

[179] Anita Solow. *Learning by Discovery: A Lab Manual for Calculus*, 1993.

[180] Staff of Project CLUME MAA and Anthony Thomas and MAA Staff and Anthony Thomas. *Cooperative Learning in Undergraduate Mathematics: Issues that Matter and Strategies that Work*, 2001.

[181] Jeffrey A. Stone. Using reflective blogs for pedagogical feedback in cs1. *SIGCSE '12 Proceedings of the 43rd ACM technical symposium on Computer Science Education*, pages 259–264, 2012.

[182] Nicholas C. Swenson, Rosalind W. Picard, and Ming-Zher Poh. A wearable sensor for unobtrusive, long-term assessment of electrodermal activity. *IEEE Transactions on Biomedical Engineering*, 57(5):1243–1252, May 2010. http://hdl.handle.net/1721.1/62149.

[183] Andrea Tartaro, Christopher Healy, and Kevin Treu. Computer science in general education: beyond quantitative reasoning. *Journal of Computing Sciences in Colleges*, 32(2):177–184, December 2016.

[184] Team-Based Learning Collaborative (TBLC). *TBLC Home Page*, 2017 (accessed June 25, 2017). https://www.teambasedlearning.org.

[185] Sheila Tobias. *They're Not Dumb, They're Different: Stalking the Second Tier*. Research Corp., Tucson, Arizona, 1990.

[186] Alan Tucker. Models that work: Case studies in effective undergraduate mathematics programs. *Notices of the AMS*, 43(11):1356–1358, November 1996.

[187] Alan C. Tucker, editor. *Models That Work: Case Studies in Effective Undergraduate Mathematics Programs*. Number 38 in MAA Notes. Mathematical Association of America, 1995.

[188] turnitin.com. *Turinitin for Higher Education*, 2017 (accessed: April 15, 2017).

[189] *United States Census Bureau*, 2017 (accessed April 15, 2017). https://www.census.gov.

[190] *U. S. House of Representatives*, 2017 (accessed April 15, 2017). http://www.house.gov.

[191] *United States Senate*, 2017 (accessed April 15, 2017). https://www.senate.gov.

[192] Union College Computer Science Department. *CSC-10X: Union College's Unique Introduction to CS*, 2017 (accessed July 8, 2017). http://cs.union.edu/intro/.

[193] United States Department of Commerce. *SOFTWARE AND INFORMATION TECHNOLOGY SPOTLIGHT: The Software and Information Technology Services Industry in the United States*, 2015 (accessed March 3, 2017). https://www.selectusa.gov/software-and-information-technology-services-industry-united-states.

[194] Dilan Ustek, Erik Opavsky, Henry M. Walker, and David Cowden. Course development through student-faculty collaboration: a case study. *ITiCSE '14: Proceedings of the 2014 conference on Innovation and technology in computer science education*, pages 189–194, June 2014.

[195] Chris van der Kuyl. Where have all the computer scientists gone? *ITiCSE '07 Proceedings of the 12th annual SIGCSE conference on Innovation and technology in computer science education*, page 2, jun 2007.

[196] World Wide Web Consortium (W3C). *Web Content Accessibility Guidelines (WCAG) 2.1*, January 2018 (accessed February 20, 2018). https://www.w3.org/TR/WCAG21/.

[197] Henry Walker. *Feedback on Class Presentations, CSC 341*, 2014. http://www.cs.grinnell.edu/~walker/courses/341.sp14/presentation-feedback-form.txt.

[198] Henry Walker. *Proofs for Class Discussion, CSC 341*, 2014. http://www.cs.grinnell.edu/~walker/courses/341.sp14/proofs-for-class-discussion.pdf.

[199] Henry Walker and Joan Krone. *Conducting Department/Program Reviews and Serving as a Reviewer*, September 2014. http://www.cs.grinnell.edu/~walker/talks/dept-reviews-mw-2014/.

[200] Henry M. Walker. *Problems for Computer Solutions Using Basic*. Winthrop Publishers, 1980.

[201] Henry M. Walker. *Introduction to Computing and Computer Science with Pascal*. Little, Brown, and Company, 1982.

[202] Henry M. Walker. *Introduction to Computing and Computer Science*. Little, Brown, and Company, 1986.

[203] Henry M. Walker. Collaborative learning: a case study for cs1 at grinnell college and austin. *SIGCSE '97 Proceedings of the twenty-eighth SIGCSE technical symposium on Computer science education*, pages 209–213, February 1997.

[204] Henry M. Walker. A racquetball or volleyball simulation. *SIGCSE Bulletin*, 29(4):22–23, December 1997.

[205] Henry M. Walker. The balance between programming and other assignments. *SIGCSE Bulletin*, 30(4):23a–25a, December 1998.

[206] Henry M. Walker. Writing within the computer science curriculum. *SIGCSE Bulletin*, 30(2):24–25, June 1998.

[207] Henry M. Walker. *Expert Systems*, January 1999. http://www.cs.grinnell.edu/~walker/courses/151.sp99/lab-placement.html.

[208] Henry M. Walker. Finding interesting examples and assignments for cs1/2 assignments. *SIGCSE Bulletin*, 31(2):24–25, June 1999.

[209] Henry M. Walker. Guided reading and seminar numbers. *SIGCSE Bulletin*, 31(4):27–28, December 1999.

[210] Henry M. Walker. Balancing the forest and the trees in courses. *SIGCSE Bulletin*, 32(4):17–18, December 2000.

[211] Henry M. Walker. Notes on grading. *SIGCSE Bulletin*, 32(2):18–19, June 2000.

[212] Henry M. Walker. Teaching and a sense of the dramatic. *SIGCSE Bulletin*, 33(4):16–17, December 2001.

[213] Henry M. Walker. Teaching and a sense of the dramatic, act ii. *SIGCSE Bulletin*, 34(4):18–19, December 2002.

[214] Henry M. Walker. Do computer games have a role in the computing classroom. *SIGCSE Bulletin*, 35(4):18–20, December 2003.

[215] Henry M. Walker. Lessons from the cupm. *SIGCSE Bulletin*, 35(2):19–21, June 2003.

[216] Henry M. Walker. Academic honesty in the classroom. *SIGCSE Bulletin*, 36(4):18–19, December 2004.

[217] Henry M. Walker. What teachers should, can, and cannot do. *SIGCSE Bulletin*, 36(2):20–21, June 2004.

[218] Henry M. Walker. Mathematics and cs topics in the classroom. *SIGCSE Bulletin*, 37(2):15–17, June 2005.

[219] Henry M. Walker. What should be in a syllabus. *SIGCSE Bulletin*, 37(4):19–21, December 2005.

[220] Henry M. Walker. Thoughts about lecturing. *SIGCSE Bulletin*, 38(2):19–21, June 2006.

[221] Henry M. Walker. Thoughts on student feedback to help teaching. *SIGCSE Bulletin*, 38(4):13–14, December 2006.

[222] Henry M. Walker. Reading and class work. *SIGCSE Bulletin*, 39(2):13–14, June 2007.

[223] Henry M. Walker. What image do cs1/cs2 present to our students. *SIGCSE Bulletin*, 39(4):18–19, December 2007.

[224] Henry M. Walker. Advertising and recruiting. *SIGCSE Bulletin*, 40(2):16–17, June 2008.

[225] Henry M. Walker. Panel: Games in the computer science classroom: good or evil? *Proceedings of the 39th ACM/SIGCSE Technical Symposium on Computer Science Education*, 2008. retrieved on 16 April 2010 from http://www.cs.grinnell.edu/~walker/talks/games-debate/games-evil.html.

[226] Henry M. Walker. *Paper Assignment: Computing Applications and Ethical Viewpoints*, September 2008. http://www.cs.grinnell.edu/~walker/courses/tutorial.fa08/writing/paper1.shtml.

[227] Henry M. Walker. Staying connected with the big picture. *SIGCSE Bulletin*, 40(4):18–20, December 2008.

[228] Henry M. Walker. Course descriptions and public relations for computer science. *SIGCSE Bulletin*, 41(2):74–75, June 2009.

[229] Henry M. Walker. Grading and the allocation of points. *SIGCSE Bulletin*, 41(4):14–16, December 2009.

[230] Henry M. Walker. Computing teaching labs can communicate negative messages. *ACM Inroads*, 1(3):13–14, September 2010.

[231] Henry M. Walker. Configurations for teaching labs. *ACM Inroads*, 1(3):26–30, September 2010.

[232] Henry M. Walker. Eight principles of an undergraduate curriculum. *ACM Inroads*, 1(1):18–20, March 2010.

[233] Henry M. Walker. Prerequisites: shaping the computing curriculum. *ACM Inroads*, 1(4):14–16, December 2010.

[234] Henry M. Walker. The role of programming in introductory computing courses. *ACM Inroads*, 1(2):12–15, June 2010.

[235] Henry M. Walker. Wellness and the classroom. *ACM Inroads*, 1(1):27–30, March 2010.

[236] Henry M. Walker. How to challenge students. *ACM Inroads*, 2(3):14–15, September 2011.

[237] Henry M. Walker. A lab-based approach for introductory computing that emphasizes collaboration. *Proceedings of CSERC '11, Computer Science Education Research Conference*, pages 21–31, 2011. Open Universiteit, Heerlen, Netherlands.

[238] Henry M. Walker. Resolved: ban 'programming' from introductory computing courses. *ACM Inroads*, 2(4):16–17, December 2011.

[239] Henry M. Walker. The role of textbooks. *ACM Inroads*, 2(1):14–16, March 2011.

[240] Henry M. Walker. When is a computing curriculum bloated? *ACM Inroads*, 2(2):18–20, June 2011.

[241] Henry M. Walker. *A Racquetball or Volleyball Simulation*, 2011 (accessed: April 19, 2017). http://www.cs.grinnell.edu/~walker/racquetball/.

[242] Henry M. Walker. Course planning: the day-to-day schedule. *ACM Inroads*, 3(3):22–24, September 2012.

[243] Henry M. Walker. Developing a useful curricular map. *ACM Inroads*, 3(4):14–16, December 2012.

[244] Henry M. Walker. How to prepare students for lifelong learning. *ACM Inroads*, 3(2):10–11, June 2012.

[245] Henry M. Walker. Mid-course corrections. *ACM Inroads*, 3(1):20–21, March 2012.

[246] Henry M. Walker. $1000_{(binary)}$ thoughts for developing and using examples. *ACM Inroads*, 4(3):40–41, September 2013.

[247] Henry M. Walker. Exercise solutions: motivations, messages sent, and possible distribution. *ACM Inroads*, 4(1):14–16, March 2013.

[248] Henry M. Walker. Homework assignments and internet sources. *ACM Inroads*, 4(4):16–17, December 2013.

[249] Henry M. Walker. Mathematics for/with computing and computational science: an opportunity for reflection and dialog. *ACM Inroads*, 4(2):32–34, June 2013.

[250] Henry M. Walker. *The Tao of Computing, Second Edition.* Taylor & Francis Group, CRC Press, 2013.

[251] Henry M. Walker. College courses of varying credit. *ACM Inroads*, 5(2):26–28, June 2014.

[252] Henry M. Walker. Encouraging student preparation for class. *ACM Inroads*, 5(1):24–25, March 2014.

[253] Henry M. Walker. Some strategies when teaching theory courses. *ACM Inroads*, 5(3):32–34, September 2014.

[254] Henry M. Walker. Structuring student work. *ACM Inroads*, 5(4):30–33, December 2014.

[255] Henry M. Walker. *Talk: A Lab-based Approach for Introductory Computing that Emphasizes Collaboration.* The University of Nebraska at Lincoln, October 2014. http://www.cs.grinnell.edu/~walker/talks/lab-based-unl/.

[256] Henry M. Walker. Beyond the cliche, mathematical fluency, in the computing curriculum. *ACM Inroads*, 6(4):24–26, December 2015.

[257] Henry M. Walker. Computational thinking in a non-majors cs course requires a programming component. *ACM Inroads*, 6(1):58–61, March 2015. Special issue on "The Role of Programming in a Non-Major, CS Course".

[258] Henry M. Walker. Priorities for the non-majors, cs course. *ACM Inroads*, 6(1):46–49, March 2015. Special issue on "The Role of Programming in a Non-Major, CS Course".

[259] Henry M. Walker. Recovering from disappointing test results. *ACM Inroads*, 6(3):38–39, September 2015.

[260] Henry M. Walker. Sorting algorithms: when the internet gives you lemons, organize a course festival. *ACM Inroads*, 6(1):28–29, March 2015.

[261] Henry M. Walker. *Talk: MyroC 3.0: A C-based Project for using Robots in CS2: Design Approaches and Techniques to Implement Robot Commands*, October 2015. http://www.cs.grinnell.edu/~walker/talks/robots-2015-williams/.

[262] Henry M. Walker. Why a required course on theory? *ACM Inroads*, 6(2):24–26, June 2015.

[263] Henry M. Walker. Basic do's and don'ts in the classroom: General environment and course suggestions. *ACM Inroads*, 7(3):20–24, September 2016.

[264] Henry M. Walker. Planning and organizing a course for the first time. *ACM Inroads*, 7(4):12–17, December 2016.

[265] Henry M. Walker. Teacher as coach, mentor, listener (part 1?). *ACM Inroads*, 7(1):18–21, March 2016.

[266] Henry M. Walker. Using the hill-climbing algorithm with curricula and courses. *ACM Inroads*, 7(2):36–38, June 2016.

[267] Henry M. Walker. *CSC 161: Imperative Problem Solving and Data Structures, Fall 2016*, 2016 (accessed: December, 2016). http://www.cs.grinnell.edu/~walker/courses/161.fa16/.

[268] Henry M. Walker. Basic do's and don'ts in the classroom: Combating bias, making presentations, and preparing slides. *ACM Inroads*, 8(1):12–15, March 2017.

[269] Henry M. Walker. Connecting computing with other disciplines and the wider community. *ACM Inroads*, 8(2):29–32, June 2017.

[270] Henry M. Walker. Lab-based courses with the 3 c's: Content, collaboration, and communication. *ACM Inroads*, 8(4):26–29, December 2017.

[271] Henry M. Walker. Lab layouts. *ACM Inroads*, 8(3):17–19, September 2017.

[272] Henry M. Walker. *Paper Assignment: The Risks Digest*, February 2017. http://www.cs.grinnell.edu/~walker/courses/101.sp17/show-path-file.php?pathFile=writing/paper-1.php.

[273] Henry M. Walker. *Research Paper Assignment*, February 2017. http://www.cs.grinnell.edu/~walker/courses/101.sp17/show-path-file.php?pathFile=writing/paper-2.php.

[274] Henry M. Walker. Retention of students in introductory computing courses: Curricular issues and approaches. *ACM Inroads*, 8(4):14–16, December 2017.

[275] Henry M. Walker. Retention of students in introductory computing courses: Preliminary plans—acm retention committee. *ACM Inroads*, 8(4):12, December 2017.

[276] Henry M. Walker. *MyroC: A Project to Support a C-based Course with Scribbler 2 Robots*, 2017 (accessed: April 15, 2017). http://www.cs.grinnell.edu/~walker/MyroC/.

[277] Henry M. Walker. Software correctness and usefulness in the classroom. *ACM Inroads*, 9(1), March 2018.

[278] Henry M. Walker and John F. Dooley. The sigcse submission and review software: 10(hexadecimal) lessons. *Journal of the Consortium for Computing Sciences in Colleges*, 25(5):196–206, May 2010. slides for the corresponding talk at CCSC Midwest 2010 are available at http://www.cs.grinnell.edu/~walker/talks/software-lessons/.

[279] Henry M. Walker and Kevin Engel. Research exercises: immersion experiences to promote information literacy. *Journal of Computing Sciences in Colleges*, 21(4):61–68, April 2006.

[280] Henry M. Walker, Daniel Kaplan, and Douglas Baldwin. Summary Report: MAA Program Study Group on Computing and Computational Science, 2015. http://www2.kenyon.edu/Depts/Math/schumacherc/public_html/Professional/CUPM/2015Guide/Program%20Reports/CompSciSummaryReport.pdf.

[281] Henry M. Walker, Daniel Kaplan, and Douglas Baldwin. Supplemental Report: MAA Program Study Group on Computing and Computational Science, 2015. http://www2.kenyon.edu/Depts/Math/schumacherc/public_html/Professional/CUPM/2015Guide/Program%20Reports/CompSciSupplementalReport.pdf.

[282] Henry M. Walker, Dorene Mboya, and Weichao Ma. Variability of referees' ratings of conference papers. *ITiCSE '02: Proceedings of the 7th annual conference on Innovation and technology in computer*, pages 178–182, 2002.

[283] Henry M. Walker and G. Michael Schneider. A revised model curriculum for a liberal arts degree in computer science. *Communications of the ACM*, 39(12), December 1996. 85–95.

[284] Tom Whaley. Using a blog for enrichment readings in cso. *Journal of Computing Sciences in Colleges*, 22(2):203–205, December 2006.

[285] Diane Whitehouse, Penny Duquenoy, Kai K. Kimppa, Oliver K. Burmeister, Don Gotterbarn, David Kreps, and Norberto Patrignani. Codes of ethics and cloud computing. *SIGCAS Computers and Society Newsletter*, 45(3):18–24, September 2015.

[286] Judith C. Williams, Bettina Bair, Jrgen Brstler, Timothy C. Lethbridge, and Ken Surendran. Client sponsored projects in software engineering courses (panel). *SIGCSE '03: Proceedings of the 34th SIGCSE technical symposium on Computer science*, pages 401–402, 2003.

[287] Williams College. *Tutorials*, 2017 (accessed May 31, 2017). https://www.williams.edu/academics/tutorials/.

[288] Graham Wilson. Building a new mythology: The coding boot-camp phenomenon. *ACM Inroads*, 9(4):66–71, December 2017.

[289] Jeannette M. Wing. Computational thinking. *Communications of the ACM*, 49(3):33–35, March 2006.

[290] William Wulf. The urgency of engineering education reform, excerpts from the litee 2002 distinguished lecture, auburn university, al. *Journal of SMET Education*, July-December, March 2002.

[291] Aharon Yadin. Reducing the dropout rate in an introductory programming course. *ACM Inroads*, 2(4):71–76, December 2011.

[292] Aharon Yadin. Using unique assignments for reducing the bimodal grade distribution. *ACM Inroads*, 4(1):38–42, March 2013.

[293] Yoav Yair. Did you let a robot check my homework? *ACM Inroads*, 5(2):33–35, June 2012.

[294] Stuart Zweben and Betsy Bizot. 2015 taulbee survey. *Computing Research News*, 28(5), May 2016.

Index

Milton Keynes UK
Ingram Content Group UK Ltd.
UKHW051929141024
449569UK00027B/1419